Pattern Recognition Applications in Engineering

Diego Alexander Tibaduiza Burgos
Universidad Nacional de Colombia, Colombia

Maribel Anaya Vejar
Universidad Sergio Arboleda, Colombia

Francesc Pozo
Universitat Politècnica de Catalunya, Spain

A volume in the Advances in Computer and
Electrical Engineering (ACEE) Book Series

Published in the United States of America by
IGI Global
Engineering Science Reference (an imprint of IGI Global)
701 E. Chocolate Avenue
Hershey PA, USA 17033
Tel: 717-533-8845
Fax: 717-533-8661
E-mail: cust@igi-global.com
Web site: http://www.igi-global.com

Library of Congress Cataloging-in-Publication Data

Names: Tibaduiza Burgos, Diego Alexander, 1980- editor. | Anaya Vejar,
    Maribel, 1981- editor. | Pozo, Francesc, 1976- editor.
Title: Pattern recognition applications in engineering / Diego Alexander
    Tibaduiza Burgos, Maribel Anaya Vejar, Francesc Pozo, editors.
Description: Hershey, PA : Engineering Science Reference, 2020. | Includes
    bibliographical references and index. | Summary: ""This book explores
    strategies of pattern recognition algorithms in industrial and research
    applications"--Provided by publisher"-- Provided by publisher.
Identifiers: LCCN 2019034145 (print) | LCCN 2019034146 (ebook) | ISBN
    9781799818397 (h/c) | ISBN 9781799818403 (s/c) | ISBN 9781799818410
    (eISBN)
Subjects: LCSH: Pattern perception. | Engineering.
Classification: LCC Q327 .P3684 2020 (print) | LCC Q327 (ebook) | DDC
    620.00285/64--dc23
LC record available at https://lccn.loc.gov/2019034145
LC ebook record available at https://lccn.loc.gov/2019034146

This book is published in the IGI Global book series Advances in Computer and Electrical Engineering (ACEE) (ISSN: 2327-039X; eISSN: 2327-0403)

British Cataloguing in Publication Data
A Cataloguing in Publication record for this book is available from the British Library.

All work contributed to this book is new, previously-unpublished material. The views expressed in this book are those of the authors, but not necessarily of the publisher.

For electronic access to this publication, please contact: eresources@igi-global.com.

Advances in Computer and Electrical Engineering (ACEE) Book Series

Srikanta Patnaik
SOA University, India

ISSN:2327-039X
EISSN:2327-0403

MISSION

The fields of computer engineering and electrical engineering encompass a broad range of interdisciplinary topics allowing for expansive research developments across multiple fields. Research in these areas continues to develop and become increasingly important as computer and electrical systems have become an integral part of everyday life.

The **Advances in Computer and Electrical Engineering (ACEE) Book Series** aims to publish research on diverse topics pertaining to computer engineering and electrical engineering. **ACEE** encourages scholarly discourse on the latest applications, tools, and methodologies being implemented in the field for the design and development of computer and electrical systems.

COVERAGE

- Digital Electronics
- Computer Science
- Applied Electromagnetics
- Power Electronics
- Sensor Technologies
- Computer Hardware
- Programming
- Microprocessor Design
- Optical Electronics
- Chip Design

IGI Global is currently accepting manuscripts for publication within this series. To submit a proposal for a volume in this series, please contact our Acquisition Editors at Acquisitions@igi-global.com or visit: http://www.igi-global.com/publish/.

Titles in this Series

For a list of additional titles in this series, please visit:
https://www.igi-global.com/book-series/advances-computer-electrical-engineering/73675

Novel Approaches to Information Systms Design
Naveen Prakash (Indraprastha Institute of Information Technology, Delhi, India) and Deepika Prakash (NIIT University, India)
Engineering Science Reference • © 2020 • 299pp • H/C (ISBN: 9781799829751) • US $215.00

IoT Architectures, Models, and Platforms for Smart City Applications
Bhawani Shankar Chowdhry (Mehran University of Engineering and Technology, Pakistan) Faisal Karim Shaikh (Mehran University of Engineering and Technology, Pakistan) and Naeem Ahmed Mahoto (Mehran University of Engineering and Technology, Pakistan)
Engineering Science Reference • © 2020 • 291pp • H/C (ISBN: 9781799812531) • US $245.00

Nature-Inspired Computing Applications in Advanced Communication Networks
Govind P. Gupta (National Institute of Technology, Raipur, India)
Engineering Science Reference • © 2020 • 319pp • H/C (ISBN: 9781799816263) • US $225.00

Tools and Technologies for the Development of Cyber-Physical Systems
Sergey Balandin (FRUCT Oy, Finland) and Ekaterina Balandina (Tampere University, Finland)
Engineering Science Reference • © 2020 • 300pp • H/C (ISBN: 9781799819745) • US $235.00

Handbook of Research on New Solutions and Technologies in Electrical Distribution Networks
Baseem Khan (Hawassa University, Hawassa, Ethiopia) Hassan Haes Alhelou (Tishreen University, Syria) and Ghassan Hayek (Tishreen University, Syria)
Engineering Science Reference • © 2020 • 439pp • H/C (ISBN: 9781799812302) • US $270.00

Major Applications of Carbon Nanotube Field-Effect Transistors (CNTFET)
Balwinder Raj (National Institute of Technical Teachers Training and Research, Chandigarh, India) Mamta Khosla (Dr. B. R. Ambedkar National Institute of Technology, Jalandhar, India) and Amandeep Singh (National Institute of Technology, Srinagar, India)
Engineering Science Reference • © 2020 • 255pp • H/C (ISBN: 9781799813934) • US $185.00

Neural Networks for Natural Language Processing
Sumathi S. (St. Joseph's College of Engineering, India) and Janani M. (St. Joseph's College of Engineering, India)
Engineering Science Reference • © 2020 • 227pp • H/C (ISBN: 9781799811596) • US $235.00

701 East Chocolate Avenue, Hershey, PA 17033, USA
Tel: 717-533-8845 x100 • Fax: 717-533-8661
E-Mail: cust@igi-global.com • www.igi-global.com

Table of Contents

Detailed Table of Contents

Chapter 1
    Julián Sierra-Pérez, Universidad Pontificia Bolivariana, Colombia
    Joham Alvarez-Montoya, Universidad Pontificia Bolivariana, Colombia

Strain field pattern recognition, also known as strain mapping, is a structural health monitoring approach based on strain measurements gathered through a network of sensors (i.e., strain gauges and fiber optic sensors such as FGBs or distributed sensing), data-driven modeling for feature extraction (i.e., PCA, nonlinear PCA, ANNs, etc.), and damage indices and thresholds for decision making (i.e., Q index, T2 scores, and so on). The aim is to study the correlations among strain readouts by means of machine learning techniques rooted in the artificial intelligence field in order to infer some change in the global behavior associated with a damage occurrence. Several case studies of real-world engineering structures both made of metallic and composite materials are presented including a wind turbine blade, a lattice spacecraft structure, a UAV wing section, a UAV aircraft under real flight operation, a concrete structure, and a soil profile prototype.

Chapter 2
    Feyzan Saruhan-Ozdag, Istanbul University-Cerrahpasa, Turkey
    Derya Yiltas-Kaplan, Istanbul University-Cerrahpasa, Turkey
    Tolga Ensari, Istanbul University-Cerrahpasa, Turkey

Intrusion detection systems are one of the most important tools used against the threats to network security in ever-evolving network structures. Along with evolving technology, it has become a necessity to design powerful intrusion detection systems and integrate them into network systems. The main purpose of this research is to develop a new method by using different techniques together to increase the attack detection rates. Negative selection algorithm, a type of artificial immune system algorithms, is used and improved at the stage of detector generation. In phase of the preparation of the data, information gain is used as feature selection and principal component analysis is used as dimensionality reduction method. The first method is the random detector generation and the other one is the method developed by combining the information gain, principal component analysis, and genetic algorithm. The methods were tested using the KDD CUP 99 data set. Different performance values are measured, and the results are compared with different machine learning algorithms.

Chapter 3

Currently, with the implementation of IoT, it is expected that medicine and health obtain a great benefit
derived from the development of portable devices and connected sensors, which allow acquiring and
communicating data on symptoms, vital signs, medicines, and activities of daily life that can affect health.
Despite the possible benefits of health services assisted by IoT, there are barriers such as the storage of
data in the cloud for analysis by physicians, the security and privacy of the data that are communicated,
the cost of communication of the data that is collected, and the manipulation and maintenance of the
sensors. This chapter intends to deploy and develop the context of the IoT platforms in the field of health
and medicine by means of the transformation of edge and fog computing, as intermediate layers that
provide interfaces between heterogeneous networks, networks inherited infrastructure, and servers in the
cloud for the ease of data analysis and connectivity in order to implement a structural health monitoring
based on IoT for application of early warning score.

Chapter 4

This chapter reviews the development of solutions related to the practical implementation of electronic
tongue sensor arrays. Some of these solutions are associated with the use of data from different
instrumentation and acquisition systems, which may vary depending on the type of data collected,
the use and development of data pre-processing strategies, and their subsequent analysis through the
development of pattern recognition methodologies. Most of the time, these methodologies for signal
processing are composed of stages for feature selection, feature extraction, and finally, classification or
regression through a machine learning algorithm.

Software-defined networks (SDN) is an emerging paradigm that has been widely explored by the research
community. At the same time, it has attracted a lot of attention from the industry. SDN breaks the integration
between control and data plane and creates the concept of a network operating system (controller). The
controller should be logically centralized, but it must comply with availability, reliability, and security
requirements, which implies that it should be physically distributed in the network. In this context, two
questions arise: How many controllers should be included? and Where should they be located? These
questions comprise the controller placement problem (CPP). The scope of this study is to solve the CPP

using the meta-heuristic Tabu search algorithm to optimize the cost of network operation, considering flow setup latency and inter-controller latency constraints. The network model presented considers both controllers and links as IT resources as a service, which allows focusing on operational cost.

Chapter 6

Pragathi Penikalapati, Vellore Institute of Technology, India
A. Nagaraja Rao, Vellore Institute of Technology, India

The compatibility issues among the characteristics of data involving numerical as well as categorical attributes (mixed) laid many challenges in pattern recognition field. Clustering is often used to group identical elements and to find structures out of data. However, clustering categorical data poses some notable challenges. Particularly clustering diversified (mixed) data constitute bigger challenges because of its range of attributes. Computations on such data are merely too complex to match the scales of numerical and categorical values due to its ranges and conversions. This chapter is intended to cover literature clustering algorithms in the context of mixed attribute unlabelled data. Further, this chapter will cover the types and state of the art methodologies that help in separating data by satisfying inter and intracluster similarity. This chapter further identifies challenges and Future research directions of state-of-the-art clustering algorithms with notable research gaps.

Chapter 7

Nancy E. Ochoa Guevara, Fundación Universitaria Panamericana, Colombia
Andres Esteban Puerto Lara, Fundación Universitaria Panamericana, Colombia
Nelson F. Rosas Jimenez, Fundación Universitaria Panamericana, Colombia
Wilmar Calderón Torres, Fundación Universitaria Panamericana, Colombia
Laura M. Grisales García, Fundación Universitaria Panamericana, Colombia
Ángela M. Sánchez Ramos, Fundación Universitaria Panamericana, Colombia
Omar R. Moreno Cubides, Fundación Universitaria Panamericana, Colombia

This chapter presents a study to identify with classification techniques and digital recognition through the construction of a prototype phase that predicts criminal behavior detected in video cameras obtained from a free platform called MOTChallenge. The qualitative and descriptive approach, which starts from individual attitudes, expresses a person in his expression, anxiety, fear, anger, sadness, and neutrality through data collection and feeding of some algorithms for assisted learning. This prototype begins with a degree higher than 40% on a scale of 1-100 of a person suspected, subjected to a two- and three-iterations training parameterized into four categories—hood, helmet, hat, anxiety, and neutrality—where through orange and green boxes it is signaled at the time of the detection and classification of a possible suspect, with a stability of the 87.33% and reliability of the 96.25% in storing information for traceability and future use.

Chapter 8

Mauricio Orozco-Alzate, Universidad Nacional de Colombia, Manizales, Colombia

The accurate identification of plant species is crucial in botanical taxonomy as well as in related fields such as ecology and biodiversity monitoring. In spite of the recent developments in DNA-based analyses for phylogeny and systematics, visual leaf recognition is still commonly applied for species identification in botany. Histograms, along with the well-known nearest neighbor rule, are often a simple but effective option for the representation and classification of leaf images. Such an option relies on the choice of a proper dissimilarity measure to compare histograms. Two state-of-the-art measures—called weighted distribution matching (WDM) and Poisson-binomial radius (PBR)—are compared here in terms of classification performance, computational cost, and non-metric/non-Euclidean behavior. They are also compared against other classical dissimilarity measures between histograms. Even though PBR gives the best performance at the highest cost, it is not significantly better than other classical measures. Non-Euclidean/non-metric nature seems to play an important role.

Chapter 9

Andres Esteban Puerto Lara, Fundacion Universitaria Panamericana, Colombia
Cesar Pedraza, Universidad Nacional de Colombia, Colombia
David A. Jamaica-Tenjo, Universidad Nacional de Colombia, Colombia

Each crop has their own weed problems. Therefore, to understand each problem, agronomists and weed scientists must be able to determine the weed abundance with the most precise method. There are several techniques to scouting, including visual counting for density or estimations for coverage of weeds. However, this technique depends by the evaluator subjectivity, performance, and training, causing errors and bias when estimating weeds abundance. This chapter introduces a methodology to process multispectral images, based on histograms of oriented gradients and support vector machines to detect weeds in lettuce crops. The method was validated by experts on weed science, and the statistical differences were calculated. There were no significant differences between expert analysis and the proposed method. Therefore, this method offers a way to analyze large areas of crops in less time and with greater precision.

Chapter 10

Rohit Rastogi, ABES Engineerig College, Ghaziabad, India
Devendra Kumar Chaturvedi, Dayalbagh Educational Institute, Agra, India
Mayank Gupta, Tata Consultancy Services, Noida, India
Heena Sirohi, ABES Engineering College, Ghaziabad, India
Muskan Gulati, ABES Engineering College, Ghaziabad, India
Pratyusha, ABES Engineerig College, Ghaziabad, India

Increasing stress levels in people is creating higher tension levels that ultimately result in chronic headaches. To get the best result, the subjects are divided into two groups. One group will be introduced under EMG, and the other will be handled under GSR. The change in the behaviour of subject (i.e., the change in the stress level) is measured at the intervals of one month, three months, six months, and twelve months. The main aim of the research is to study the effects of tension type headache using biofeedback therapies on various modes such as audio modes, visual modes, and audio-visual modes. The groups were randomly allocated for galvanic skin resistance (GSR) therapies, and the other one was control group (the group that was not under any type of allopathic or other medications). Except for the control group, the groups were treated in a session for 20 minutes in isolated chambers. The results were recorded over a specific period of time.

Chapter 11

Santiago Morales, Universidad Nacional de Colombia, Colombia
César Pedraza Bonilla, Universidad Nacional de Colombia, Colombia
Felix Vega, Universidad Nacional de Colombia, Colombia

Traffic volume is an important measurement to design mobility strategies in cities such as traffic light configuration, civil engineering works, and others. This variable can be determined through different manual and automatic strategies. However, some street intersections, such as traffic circles, are difficult to determine their traffic volume and origin-destination matrices. In the case of manual strategies, it is difficult to count every single car in a mid to large-size traffic circle. On the other hand, automatic strategies can be difficult to develop because it is necessary to detect, track, and count vehicles that change position inside an intersection. This chapter presents a vehicle counting method to determine traffic volume and origin-destination matrix for traffic circle intersections using two main algorithms, Viola-Jones for detection and on-line boosting for tracking. The method is validated with an implementation applied to a top view video of a large-size traffic circle. The video is processed manually, and a comparison is presented.

Chapter 12

Jessica Gissella Maradey Lázaro, Universidad Autónoma de Bucaramanga, Colombia
Carlos Borrás Pinilla, Universidad Industrial de Santander, Colombia

Variable displacement axial piston hydraulic pumps (VDAP) are the heart of any hydraulic system and are commonly used in the industrial sector for its high load capacity, efficiency, and good performance in the handling of high pressures and speeds. Due to this configuration, the most common faults are related to the wear and tear of internal components, which decrease the operational performance of the hydraulic system and increase maintenance costs. So, through data acquisition such as signals of pressure and the digital processing of them, it is possible to detect, classify, and identify faults or symptoms in hydraulic machinery. These activities form the basis of a condition-based maintenance (CBM) program. This chapter shows the developed methodology to detect and classify a wear fault of valve plate taking into account six conditions and the facilities providing by wavelet analysis and ANNs.

Preface

.

Over the years, engineering has become a very useful tool for bringing ideas to the real world. Scientifics from the fundamental sciences together with engineers have worked in the development of several inventions which has changed the world and the way as humans interact all days, as one example, it is possible to cite the invention of the transistors in 1947 (Lojek Bo, 2007; Brinkman, 1997) which result in the necessary step to accelerate the development of electronics applications and the use of gadgets and devices in the daily life. In fact, an electronic device consists in passive and active elements and memories which are made of semiconductors and includes a big number of transistors in ICs with a very large scale of integration.

Currently, humanity depends significantly on electronic devices and processor-based systems, some examples are the modern medicine, on-line monitoring systems, home appliances, office gadgets, storage systems, audio and video systems, industrial automation and automatic control applications, surveillance systems, meteorological and oceanographic monitoring systems and precision farming devices, among others.

The development of this simple device (transistor) has brought revolutionary changes affecting the human interaction and relegating some activities to the machines and automatic systems. Of course, electronic and computing capabilities developments are hand in hand and requires together to answer the needs of the industry and in general of the people. One of these activities that human can perform in an inherently way is pattern recognition, however in some cases, machines have proven better performance due to the human nature which cannot preserve his concentration for long periods of time, are limited by its senses and can avoid relevant information due to the high number of variables and sources to process by the human brain. In this sense, artificial pattern recognition is a need and can help to the humans with some of these difficult tasks. Here, engineering appear again to provide solutions to apply pattern recognition tasks by using and developing new resources for improving this kind of work. In fact, pattern recognition comes from engineering and has proved to work in many different fields where classification is a need.

Pattern recognition can be understood as a discipline where the goal is to classify data/objects into several categories or classes (Theodoridis, 2003). These data/objects are in most cases electrical signals collected by data acquisition systems through sensors and provided as data in digital or analog formats, however, its processing requires in most of the cases to be developed digitally, this means that analog signals from sensors require a data conversion from analogic to digital to be processed by a micro-controlled or a micro-processed system. Some examples of sensors used in applications for pattern recognition includes cameras for image processing (Fu, 1976; Hu. 1962), piezoelectric sensors (Vitola, 2017), fiber Bragg gratings (FBGs) and strain sensors (Sierra, 2013), accelerometers (Pires, 2019), electronic tongues (León, 2019), electronic noses (Fu, 2007) for waveform analysis (Dagamseh, 2010; Nixon, 1995; Kuhl, 2007).

To analyze these data, it is necessary to consider if the data come from different sources or a single source to define the normalization to apply (Anaya, 2015). At this point a multivariable analysis can be required to provide the same relevance to all data and the use of sensor data fusion strategies (Tibaduiza, 2013). Once data are organized and normalized, data reduction and feature extraction are basic for obtaining the most relevant information in a reduced version of the data (Anaya, 2016). These steps allow to the classification techniques to obtain the patterns and perform the classification in a better way by avoiding noise and unnecessary data that can increase the computational cost of the solution. The last step in the analysis of the data is concerning with automatic discovery of regularities in data by using computer algorithms (Bishop, 2006) and the use of this patterns for the classification tasks. In general the process of pattern recognition includes two steps well known as training and testing. The first is required to provide information about the patterns, and the second is used in practice to detect and classify the patterns detected in the new inputs according to the training process. According to training process, it is possible to find supervised and unsupervised learning approaches.

There are many applications of the pattern recognition in multiple areas, following are some common areas were pattern recognition has been used:

- Structural health monitoring.
- Speech recognition.
- Data Mining.
- Industrial automation.
- Cell classification.
- Visual pattern recognition.
- Handwritten digits and words.

In general, it is possible to use pattern recognition anywhere classification task is required.

THE CHALLENGES

Data is now in all the applications, and its proper analysis can bring multiple benefits in the decision-making process. It is very common that data from the sensors require in most of the cases the use or development of multiple strategies to analyze this information, however, real implementations require that these strategies can be developed to be implemented as portable solutions with the limitations of the current hardware/software. In this sense, the use of sensor data fusion, pre-processing strategies, data reduction, feature extraction, and pattern recognition approaches can provide very useful tools for data analysis and the decision-making process.

In general terms, pattern recognition is based on the analysis of data by defining a baseline or a pattern that can be used as an element to compare changes in the new data and define some characteristics or for obtaining information about the current state of a process. This is based on the use of data from experiments and presents high challenges in its application due to the problems associated with the acquisition, noise and other problems that can be solved by different strategies and are currently a focus of research.

It goes without saying that, continuous development of the technology brings further new challenges to the pattern recognition task. This is that the development of new sensors provides new information to process which is often added to big multisensory networks adding variables to analyze which requires to increase the computing capabilities. In this sense, is now open among others the following issues:

- Development of Bioinspired algorithms to imitate the way as the nature process the information. These approaches could allow reducing the computational cost by optimizing the features to be analyzed by the pattern recognition strategies.
- Development of pre-processing techniques to improve the accuracy of the current approaches for classification.
- Development of more light strategies that can be used in embedded systems.

-Increasing the computing capabilities, this means, improves the way as the computers process the information, some examples and new directions the following elements have been used or still in development:

- Parallel computing.
- Big data analysis.
- Quantum computing.
- Cluster computing.
- Massive storage.
- Development of more complex embedded systems.

Although these elements have some current solutions, the economic cost still high for its implementation and use. These challenges open the door to researchers from different fields to carry out further research on these topics to provide better tools for the growth of pattern recognition in real applications.

ORGANIZATION OF THE BOOK

This book is aimed to provide concepts on different strategies of pattern recognition algorithms in industrial and research applications. In the same way to provide examples of applications and results in different areas such as electronics, structural Health Monitoring, automation, Computation, condition monitoring, IoT, etc. It is expected that the book can be used for students of courses in the area of data analysis and pattern recognition as well as researchers and engineers that are developing applications where a proper treatment of the data is a need to ensure an adequate analysis and result in classification tasks.

The book is organized into 12 chapters that cover different engineering areas and their applications to different disciplines. A brief description of each of the chapters follows:

Chapter 1 is aimed to study the correlations among strain readouts by using machine learning approaches for damage detection applications. These are evaluated in some real cases of studies including a wind turbine blade, a lattice spacecraft structure, a UAV wing section, a UAV aircraft under real flight operation, a concrete structure and a soil profile prototype.

Chapter 2 presents the development of a methodology for intrusion detection to increase the attack detection rates. The methodology includes the use of negative selection in artificial immune systems, information gain, principal component analysis and a genetic algorithm to propose different techniques. To validate the developed methodology, different machine algorithms are evaluated by using the KDD CUP 99 dataset.

Chapter 3 is aimed to present a brief review of the Internet of Things applied to health monitoring, including the current challenges of IoT, a description of the infrastructure and architecture and some applications.

Chapter 4 is devoted to exploring different pattern recognition techniques used to process information from electronic tongues to classify liquid substances. This start by introducing different kind of electronic tongues and the way as they interact with the substances to continue with the pre-processing data strategies and finishing with the pattern recognition methods applied for the classification of the liquid substances.

Chapter 5 is oriented to propose an approach to find a solution to the Controller Placement Problem in a Software-Defined Network with the Tabu Search heuristic optimization method. The algorithm considers OPEX cost as the optimization objective, as well as flow setup and inter-controller latency, and feasibility constraints. This is evaluated using real-life topologies from the Topology Zoo which is an ongoing project to collect data network topologies from around the world.

Chapter 6 shows the problem of analyses mixed data from different sources and shows some solutions in this direction and means of a review of the different solutions in the literature.

Chapter 7 presents a system for gesture recognition, to criminal behavior called MOTChallenge. This considers individual attitudes for each person, as his expression, anxiety, fear, anger, sadness and neutrality, through data collection, and feeding of some algorithms detection by using information from video cameras obtained from a free platform for assisted learning.

Chapter 8 is oriented to present an experimental comparison of the weighted distribution matching similarity and the Poison-Binomial Radius measures between histograms for recognizing many classes of plant leaves.

Chapter 9 presents a methodology to process multispectral images, based on a combination of a histogram of oriented gradients as feature extractor and support vector machines as a classifier in order to detect weed and consequently get its coverage inside a determined crop grid.

Chapter 10 is devoted to studying the effects of tension-type headache using biofeedback therapies on various modes such as audio modes, visual modes, and audiovisual modes, to do this, a study applied to 90 people was performed and correlations between the variables of the experiment are analyzed.

Chapter 11 is focused to present a vehicle counting method to determine traffic volume and origin-destination matrix for traffic circle intersections. To do that, authors use Viola-Jones algorithm for detection and on-line boosting algorithm for tracking. Methodology is validated with data from of a large-size traffic circle.

Chapter 12 is oriented to present a methodology to detect, classify and analyze off-line faults of a hydraulic pump with axial pistons corresponding to different valve plate wear conditions. The methodology uses the pressure signals and includes the use of spectral analysis and artificial neural networks.

Diego Alexander Tibaduiza Burgos
Universidad Nacional de Colombia, Colombia

Maribel Anaya Vejar
Universidad Santo Tomás, Colombia

Francesc Pozo
Universitat Politècnica de Catalunya, Spain

REFERENCES

Anaya, M. (2016). *Design and validation of a structural health monitoring system based on bio-inspired algorithms* (PhD. Thesis). Universitat Politecnica de Catalunya.

Anaya, M., Tibaduiza, D. A., Forero, E., Castro, R., & Pozo, F. (2015). An acousto-ultrasonics pattern recognition approach for damage detection in wind turbine structures. *20th Symposium on Signal Processing, Images and Computer Vision (STSIVA)*. 10.1109/STSIVA.2015.7330419

Bishop, M. (2006). *Pattern recognition and machine learning*. Springer.

Brinkman, W. F., Haggan, D. E., & Troutman, W. W. (1997). A history of the invention of the transistor and where it will lead us. *IEEE Journal of Solid-State Circuits, 32*(12), 1858–1865. doi:10.1109/4.643644

Dagamseh, A., Bruinink, C., Kolster, M., Wiegerink, R., Lammerink, T., & Krijnen, G. (2010). Array of biomimetic hair sensor dedicated for flow pattern recognition. *2010 Symposium on Design Test Integration and Packaging of MEMS/MOEMS (DTIP)*, 48-50.

Fu, J., Li, G., Qin, Y., & Freeman, W. (2007). A pattern recognition method for electronic noses based on an olfactory neural network. *Sensors and Actuators. B, Chemical, 125*(2), 489–497. doi:10.1016/j.snb.2007.02.058

Fu, K.-S. (1976). Pattern Recognition and Image Processing. *IEEE Transactions on Computers, C-25*(12), 1336–1346. doi:10.1109/TC.1976.1674602

Hu, M. (1962). Visual pattern recognition by moment invariants. *I.R.E. Transactions on Information Theory, 8*(2), 179–187. doi:10.1109/TIT.1962.1057692

Kuhl, M., Neugebauer, R., & Mickel, P. (2007). Methods for a multisensorsystem for in-line process- and quality monitoring of welding seams using fuzzy pattern recognition. *2007 IEEE Conference on Emerging Technologies and Factory Automation (EFTA 2007)*, 908-911. 10.1109/EFTA.2007.4416879

Leon-Medina, J. X., Cardenas-Flechas, L. J., & Tibaduiza, D. A. (2019). A data-driven methodology for the classification of different liquids in artificial taste recognition applications with a pulse voltammetric electronic tongue. *International Journal of Distributed Sensor Networks, 15*(10). doi:10.1177/1550147719881601

Lojek, B. (2007). Shockley Semiconductor Laboratories. In History of Semiconductor Engineering. Springer.

Nixon, O., & Nathan, A. (1995). Magnetic Pattern Recognition Sensor Arrays using CCD Readout. *ESSDERC '95: Proceedings of the 25th European Solid State Device Research Conference*, 273-276.

Pires, I.M., Garcia, N.M., Pombo, N., Flórez-Revuelta, F., Spinsante, S., Canavarro Teixeira, M., & Zdravevski, E. (2019). *Pattern recognition techniques for the identification of Activities of Daily Living using mobile device accelerometer*. PeerJ Preprints 7:e27225v2.

Sierra, J., Güemes, A., & Mujica, E. (2013). Damage detection by using FBGs and strain field pattern recognition techniques. *Smart Materials and Structures, 22*(2).

Theodoridis, S., & Koutroumbas, K. (2003). *Pattern Recognition* (2nd ed.). Elsevier Acedemic press.

Tibaduiza, D. A. (2013). *Design and validation of a structural health monitoring system for aeronautical structures* (PhD thesis). Universitat Politecnica de Catalunya.

Vitola, J., Pozo, F., Tibaduiza, D. A., & Anaya, M. (2017). Distributed Piezoelectric Sensor System for Damage Identification in Structures Subjected to Temperature Changes. *Sensors*, *17*(6), 1252. doi:10.339017061252 PMID:28561786

Acknowledgment

The editors would like to acknowledge the help of all the people involved in this project and, more specifically, to the authors and reviewers that took part in the review process. Without their support, this book would not have become a reality.

First, the editors would like to thank each one of the authors for their contributions. Our sincere gratitude goes to the chapters' authors who contributed their time and expertise to this book.

Second, the editors wish to acknowledge the valuable contributions of the reviewers regarding the improvement of quality, coherence, and content presentation of chapters. Most of the authors also served as referees; we highly appreciate their double task.

In addition, Maribel Anaya and Diego Tibaduiza would like to acknowledge to their sons Juan Diego and Liam Alexander for their patient and endless love.

Diego Alexander Tibaduiza Burgos
Universidad Nacional de Colombia, Colombia

Maribel Anaya Vejar
Universidad Santo Tomás, Colombia

Francesc Pozo
Universitat Politècnica de Catalunya, Spain

Chapter 1
Strain Field Pattern Recognition for Structural Health Monitoring Applications

Julián Sierra-Pérez
Universidad Pontificia Bolivariana, Colombia

Joham Alvarez-Montoya
https://orcid.org/0000-0003-3174-3977
Universidad Pontificia Bolivariana, Colombia

ABSTRACT

Strain field pattern recognition, also known as strain mapping, is a structural health monitoring approach based on strain measurements gathered through a network of sensors (i.e., strain gauges and fiber optic sensors such as FGBs or distributed sensing), data-driven modeling for feature extraction (i.e., PCA, nonlinear PCA, ANNs, etc.), and damage indices and thresholds for decision making (i.e., Q index, T2 scores, and so on). The aim is to study the correlations among strain readouts by means of machine learning techniques rooted in the artificial intelligence field in order to infer some change in the global behavior associated with a damage occurrence. Several case studies of real-world engineering structures both made of metallic and composite materials are presented including a wind turbine blade, a lattice spacecraft structure, a UAV wing section, a UAV aircraft under real flight operation, a concrete structure, and a soil profile prototype.

INTRODUCTION

Structural health monitoring (SHM), as a discipline named in this way, started about four decades ago (Boller, 2008). Since then, several definitions have been proposed being the one by Boller (2008), the most accepted one. According to this definition, SHM consists in the integration of sensing and/or actuating devices in order to record, analyze, localize and predict the damage and infer the load conditions of a structure in such a way that nondestructive testing (NDT) becomes an integral part of the structure (Boller, 2008).

DOI: 10.4018/978-1-7998-1839-7.ch001

Several classification schemes have also been stated, however, the most complete includes seven basic levels serving as the objectives to achieve by an SHM system (Farrar & Worden, 2007). In the following proposed classification, load monitoring has been included in order to match with the broad definition of SHM proposed by Boller (2008).

Level 1: Monitor and record the loads.
Level 2: Determine the existence of the damage.
Level 3: Determine the location of the damage.
Level 4: Determine the kind of damage.
Level 5: Quantify the severity of the damage.
Level 6: Estimate the remaining lifetime.
Level 7: Develop capacities for self-diagnosis and self-healing.

Up to date, the scientific community has provided laboratory solutions under simulated operational conditions mainly for the first five levels of this damage assessment scheme. Among the reasons for a slow development of real-world SHM applications is the uncertainty involving damage assessment. The major progression in this aspect has been accomplished in the context of condition monitoring in rotating machinery due to the large amount of available data and the predictable nature of the damages typically presented in this type of structures (Lopez & Sarigul-Klijn, 2010). This chapter is mainly focused on the first two levels of the scheme, namely, load monitoring and damage detection, where most of the studies have been focused on.

The importance of damage detection in structures is related to the fact that materials deteriorate with time and usage due to several types of loads. It is paramount to guarantee in all engineering structures their serviceability and reliability during their lifetime. In addition, damage-tolerant designs, specifically used in the aerospace field, require the proper identification of damages since they can grow until a state where the integrity of the structure is not at risk. This led to lighter structures and more efficient designs (Rocha et al., 2013).

Historically, early studies within the context of damage detection focused on physical models such as stiffness and modal parameter determination. Such approaches rely on deterministic modeling where all variables are assumed to be measurable and several uncertainties are not directly included in the model. Consequently, it is complex to determine the reliability of such estimation. This is even more difficult when considering the tendency in several industries to use sophisticated designs and materials in order to increase efficiency and performance. Thus, the physical modeling of the actions and responses in the state-of-the-art engineering structures is difficult to accomplish.

Here, is where statistical pattern recognition comes to play, since it is possible to use approaches that develop data-driven models through machine learning algorithms instead of using complex physical models. Data-driven models rely on experimental data for *training* or *learning* the current state of a structure. These approaches have proved to be robust in the damage assessment tasks. However, some difficulties are posed such as variability related to the physical system and the environment where it operates and the management of a large amount of data (Figueiredo & Santos, 2018).

In the damage detection process in an online and automated way using statistical pattern recognition, it is common that every SHM system addresses four stages: operational evaluation, data acquisition, feature extraction and statistical modeling for decision-making. Operational evaluation aims to customize the damage detection process by considering the limitations based on the characteristics of the structure to be monitored. Data acquisition refers to the selection of the excitation methods, sensor architecture and, in general, the design of the hardware involved in the process of gathering information about the structure. Feature extraction corresponds to obtain damage-sensitive features from the gathered information in order to ease the analysis process. Finally, statistical modeling is performed for decision-making about whether the structure is undamaged or damaged (Farrar & Worden, 2012).

This chapter lays in the context of feature extraction and statistical modeling where pattern recognition is mostly used. Among the available options for data acquisition and information, gathering strain measurements point out due to their intrinsic relationship with damage occurrence, sensitivity to small-scale damages and profuse available type of sensors.

The aim of this chapter is to present an approach that has been called *strain field pattern recognition*, also known as *strain mapping*. This approach is based on strain measurements gathered through a network of sensors, data-driven modeling for feature extraction and damage indices and thresholds for decision-making. Several case studies have been successfully applied from engineering structures under simulated operational conditions to complex structures under real operational conditions. Additionally, this concept has been applied to several fields including wind power, aerospace and civil engineering.

This chapter is divided into four sections. After introduction, the motivation for using strain data for damage detection is reviewed discussing the advantages and drawbacks of strain as a feature for damage detection. This discussion includes strain fundamentals, a brief definition of the strain field pattern recognition and a review of the state-of-the-art strain sensors. Then, strain field pattern recognition framework is presented as a set of steps to process strain data for damage detection. Finally, case studies consisting of real-world engineering structures are shown and compared using several performance metrics.

MOTIVATION FOR USING STRAIN DATA FOR DAMAGE DETECTION

Different SHM techniques offer promising perspectives, indistinctly of the technique, the fundamental task is to determine changes in a physical parameter of the structure as a result of a possible anomaly in this. This is where the data acquisition stage comes to play by providing methods for gathering information about the structure.

Data acquisition stage is one of the most critical ones, select the type of sensor, the number and the location is a non-trivial task, which requires much knowledge as possible from the operational evaluation stage. Currently, due to the advances in microelectronics and smart materials, there is a surplus in the type of sensors that can be used as a basis for SHM systems. According to this, SHM techniques can be classified into (Ostachowicz et al., 2019):

- Vibration-based monitoring.
- Strain monitoring.
- Elastic waves-based monitoring.
- Electromechanical impedance-based monitoring.
- Comparative vacuum monitoring (CVM).

Regardless of the used technique, all have their own advantages and limitations and, therefore, it is required to have knowledge about the system or structure under study in order to select the suitable technique. Such techniques can also be classified according to whether an external stimulus is required.

Active sensing approaches consider the use of physical variables that requires actively interrogating the structure like the most-common NDT techniques such as ultrasonic testing or electromagnetic waves. On the other hand, passive sensing approaches bank on monitoring a structural/material variable without imparting energy to the structure. Vibration-based monitoring and strain monitoring can be classified mainly as passive while electromechanical impedance-based monitoring and CVM as active. Elastic waves-based monitoring can be passive as in acoustic emission method or active as in acousto-ultrasonics (Giurgiutiu, 2016).

Vibration-based monitoring, among the earliest proposed techniques, lacks providing reliable solutions for SHM in real-world operating structures due to the uncertainty associated with noise, operational and environmental variability and excitation mechanisms. In general, the high noise and harsh conditions under operation in conjunction with the large amount of time required for the measurements hinder the damage assessment tasks.

Indeed, the issue associated with excitation mechanisms is also a common disadvantage in the active sensing approaches, the fact that it is required to interrogate the structure lead to constraint these inspections to an offline way (not during operation). However, it is important to highlight that in the first stage, operational evaluation, it should be assessed whether an online damage detection is required. Offline damage detection is suitable in most cases due to inspections are performed on-demand and real-time damage detection is not the priority. Indeed, NDT procedures, which have been the way for detecting damages over years is performed offline and on-demand.

In addition, active sensing approaches often are more invasive than passive sensing since they may alter the performance of the structure, several sensors like piezoelectric and piezoresistive sensors attached to the structure are required, which increases the weight and may affect the aerodynamic performance in case of aerospace structures. Active approaches require the application of energy; therefore, in the light of power consumption, these techniques are less favorable than passive. Although these limitations are being addressed by the continuous development of new advanced sensors (Qing et al., 2019).

However, passive sensing approaches are not a panacea. They often rely on excitation provided by the operation since they lack their own. Usually, active sensing can detect minor damages in a more accurate way. There is no guarantee that the excitations are persistent or at the level required to allow the detection of several types of damages in an automated way. Operational and environmental variability is also an important concern because they may change the response of the structure under the same excitation or input. This is quantified or counteracted by measuring the variables that can change the response of the structure, however, there are always unmeasurable variables that may have an impact (Haroon, 2009).

One of the most researched passive sensing techniques is strain monitoring. Strain is a variable that has played an important role in the mechanics of materials field, that is why, along the years, several ways of measuring strain, mathematical models and experience have been acquired. Over other SHM techniques, strain-based SHM offers several advantages such as the different types of sensors available (e.g. resistance strain gauges, fiber optic sensor, etc.), which provide interesting capabilities with respect to other techniques, the high sensitivity to small-scale damages, the sensitivity to almost all the damages and the capability of performing load monitoring, fatigue and life time management.

Strain-based damage detection relies on the redistribution of loads and strains in a structure because of a damage occurrence. The global strain field in a structure is only modified by large-scale damage. That is why strain-based SHM has been considered as a local approach for damage detection, that is to say, only damages in the vicinity of the sensor are possible to be detected.

Usually, strain-based methods have been used only for monitoring structural *hotspots,* namely, where the probability of a damage occurrence is higher. However, this leads to uncertainty about how these hotspots should be selected accordingly, which is even more complex when dealing with composite materials due to their anisotropic nature. Additionally, strain changes fade away quickly; therefore, they can be easily masked by temperature, load or any other environmental change.

An example of this local approach of strain monitoring is the so-called technique *damage-induced strains* (Güemes et al., 2018). It has been demonstrated that new internal strains appear in a structure due to damages and it is possible to detect such induced strains by means of sensors, specifically by using distributed sensing. This technique can be applied to unloaded structures since strains are zero everywhere except in the damaged region. It has been proved to be robust in the detection of delamination caused by impacts and delamination at laminate edges in composite materials (Güemes et al., 2016).

However, global damage detection can be performed by using strain data and pattern recognition techniques. Local damages produce changes in the local stiffness that can be detected by using multiple strain sensors distributed along the structure. Through the study of the correlations by means of robust techniques rooted in the machine learning field, it is possible to reconstruct the strain field of the structure and infer some change in the global behavior associated with damage occurrence. This approach is known as strain field pattern recognition (Güemes et al., 2018; Sierra-Pérez, 2014).

In the following subsections the theoretical fundamental of strains will be reviewed as a basis for then introducing the principles of the strain field pattern recognition technique to detect damages and, finally, providing an overview of the state-of-the-art sensors to measure strains.

Strain Fundamentals

Deformation occurs when a solid body is subjected to external loads (e.g. mechanical or thermal loads). It implies a change in the dimensions of a solid body and, therefore, a strain. Strain field, on the other hand, refers to the distribution of strains in a region of a body. From a more formal point of view, the strain field is a tensor field that assigns a particular strain tensor to each point of the solid.

In engineering, two different types of strains are considered: normal and shear strains, denoted by ε and γ, respectively. The normal strain is formally defined as the rate of change of the length of a body subjected to stress in a defined direction and the shear strain refers as the angular distortion of a body (Popov, 1998).

$$\varepsilon = \frac{L - L_0}{L_0} = \frac{\Delta L}{L_0}.$$

Additionally, engineering shear stress is expressed as:

$$\gamma_{xy} = \theta + \beta,$$

where θ and β are defined as the change in the angles between the deformed lines of the body and the original, horizontal and vertical lines, respectively. In some cases, where strain is much larger, other definitions of strains are used. True strain, $\bar{\varepsilon}$, is one of them. Other theories for large strain include Green's strain, Almansi's strain, etc (Boresi & Schmidt, 2003).

However, this general definition may seem *superficial* from the point of view of continuum mechanics. The same region of a material regardless size can be simultaneously stressed and deformed in different directions at the same time; therefore, a single vector cannot describe the strain state in some point of a body. To take into consideration this state, the state of the medium around a particular point or region, must be represented by the strain tensor, both for infinitesimal theories as to the finite strain theories.

The theory of stress in a continuous medium is based on Newton's laws and the theory of strain is based only on geometric concepts. Therefore, both theories are independent of the material behavior and, consequently, can be applied to the study of any material. Both theories are equivalent from a mathematical point of view regardless of their different basis derivation.

Since the strains vary from point to point in a medium, creating in that way a strain field, the mathematical definition of strain must be related to an infinitesimal element. When a fixed point inside a body suffers a displacement, this may be a consequence of deformation or rigid body motion. The deformation is a consequence of internal stresses, external forces or other causes like thermal expansion or contraction, humidification, chemical reactions, etc. On the other hand, for a rigid body motion, forces are not necessarily involved, unless, there exist acceleration.

In a general way, the considerations about the deformation associated with the displacement must be considered since the way the material deforms influences the way in which, internal forces are distributed inside the body. There are some exceptions for this general rule, for instance, if the problem under study is a fully statically determinate problem.

If the continuum hypotheses are accepted, namely, the macroscopic behavior of the material does not change with the size, the internal forces can be fully defined by the stress tensor and, in the same way, deformations can be defined by the strain tensor. The strain tensor is mathematically equivalent to the stress tensor. The strain tensor is a symmetric tensor used to characterize the changes in the shape of a body. In three dimensions, the strain tensor of second order can be expressed, in general form, as:

$$\varepsilon = \begin{bmatrix} \varepsilon_{xx} & \varepsilon_{xy} & \varepsilon_{xz} \\ \varepsilon_{yx} & \varepsilon_{yy} & \varepsilon_{yz} \\ \varepsilon_{zx} & \varepsilon_{zy} & \varepsilon_{zz} \end{bmatrix}.$$

In this tensor, each one of the components is a function whose domain is the set of points of the body where it is desired to characterize the strain. The normal strain referred in is a general expression for an arbitrary direction and can be defined for the three principal directions as:

$$\varepsilon_{xx} = \frac{\partial u_x}{\partial x}, \varepsilon_{yy} = \frac{\partial u_y}{\partial y}, \varepsilon_{zz} = \frac{\partial u_z}{\partial z}$$

The shear strain defined in can be expressed for the three main planes as:

$$\gamma_{xy} = \gamma_{yx} = \frac{\partial u_y}{\partial x} + \frac{\partial u_x}{\partial y}, \gamma_{yz} = \gamma_{zy} = \frac{\partial u_y}{\partial z} + \frac{\partial u_z}{\partial y}, \gamma_{zx} = \gamma_{xz} = \frac{\partial u_z}{\partial x} + \frac{\partial u_x}{\partial z}.$$

In order to correlate external forces and stresses with displacements and strain, a constitutive law must be used. In constitutive laws, the rheological behavior of the material is preponderant due to the relations between the forces and deformations must be established. However, the phenomena that influence the material's behavior are very complex and occur at the level of atoms, molecules, crystals, etc. Therefore, quantitative descriptions based on elementary interactions are still to date a relatively open field in material science.

The most usual constitutive law for elastic materials is the generalized Hooke's law. According to the stated by this formulation, there exists a linear relationship between the applied stress and resulting strain. In the simplest case, Hooke's law can be written as a set of six basic relationships between a general state of stress and strain. In the case of an orthotropic material, which has different properties in the orthogonal directions, nine independent material constants are required, while, for an isotropic material only two. For a total anisotropic material, up to 21 independent constants are needed.

In the case of a continuous material, the Hooke's law can be expressed as in terms of the strain sensor:

$$\varepsilon = -s\sigma,$$

where s is the compliance tensor, which represents the inverse of the stiffness tensor. In the case of isotropic materials, Hooke's law can be expressed in matrix form as:

$$
\begin{bmatrix} \varepsilon_{xx} \\ \varepsilon_{yy} \\ \varepsilon_{zz} \\ \gamma_{xy} \\ \gamma_{xz} \\ \gamma_{yz} \end{bmatrix} =
\begin{bmatrix}
1/E & -v/E & -v/E & 0 & 0 & 0 \\
-v/E & 1/E & -v/E & 0 & 0 & 0 \\
-v/E & -v/E & 1/E & 0 & 0 & 0 \\
0 & 0 & 0 & 2(1+v)/E & 0 & 0 \\
0 & 0 & 0 & 0 & 2(1+v)/E & 0 \\
0 & 0 & 0 & 0 & 0 & 2(1+v)/E
\end{bmatrix}
\begin{bmatrix} \sigma_{xx} \\ \sigma_{yy} \\ \sigma_{zz} \\ \tau_{xy} \\ \tau_{xz} \\ \tau_{yz} \end{bmatrix},
$$

where E is the Young's modulus, v the Poisson's ratio, σ represent normal stresses in different directions and τ shear stress in different main planes.

The distribution of the elastic stress/strain across a section or a region body may be uniform or may vary in a regular or nonregular way. In cases where the variation occurs suddenly, in the very short distance, the intensity of stress/strain increases abruptly. This condition is known in the literature as *stress concentration*, however, as mention before, regardless of the constitutive law used in order to correlate stress and strain responses in a material, always exist a relationship between these. For this reason, *stress concentration* and *strain concentration* can be considered as equivalent terms.

Commonly, the strain concentration is due to some sort of *strain raiser* like holes, small contact points for applied loads, screw threads, abrupt changes in the cross-section or, in general, any type of local irregularity or discontinuity such as inclusions, corrosion, cracks, dents, etc. The term stress/strain gradient is used to indicate the rate of increase as a stress raiser approached. As an example, in Figure 1 is shown the strain field change produced by a hole in an aluminum plate under tension load.

The maximum intensity of stress/strain concentration can be calculated analytically, numerically (e.g. finite element modeling, FEM) or by means of experimental techniques such as photoelasticity or direct strain measurements. In practical problems involving strain concentration, the state of strain

Figure 1. Strain distribution from a finite-element simulation of an aluminum plate with a hole under tension load

near to a defect has a complex nature and few and limited analytical solutions exist. Usually, the stress concentration is expressed in terms of the stress concentration factor K_t, which is explained as the ratio between the actual stress divided the nominal stress. An analogous factor may be defined for the strain concentration factor; however, it is not common in the literature.

As shown before, it is possible to estimate the stress/strain concentration in the region close to a discontinuity, in other words, the stress/strain concentration in the near field. However, when moving away from the discontinuity to the far field the scenario changes. The most famous study about far field stress/strain concentration was developed by Saint-Venant. The Saint-Venant principle stated that the way on which a force is applied to a body only influences the stresses/strains near the region where that force is applied. More important, this principle is also valid for the disturbances caused by changes in cross-section. Over the years other authors like Von Mises have demonstrated that there are some exceptions for particular cases where Saint-Venant's principle is not strictly true (Boresi & Schmidt, 2003).

Strain Field Pattern Recognition

Strain field pattern recognition focuses on studying changes in the local strain field. For example, by means of analyzing the response of two sensors, it is possible to detect small changes in the local stiffness as can be seen in Figure 2.

Some researchers have used this approach as an SHM methodology. Menendez & Güemes (2006) demonstrated in a composite structure provided with several fiber Bragg Grating (FBG) sensors attached along a bonded stiffener. As the bond breaks, local strains around the FBG sensors were disturbed. By studying the *differential strains* between different couples of sensors, they were able to assess debonding.

Li et al. demonstrated the application of differential strains in health assessment of bonded composite repairs in several experimental cases (Li, et al., 2006a; Li et al., 2006b). Similarly, Fernández-Lopez et al. (2007) used differential strains for detecting debonding in stiffened plates made of composite materials by detecting changes in the strain field distribution. The authors studied the difference of strains between neighbor sensors.

Figure 2. Differences in slope between two neighboring sensors representing a change in the strain field due to different damage conditions (five cumulatively induced holes and one artificial crack) for a same operational condition of an aluminum beam under dynamic loading (Sierra-Pérez et al., 2018)

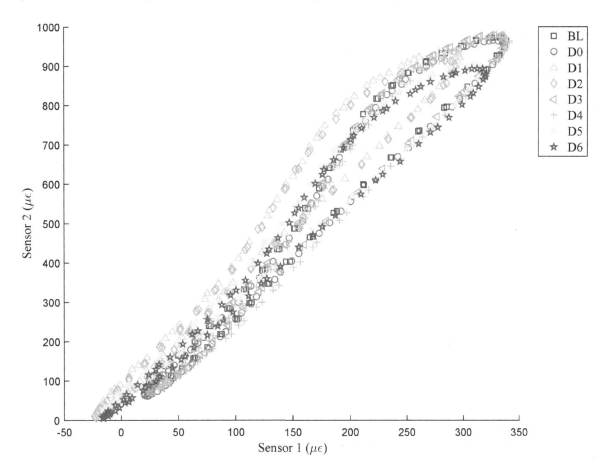

The authors of this chapter have worked in developing techniques to perform damage detection by studying all the possible correlations between all the possible couples of sensors to rebuild the strain field in the structure for a determined state. By means of this information and pattern recognition is possible to infer subtle changes in the global strain field that allow global damage detection even in case of small-scale damages.

Strain Sensors

In order to perform damage detection by using strain-based techniques, it is required to have multiple accurate strain sensors that do not affect the structure's performance to effectively detect changes in the strain field even when produced by small damages. In fact, one of the discussed advantages of strain monitoring compared with other techniques is related to the large quantity of available strain sensors. Three types of sensors are basically employed: resistance strain gauges and fiber optic sensors (FOS), both distributed and discrete sensing.

As mentioned before, the changes in the strain field are intense near to a defect but vanish quickly as one moves away from de defect. It is paramount to try to locate sensors in areas susceptible to damages and in areas where, indeed, *significant strain* is present. The noise is used to describe undesirable component in measured signals, therefore, the signal-to-noise ratio (SNR) may be an adequate way of measuring what is a sufficient significant strain. SNRs greater than 15 are acceptable for the proposed methodology when the resolution for the acquisition equipment is assumed to be 1 με.

Another issue to take into consideration when performing damage detection in engineering structures by means of strain sensors is the installation of the sensor networks into them. This can be done during the manufacturing process or after the structure has been built by attaching or bonding the sensors on the surface. The main limitation of sensors being adhered to the structure is the maximum temperature operation of the adhesive. The more used adhesives include cyanoacrylate and epoxy due to their low cost and effectivity. However, their maximum temperature operation is about 120 °C for cyanoacrylate and 180 °C for epoxy. Moreover, not only the operation temperature is a drawback but degradation and aging due to the operational environment is also a concern.

Resistance strain gauges are a conventional way of measuring strains. They are based on the fact that a relative strain change represents a relative change in the resistance due to geometric effects or by piezoresistive effects showed by some smart materials. Resistance strain gauges have several disadvantages such as strain drifting due to the effect on the environment in the electrical connection and their large size and high-power consumption.

Without doubt, the advent in FOS has facilitated the development and applications of strain-based SHM in different kinds of structures. They represent several advantages over traditional sensors that can be summarized in:

- Low size and weight (diameters about 150 micrometers), which allow their use without affecting severely the structural performance.
- Electromagnetic inference immunity due to their non-electrical nature.
- Multiplexing capability, several sensors can be set in a single optical fiber.
- High sensitivity and accuracy.
- Long-term monitoring capabilities due to their high fatigue resistance and corrosion resistance.

Additionally, according to Güemes et al. (2018), FOS offer three different topologies: point sensors, (e.g. micromirrors at fiber tips), discrete sensors, which are placed at intervals along the fiber length (e.g. FBGs) and distributed sensors, where sensing is distributed along the fiber without having locally engraved sensors (e.g. Rayleigh, Raman and Brillouin distributed sensing). For a more detailed view of FOS sensing techniques, the reader is directed to (Güemes & Menendez, 2010; Guo et al., 2011; Inaudi, 2010).

Although FOS distributed sensing development is evolving in an exponential rate and with the promise of offering significant advantages over the discrete sensors, distributed sensing has some accuracy and spatial resolution limitations that restraints their application in strain field pattern recognition. In a real experimental setup for damage detection in a wind turbine blade, the authors compared the effectivity of resistance strain gauges, FBG sensors (discrete sensors) and distributed sensing using an optical backscatter reflectometer (OBR) based on Rayleigh backscattered radiation. The results are depicted in Figure 3, comparing strain profiles obtained with strain gauges, FBGs and OBR for seven different load magnitudes (pristine structure). The profiles were gathered for the optical fiber located at the trailing edge-extrados of the wind turbine blade under pressure to suction test (see Figure 4). For more details about the experimental setup see the wind power case study later in this chapter.

The employed distributed technique demonstrated difficulties with accuracy since the accuracy was strongly affected by increasing the number of defined sensors. Strain gauges and FBG sensors had similar behavior in terms of accuracy; however, it is a fact that FBGs offers advantages towards a practical application in real-world operating structures (Sierra-Pérez et al., 2016).

Therefore, FBGs have been the most used for such applications, their accuracy of about $\pm$ 1 $\mu\varepsilon$ and $\pm$ 0.1 °C for measuring strains and temperature, respectively, and the ability to engrave several sensors in an optical fiber line. However, they have also the inherent limitations of optical fibers such as sensitivity of refractive index to both ultraviolet and thermal radiation and temperature sensitivity which creates the need for thermal compensation.

FBGs are one type of intensity-modulated FOS consisting of a periodic modulation of the refraction index at a specific location of a single-mode fiber. When an FBG is subjected to uniform uniaxial tension, the photoelastic equations afford a linear relationship among the wavelength drifting and the axial strain, the proportionality coefficient is known as sensitivity factor to the axial strain S_e:

$$S_e = \frac{\Delta\lambda_b}{\varepsilon_1} = \lambda_{b,0}\left[1 - \overline{n_e^2}(p_{12} - v_f(p_{11} + p_{12}))\right],$$

where v_f is the Poisson's coefficient of the fiber, $\Delta\lambda_b$ the Bragg wavelength drift, $\Delta\lambda_{b,0}$ the initial Bragg wavelength of the FBG and p_{11} and p_{12} its photoelastic coefficients. For 1550 nm, the sensitivity is approximately 1.2 pm for a strain of 1 $\mu\varepsilon$, namely, 833 $\mu\varepsilon$/nm.

However, temperature also influences the FBGs since occurs a linear thermal expansion of the grating and there are changes in their refractive index. Therefore, the combined mechanical and thermal effects can be aggregated in the following equation:

$$\varepsilon = \frac{\Delta\lambda_b}{\lambda_b} = (1 - \rho_\alpha)\Delta\varepsilon + (\alpha_T + \xi)\Delta T = k_\varepsilon \Delta\varepsilon + k_T \Delta T,$$

where ρ_α is the photoelastic coefficient of the optical fiber. Commonly, the values for such coefficients are $k_\varepsilon \approx 0.8 \times 10^{-6}$ $\mu\varepsilon$ and $k_T \approx 6.4 \times 10^{-6}$ K^{-1} (Güemes & Menendez, 2010).

As mention before, FBGs are both sensitive to strain and temperature, an increment in temperature also causes the center wavelength to shift. There are several methods to perform a thermal compensation of FBG strain measurements based on the estimation of the thermal output. Having the thermal output, the strain can be calculated by subtracting the thermal output from the strain measured by the FBG.

One of the most used methods is the so-called *dummy sensor*. This method involves the use of a second compensating dummy sensor identical to the strain FBG. The sensor should be embedded into a small brass tube sealed with adhesive in order to be isolated from any mechanical effect. Then, the strain can be calculated as:

$$\varepsilon = k_\varepsilon \left[\underbrace{\frac{(\lambda_b - \lambda_{b,0})}{\lambda_{b,0}}}_{Strain} - \underbrace{\frac{(\lambda_b - \lambda_{b,0})}{\lambda_{b,0}}}_{Temperature} \right].$$

Figure 3. Strain profiles for seven different load magnitudes acquired with strain gauges, FBG sensors and OBR (distributed sensing)

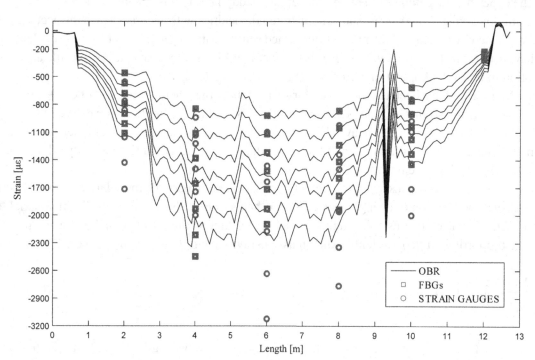

STRAIN FIELD PATTERN RECOGNITION FRAMEWORK

In this section, the theoretical background and the procedure involving the strain field pattern recognition technique is presented. The section is divided into the natural steps that a typical SHM methodology follows: preprocessing, dealing with operational environmental variability, feature extraction and modeling and decision-making.

Preprocessing

All the applications dealing with data need to implement solutions in order to assure that the quality of the datasets is adequate and organized in a way that pattern recognition can be applied. The procedures that are commonly followed are data fusion, data standardization, data cleaning and data compression (Farrar & Worden, 2012). In this case, the data consist of matrices having continuous dynamic signals of strains as a function of time, normally, a row referred to a sample and a column to each of the sensors.

Data fusion is the first step in the preprocessing stage, all the sensor readouts need to be aggregated in vectors so that a pattern of the structure strains can be obtained for each time. Other information about the dynamic and performance on the structure can be added in order to represent in a better way its behavior. For example, in the case of aircraft structures, the flight parameters should enhance the damage detection procedure (Sierra-Pérez & Alvarez-Montoya, 2018). In the data fusion stage is also important to take into consideration thermal compensation, since changes produced by damage are small, temperature must be considered in order to avoid masking the changes related to damages.

In addition, data must be organized to ease the pattern recognition process, the authors have called this step *data organizing and unfolding* which can be seen as a substage within data fusion. In the normal setup, a vector containing all the measurements for a given time is arranged in a matrix X which is $n \times m$, where n is the number of experimental trials and m is the number of sensors. However, in some cases, each sensor can be a set of dynamic measurements and, therefore, the dataset must be a 3D-matrix composed of J sensors or variables, K samples or time intervals and I experiments or trials. This raises a problem because such datasets cannot be used when using multivariable analysis techniques and proper unfolding is required.

Several techniques for unfolding 3D matrices have been proposed being the possible outputs six different 2D matrices: matrix A ($KI \times J$), matrix B ($JI \times K$), matrix C ($IJ \times K$), matrix D ($I \times KJ$), matrix E ($I \times JK$) and matrix F ($J \times IK$). There is also information in the literature about the advantages and limitations of each technique, according to Nomikos & MacGregor (1994), matrix D form is the most significant for analyzing and monitoring batch data. Through this technique all the trajectories of process variables (obtained in each test group) are considered as a single object. On the other hand, other authors have argued that by unfolding in matrix A, which implies that each point of the trajectory of each batch of data is considered an object, preserves the nonlinear trajectories in the time-dependent data array (Wold et al., 1998).

Data standardization is a paramount stage in the pattern recognition application. Different physical variables having different scales and magnitudes can be employed which often biased the *learning* process. In order to accomplish this, exist several techniques reported in the literature including continuous scaling, variable scaling, group scaling, autoscaling, etc. Autoscale, perhaps the most used technique, uses the mean and standard deviation, however, in the case of pattern recognition have not demonstrated an adequate performance since sample gathered are not usually drawn from the same distribution. Group scaling partially solves this problem by using a separate mean for each sensor (Mujica et al., 2010).

Load magnitude effects are undesirable, imagine for example a case where you have acquired strain patterns from a structure in a pristine state under a load W, if a load of $10W$ is applied, it is expected that such patterns have a similar form but different magnitude from the pristine data (assuming a linear structural behavior). The result will be a false alarm; this can be avoided if an appropriate standardization technique is used. However, the authors have demonstrated that unfolding and standardization of strain data are better performed, first, when different load cases (e.g. load magnitudes) and data from each sensor are grouped in columns. In that way, a new dimension must be added to the original 3D matrix so that each data point x_{ijkl} represents the i-th experiment trial, the j-th sensor, the k-th instant time and the lth load.

Then, the standardization is done by using the mean of each sensor at each instant of time for the total number of experiments. The standard deviation used is the corresponding to each sensor for all measurements and all load cases simultaneously. As a result, all the acquired information is seen together with all the other sensors and all the possible correlations among sensors under all the load cases are analyzed simultaneously (Sierra-Pérez et al., 2015).

Data cleaning and data compression are the last stages in preprocessing. In the former is important to identify and extract evident outliers from data (i.e. acquisition errors or spikes). On the other hand, data compression deals with reducing the quantity of data that, within this context, is aimed to filter low SNR data and discretize continuous measurements. For example, in dynamic measurements with damping phenomenon, some sensors, especially located at the cantilever tip of the beam, read strains near to zero. This implies a loss in sensitivity since the data is near the noise level and SNR becomes critical.

To solve this problem, the authors have used several data filtering techniques such as low-pass filter, local maximum extraction based on peak detection, kernel density estimation (KDE) and Hampel filtering. The first two have demonstrated suitable results for the experiments showing perfectly periodic load spectra while KDE and Hampel filtering in applications involving more random load spectra (Alvarez-Montoya et al., 2020).

Dealing with Operational and Environmental Variability

Operational and environmental variability is a serious concern in real-world SHM applications. Since the idea is to study changes in the strain field, the phenomena and external stimuli producing changes in the strain field should be isolated from those produced particularly for a damage occurrence. These phenomena could be load magnitude changes, temperature and load conditions variations, sudden environment changes and so on.

Temperature changes can be solved by means of thermal compensation as explained before. Load magnitude variations can be dealt with data standardization as discussed in the previous section since patterns are expected to change linearly. However, sometimes nonlinearities can be caused by manufacturing defects such as wrinkles, delamination, debonding, kiss bonding, reinforcements fiber misalignment, local buckling, etc. In such cases and when dealing with other changes associated with operational condition variations, more robust techniques are required.

In order to explain this concept clearly, think of an aircraft beam. It is a fact that there is a variety of different load conditions in normal operation. For example, the load distribution in the wing follows a parabolic shape, that is to say, it will produce patterns associated with such normal operation. When, for a change in wind speed, air density or another condition, the load magnitudes vary proportional, the patterns will continue being the same after standardization in the absence of nonlinearities. However, if a deflection in the control surfaces (e.g. aileron, flaps, spoilers, etc.) occurs, the load distribution will be changed in an unpredictable way, changing the strain field. Or, if the angle of attack is modified producing a change in the beam's second moment of area, the strain field is also modified.

The reader can also think in a wind turbine blade as depicted in Figure 4. Four tests were performed changing in each one 90° the blade pitch angle: suction side to pressure side (STP), pressure side to suction side (PTS), trailing edge to leading edge (TTL) and leading edge to trailing edge (LTT). The experimental setup is discussed in detail in the wind power case study. As can be seen in Figure 5, the response of the same couple of strain sensors for one load condition (i.e. fixed pitch angle) increases linearly as a function of load magnitude. However, for a different pitch angle (see Figure 4), the same linear behavior as a function of the slope is depicted but with a different slope.

Both examples represent cases when a change in the operational conditions produces changes in the strain field that are not related to a damage occurrence (i.e. different distribution of strains in the whole structure), usually, those changes mask the ones produce by damages. One approach is to try to include in the baseline (i.e. pristine data for *training* the pattern recognition techniques) as much as operational conditions as possible. However, this task may be difficult since, first, it is demanding to obtain all the data related to the possible operational and environmental conditions because a lot of experiments are required. Secondly, if all the conditions are considered, the model may be too *broad* so that damage sensitivity may decrease considerably.

Figure 4. Changes in pitch angles in different tests for a 13.5 m long wing turbine blade: (a) flapwise: suction side to pressure side, (b) flapwise: pressure side to suction side (PTS), (c) edgewise: trailing edge to leading edge (TTL) and (d) edgewise: leading edge to trailing edge

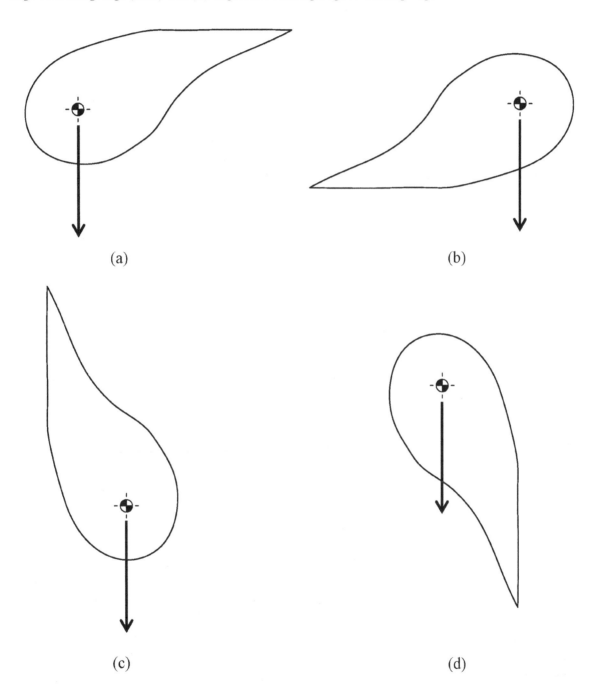

(a)

(b)

(c)

(d)

Figure 5. Differences in slope between the same couple of sensors for a 13.5 m long wing turbine blade in four different tests varying the pitch angle (Sierra-Pérez et al., 2016)

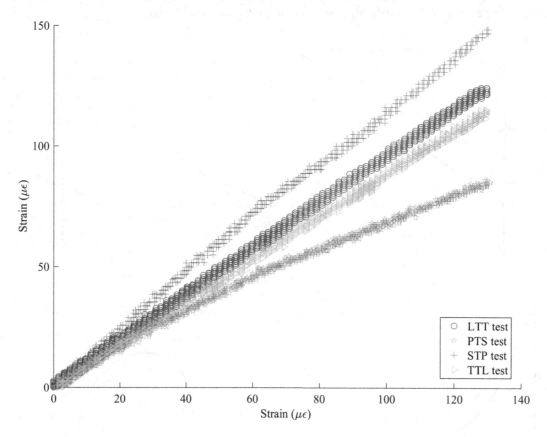

A methodology based on a procedure so-called *optimal baseline selection* (OBS) was developed by the authors in order to isolate different operational conditions. The natural approach in real structures consists of classifying the information according to the operational variables when they are known a priori. Several data groups are classified according to each specific operational case, specific baselines models for each group can be built. Then, new data for an unknown condition should be first classified according to their belonging operational case and, after that, compared with their corresponding baseline model for deciding whether it is a damaged or undamaged case.

However, in some systems it is not possible to have access to the operation variables, in such cases, OBS has been performed by using automatic clustering techniques such as standard self-organizing maps (SOMs), local density-based simultaneous two-level clustering (DS2L-SOM), density-based spatial clustering of applications with noise (DBSCAN) and fuzzy clustering (Alvarez-Montoya & Sierra-Pérez, 2018; Perafan, 2018; Sierra-Pérez et al., 2018). Indistinctly of the technique used, the approach consists of the same mentioned procedure; however, the use of unsupervised classification or clustering is paramount since the number of clusters (i.e. number of possible different operational cases) is unknown and difficult to estimate.

Feature Extraction and Modeling

Time-domain signals obtained from several experiments can be seen as patterns whose study leads to damage detection based on pattern recognition. Feature extraction and modeling refer to the step where gathered information is used to build a model yielding damage-sensitive parameters that can be used for the next step of decision-making, usually, model parameters or predictive errors associated with them are used as such damage-sensitive feature (Figueiredo & Santos, 2018). This process usually results in some form of data reduction (Kerschen & Golinval, 2009).

In the case of strain field pattern recognition, the idea is to correlate the strain measurements in order to infer if changes have been occurred, in particular, the global stiffness and strain field between different sensors. To accomplish this, several approaches have been used including so-called statistical methods and syntactic methods. Syntactic methods deal with classifying data according to its structural description; on the other hand, statistical methods employ statistical density functions. Statistical modeling, the most utilized in SHM, requires characterizing the data prior to statistical inference. This lead to a problem known as *curse of dimensionality* since required sample size (number of experimental trials) grows exponentially with respect to the dimension of the sample space (number of sensor readouts) (Sohn & Oh, 2009).

One of the reasons for the profuse use of statistical modeling methods in SHM is related to the advent of machine learning in the last decades, where rules are learned based on evidence or data. According to Figueiredo & Santos (2018), machine learning techniques used in statistical modeling include classification, regression and anomaly detection. Data classification can be defined as the association of a class (label) with a dataset. These techniques are often applied to damage localization tasks. Data classification usually requires supervised learning, that is to say, samples of excitation and responses are required for a specific postulated relationship. On the other hand, data regression can be defined as the construction of a map between a group of continuous input variables and a continuous output variable on the basis of a set of samples, which also requires supervised learning.

Without doubt, anomaly detection, or outlier detection, outperforms the other approaches since only data belonging to the pristine structure is required due to its unsupervised-learning nature. Let consider again the example of an aircraft structure under operation, if a classification algorithm is used as a basis for damage detection, in the training data should be assured that all the possible damage cases are included. In the case of non-included damage cases, the output of the algorithm may be uncertain.

Several machine learning algorithms, supervised and unsupervised, have been proposed for accomplishing SHM tasks including artificial neural networks (ANN), principal component analysis (PCA), multiway PCA (MPCA), Mahalanobis square distance, Markov models, singular value decomposition (SVD), parallel factor analysis (PARAFAC), Gaussian mixture models, partial least squares (PLS), multiway partial least square (MPLS), k-th-nearest neighbor rules and recently deep learning (Figueiredo & Santos, 2018; Yan et al, 2017).

Indistinctly of the algorithm used, within the context of anomaly detection, the problem is reduced to deal with providing output residual errors or distance metrics as damage index. In order to avoid the *curse of dimensionality*, data are often projected onto a lower-dimensional feature space using mapping functions. This concept can be understood better with an example presented in Figure 6. Assume that feature 1, 2 and 3 represent two arbitrary physical magnitudes obtained from pristine case (BL) and four damage cases (D1-D4). These three-dimensional inputs can be projected onto a one-dimensional feature space by using several transformations obtaining simultaneously the extraction of a damage sensitive feature and a reduction of data space to one-dimensional feature space (Sohn & Oh, 2009).

Among the most used projection methods are PCA, MPCA, MPLS, PARAFAC and Tucker models. Regardless of the algorithm used, the idea is to transform highly correlated, noisy and redundant data on a statistical model whose elements can provide an overview of occult phenomena representing the system. In general, a projection method deals with approximating a dataset by means of two functions f and g:

$$X = \overline{X} + \widetilde{X} = g(f(X(t))) + \widetilde{X},$$

where X is the input data that is projected by means on f to the original m-dimensional space onto a r-dimensional subspace. On the other hand, g represents the mapping function from an r-dimensional subspace back to the original m-dimensional space having $\widetilde{X}$ as a residue.

Figure 6. Feature extraction and data reduction for data gathered from an aluminum beam under dynamic loading with several damage cases (Sierra-Pérez et al., 2018)

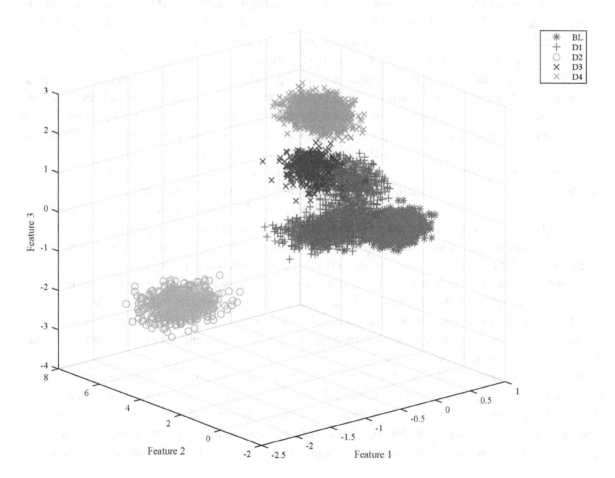

The authors have focused their attention on PCA and its nonlinear variations, the most widely used techniques by far for dimensionality reduction and characteristics identification. PCA aims to discern which data represent the most important dynamics of a particular system by finding a new space where the data is expressed based on the original covariance structure (Mujica et al., 2011). At this point, it is important to highlight that MPCA is equivalent to PCA applied to a dataset displayed as a 3D matrix. Therefore, PCA will be discussed indistinctly of the storing way.

PCA study is based on the covariance matrix, which is able to quantify the linearity degree between a couple of variables. Once the covariance matrix of X (input data) is obtained, X can be transformed by means of the linear transformation matrix P:

$$T = XP,$$

where T is the transformed matrix onto the new space and P is the principal components matrix. The columns of P possess the eigenvectors of the covariance matrix, Σ_X, sorted in descending order according to their associated eigenvalues λ_i. Usually, eigenvectors are known as principal components (PCs) and the highest their associated eigenvalue the highest importance represents. These matrices satisfy the following condition:

$$\Sigma_X P = A\Lambda,$$

where $\Lambda = diag(\lambda_1, ..., \lambda_i ... \lambda_m)$. Usually, the dimension of X is reduced by retaining only the r PCs having the most significant variance, $\overline{P} = (p_1, ..., p_r)$. Thus, the original data X can be projected onto the space contained by this matrix:

$$\overline{T} = X\overline{P},$$

where $\overline{T}$ is the matrix containing the data projected, this matrix is also known as score matrix and it will be used later for calculating damage indices as discussed in decision-making.

In the case of nonlinear dependencies among the variables, PCA may fail in providing an efficient description of the dataset. That is why several studies have focused on developing variations and improvements to the classical PCA. One of the most common approaches is nonlinear PCA (NLPCA) that uses surfaces for data projection (instead of planes as in PCA) in order to allow higher accuracies when dealing with nonlinear problems.

As stated by Scholz et al. (2008) NLPCA can be subdivided into normal NLPCA and hierarchical NLPCA (h-NLPCA), the later allows performing dimensionality reduction by using a hierarchical process similar to the PCA decomposition way.

Let consider again the expression in, in h-NLPCA f and g are generalized in order to allow nonlinear mapping functions so that:

$$T = f(X(t)),$$

where f is the vector of projection functions composed of r individual nonlinear functions. In the same way:

$$\overline{X} = g(T),$$

being g the inverse transformation needed to come back to the original dimensionality system. The task is to select f and g so that the Euclidean norm of $\widetilde{X}$ is minimized.

Commonly these functions are found by means of a feed-forward ANN, according to Kramer (1991), a three-layer neural network with m input neurons, having nonlinear transfer functions in the second layer and linear transfer functions in the third layer, is capable of approximating any continuous function from an m-dimensional space to an r-dimensional one. This statement is true provided that the number of neurons in the second layer is large enough.

Therefore, h-NLPCA implements a five-layer ANN comprising an input layer (X), three hidden layers (i.e. mapping layer, middle layer and demapping layer) and the output layer ($\overline{X}$). The middle layer, also known as bottleneck layer, outputs features T. The aim is to find the neuron weights so that the error between X and $\overline{X}$ is minimized, this error is often represented by the squared reconstruction error (SRE).

Decision-Making

As discussed above, all the aforementioned algorithms should provide output residual errors or distance metrics as damage indices (also known as statistics) in order to have a one-dimensional variable so that decision-making is straightforward. These quantitative indices seek to consider whether or not data for different experimental samples are homogeneous.

Within the context of PCA, a batch of data may exhibit abnormal behavior in two ways. First, the scores, $\overline{T}$, can move out of an allowable range defined by a control region and, second, the residual may become larger and the entire batch of data is located outside and perpendicular to the reduced space (Nomikos & MacGregor, 1994; Tibaduiza et al., 2011).

Mujica et al. (2011) proposed the use of PCA statistical tools such as the Q index, also known as squared prediction error (SPE), and the T^2 index, also known as D-statistic or Hotelling's T^2-index. Another index, although less used, include φ index, also known in the literature as combined index, which is a combination of T^2 and Q indices used to monitor the behavior of a process.

Q index analyses the residual data matrix $\widetilde{X}$ to represent the variability of the data projection in the residual subspace, in other words, it is a measure of the difference between a sample and its projection into the model. The Q index for the i-th experiment is given by:

$$Q_i = x_i(I - PP^T)x_i^T,$$

where x_i is the row vector of the original matrix X.

On the other hand, T^2 index is commonly used in multivariate hypothesis testing. It can be seen as a generalization of Student's t-statistic based on analyzing the score matrix $\overline{T}$ to check the variability of the projected data in new space of the principal components. The T^2 index for the i-th experiment is given by (Mujica et al., 2011):

$$T^2 = x_i^T (P\Lambda^{-1}P^T)x_i.$$

Normally, Q index is more sensitive to damage than T^2 index since Q index is smaller and any minor change in the characteristics of the system is observable. By contrast, T^2 index has a large variance and, therefore, it requires a higher change in the system characteristics to be detectable. However, Q index has the problem of having less power to detect heterogeneities when a small number of samples is used.

Then, it is necessary to define control limits, detection thresholds or also called confidence intervals, the aim is to take into account variability and uncertainty presented in the normal operational region to make a decision. One approach is to estimate thresholds based on values representing a specific percentage of confidence over the training data. In this manner, outliers can then be defined as data having damage indices beyond defined thresholds.

Control limits can be seen as warning and action limits. A warning limit is usually defined as a 95% of confidence bound and serves as a warning that the system or structure is approaching to an abnormal condition. On the other hand, an action limit is commonly defined as a 99% of confidence bound and serves as an indicator that an abnormal condition has occurred and an action must be performed.

It is important to highlight that using these damage thresholds requires the assumption that scores and residual follow a multivariate normal distribution. This is true if the input data X follows a multivariate normal distribution, thus, it is possible to define control limits for the damage indices.

In the case of the Q index, due to its nature rooted in a quadratic form of errors, the errors are adequately approximated by a multinormal distribution; therefore, the upper control limits (UCL) can be defined as the following expression based on the Box's equation:

$$UCL_Q = \frac{v}{2\omega}\chi^2_{\frac{2\omega^2}{v}}(\alpha),$$

where $\chi^2_{\frac{2\omega^2}{v}}(\alpha)$ is the upper 100-th percentile of a chi-square distribution with $2\omega^2/v$ degrees of freedom at a significance level α.

It is possible to define an upper control limit for T^2 index, which involves the estimated eigenvalues and eigenvectors, by using this expression:

$$UCL_{T^2} = \chi^2_{r-2}(\alpha),$$

where $\chi^2_{r-2}(\alpha)$ is the upper 100-th percentile of a chi-square distribution with $r-2$ degrees of freedom at significance level α (Johnson & Wichern, 2008).

It is important to notice that the aforementioned definitions hold true for the PCA modeling case. If NLPCA is considered, the assumption of normality may not true, therefore, new definitions are required. For Q index, the expressions in and are valid since the residuals of the NLPCA model follow a normal distribution due to they are not encapsulated in the nonlinear scores. However, the definitions for the T^2 index are not the same because they are rooted in the nonlinear scores. According to Wang et al. (2008), an improved residual vector can be defined as follows:

$$Z(\nabla \left\| \widetilde{X} \right\|^2), k) = \frac{1}{k} \sum_{j=1}^{k} \frac{\partial \widetilde{x}_j}{\partial \zeta}\bigg|_{\zeta=\zeta_0},$$

where $\nabla(\bullet)$ is the gradient vector, k is a vector of data for a discrete time instant and, finally, ζ and ζ_0 are defined as parameter vectors describing the abnormal and normal behavior, respectively. From this definition, an improved T^2 index is defined:

$$T^2(Z,k) = Z^T(\nabla \left\| \widetilde{X} \right\|^2), k) \sum_{ZZ}^{-1}(\zeta_0) Z(\nabla \left\| \widetilde{X} \right\|^2), k).$$

However, singular value decomposition can be used in order to avoid ill-conditioned T^2 indices, resulting in the approximate definition $Z \approx U_1 \Sigma_1 V_1^T$. Therefore, T^2 indices can be calculated as $T^2(v,k) = v^T(k)v(k)$, being $v(k)$ defined by:

$$v(k) = \Sigma_1 V_1^T Z(\nabla \left\| \widetilde{X} \right\|^2) \sqrt{k} = PZ(\nabla \left\| \widetilde{X} \right\|^2)$$

Dimensions of Σ_1 determine the degrees of freedom for the T^2 indices so that the threshold for a defined confidence level is given by:

$$UCL_{T^2} = \chi^2_{\dim(\Sigma_1)}(\alpha),$$

For a detailed definition of such expressions and for an overview of the methodology, the reader is referred to Sierra-Pérez et al. (2018).

CASE STUDIES

Different experiments have been performed in order to test the suitability of strain field pattern recognition in different real-world engineering structures under realistic operational conditions. These applications can be divided into three main fields: wind power, aerospace and civil engineering.

As a way of comparing the performance of the methodology in different cases, receiver operating characteristic (ROC) analyses were carried out based on true positive rate (TPR), true negative rate (TPR), false positive rate (FPR) and false negative rate (FNR). From definition, a ROC curve is a visualization of the TPR as a function of the FPR of a classifier, in this case, Q index or T^2 index. Within this context, several metrics can be defined to quantitatively assess the performance of the classifier such as accuracy, precision, recall and F1 score. Accuracy is defined as the quotient of the sum of TP and TN and the sum of TP, TN, FP and FN.

Wind Power

FBGs, distributed sensing (OBR) and resistance strain gauges were embedded and bonded to a 13.5-m long wind turbine blade which was monitored during several experimental tests until reaching the failure. The wind turbine was made of glass fiber and vynilester resin doped with carbon nanofibers through light resin transfer molding as a monocoque structure with a PVC foam core. This blade was manufactured and designed by CENER Spain (*Centro Nacional de Energías Renovables*) as part of the Spanish National Project *NeWind*.

Four optical fibers (two in the intrados and two in the extrados) were used each one containing six FBGs, so that the structure contained a total of 24 embedded FBGs. Resistance strain gauges were bonded at the same locations that the FBGs and in other locations. In addition, four plain optical fibers were installed at the surface of the blade coinciding with the optical fiber containing the FBGs with the aim of obtaining distributed strain measurement by means of the OBR (see Figure 3 for comparison of the sensor readouts by the different sensing devices used). The location of the different sensors and three damage conditions are shown in Figure 7.

Consequently, the wind turbine blade was loaded in a test bench in a cantilever configuration by means of a water deposit system provided with an automatic differential filling system to emulate the real load distribution under operation. Different positions were tested in order to accomplish the load envelope certification of the wind turbine blade, these positions include, STP, PTS, TTL and LTT as detailed in Figure 4. In the final tests, the blade was loaded in a scaled way (from 40%, 70%, 80%, 90%, 100% until 130% of the ultimate load) in the PTS position with the aim of inducing failure. The experimental setup is depicted in Figure 8.

In the first trial, the blade finished without any visible damage, therefore, an artificial damage (D1) was induced by cutting in the trailing edge (simulating a real damage such as debonding). This damage condition consisted in a 100 mm long, 3 mm wide, 300 mm deep cutting located at 2.5 m from the blade root and in a 130 mm long, 3 mm wide, 13 mm deep transversal cutting located at 7 m from the root. For the second trial, visible cracks appeared and they were treated as a different damage condition (D2). Finally, in the last trial (D3), the size of the transversal cutting of the D1 was increased so that both depth and width had 30 mm and 15 mm, respectively.

Figure 7. Location of sensors and damages in the wind turbine blade (all dimensions are in mm)

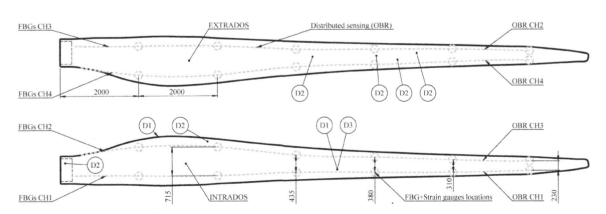

Figure 8. Experimental setup for TTL (trailing edge to leading edge) position test

An example of the Q indices obtained for the FBG sensors for the pristine and damage cases are presented in Figure 9 for both PCA and h-NLPCA. It can be seen that the damage detection performance considering this index is highly improved when the h-NLPCA model is used. Although h-NLPCA implies higher computational cost, its accuracy is 98.5% compared with an accuracy of 51.22% for the PCA model. This can be explained due to this structure presents nonlinearities associated with its manufacturing process and its nature (i.e. thin skins with stiffeners under compressive loads).

However, all the damage cases were successfully detected by means of strain field pattern recognition based on h-NLPCA using the three types of sensing techniques, namely, FBGs, OBR and strain gauges. It is important to highlight that, in order to obtain a fair comparison, 2262 sensors were defined regarding the OBR-based distributed sensing since this number demonstrated to be the compromise solution between number of sensor and resolution when tested by using ROC analysis.

An error analysis of the different sensing techniques is presented in Figure 10 only for the model h-NLPCA in order to compare them. The results showed a higher sensitivity for the OBR measurements (accuracy of 99.15%) compared with FBGs (accuracy of 98.5%) and strain gauges (accuracy of 83.3%) due to the fact that the large amount of sensor in the distributed sensing technique permits having small distances to the damages and, therefore, capture the effects in the strain field which smooth steeply. Strain gauges showed less sensitivity since their measurement accuracy and uncertainty is not as good as in FOS (Sierra-Pérez et al., 2016).

Figure 9. Q indices for the wind turbine blade case. BL corresponds to baseline, dashed line and solid line represent damage thresholds associated with 95% and 99% confidence level: (a) PCA model and (b) h-NLPCA model

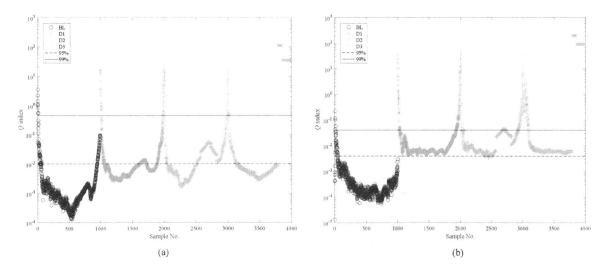

Aerospace

In the case of aerospace engineering, several real-world structures have been tested including a lattice spacecraft structure under compression test, a UAV wing section under bending test and a UAV under real flight operation.

Lattice Spacecraft Structure

A high-stiffness, low-weight open isogrid structure for launcher applications, also known as lattice structure, was manufactured by EADS CASA using an out of autoclave epoxy reinforced with carbon fibers as part of the project *ICARO* in conjunction with INTA. The lattice structure dimensions were 0.8 m of diameter and 1.1 m of height. In the upper half of the cylinder, four optical fibers (each one containing 9 FBGs) were bonded to each bar of the grid in the inner side of the cylinder. The structure under consideration is depicted in Figure 11.

The lattice structure was then screwed to aluminum inserts for assuring proper loading under compression test using a testing machine. Three tests consisting of increasing the load up to 330 kN were performed until the structure suddenly failed during the third test (D1). It is important to highlight that since it was uncertain the time of failure, all this experiment was labeled as damage; this is conservative because not all the data correspond to the damaged structure. As a consequence of failure, several bars and nodes broke instantaneously in the center region of the cylinder. Finally, three additional tests were performed with the damaged structure (D2 to D4) in order to be used for damage detection based on PCA.

Figure 10. Error analysis of different sensing technique, namely, FBG sensors, OBR-based distributed sensing and strain gauges for the wind turbine blade case (95% confidence level)

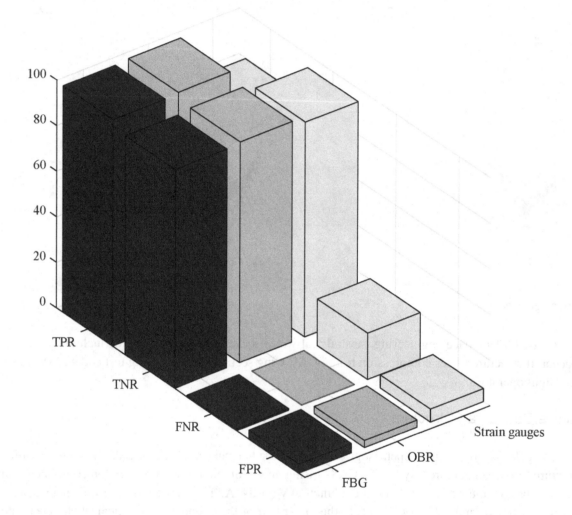

In the test campaign it was possible to detect damages by means of the Q index (see Figure 12a and Figure 12b) which demonstrated more adequate performance than T^2 (see Figure 12c and Figure 12d). Similar results were obtained with PCA and h-NLPCA with an accuracy (95% confidence level) of 90.58% and 93.01%, respectively. However, the h-NLPCA showed the most adequate results when dealing with nonlinearities caused particularly with load magnitudes associated with the start and finish of loading (Sierra et al., 2014).

This case study demonstrated the feasibility of the technique to be used in aerospace structures made of composite materials. The unique advantages of FBGs like small size and weight, embedment capability in composites and immunity to electromagnetic interference shows their suitability as SHM technique for this kind of structures.

Figure 11. Lattice spacecraft structure (all dimensions in mm)

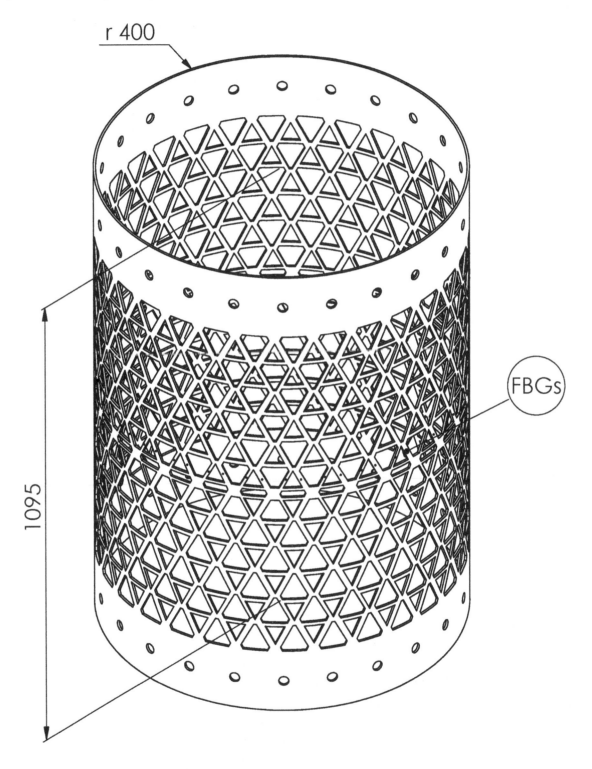

Figure 12. Damage indices for the lattice spacecraft structure. BL corresponds to baseline, D0 to validation data and dashed line and solid line represent damage thresholds associated with 95% and 99% confidence level: (a) Q indices for PCA model, (b) Q indices for h-NLPCA model, (c) T^2 indices for PCA model and (d) T^2 indices for h-NLPCA model

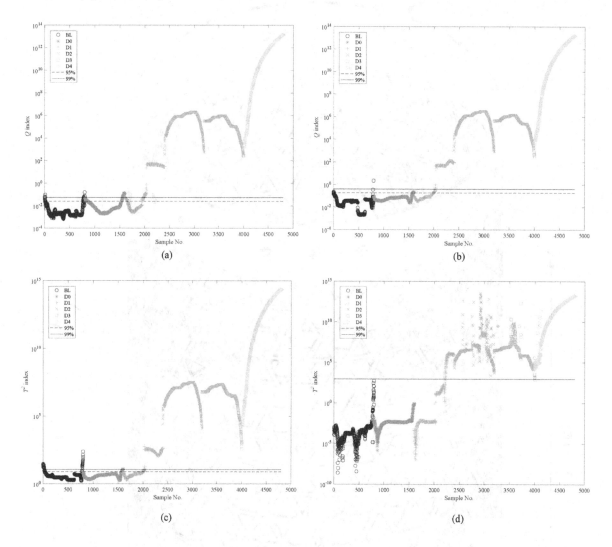

UAV Wing Section

A 1.5 m long wing section of an UAV's wing made of carbon fiber reinforced polymer (CFRP), glass fiber reinforced polymer (GFRP), PVC foam and balsa was designed by Technical University of Munich for Boeing Research and Technology Europe as part of the project SINTONIA (Röbler, 2008). Drawings of the UAV are depicted in Figure 13.

In this case, four optical fibers, each one having eight FBGs, were bonded to the structure at the wing skin. Two optical fibers were placed at the intrados (one close to the main spar and the other close to the secondary spar) and two optical fibers were placed at the extrados (in the same configuration that the located at the intrados). The wing section was attached in a cantilever configuration by means of

Figure 13. Drawings of the tested UAV (all dimensions are in mm)

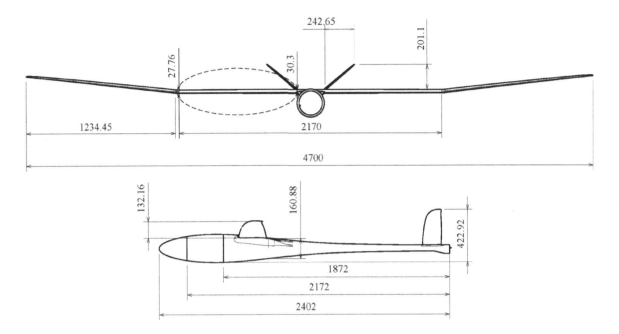

screws, an extruded aluminum profile and C-clamps to secure the fixed condition and simulate a real operational condition. After that, several tests consisting of loading the structure by means of different masses suspended from the main spar at the free end were performed. The masses used were 3.25 kg, 4.75 kg, 6.25 kg and 7.25 kg.

A total of 20 tests with the pristine structure were performed for each load magnitude, of which 50% were used to create the baseline dataset and the rest to validate the model. Then, six damage cases were induced in the structure by means of longitudinal and transversal cutting in a cumulative way with a 0.17 mm thick cutting disk. All damage cases were performed in the intrados aiming to represent the less severe case in order to test methodology's sensitivity. Wings are commonly in bending loading; therefore, extrados is under compression and intrados under tension, this means that damages in extrados are more critical since promote buckling failure, which is the most typical failure in this type of structures.

Damage case 1 (D1) consisted of 10 mm long longitudinal cutting near spar that then was expanded to 30 mm to form damage case 2 (D2). From damage cases 3 to 6 (D3-D6), transversal cuttings were performed with a length of 10, 20, 30 and 40 mm, respectively. In damage case 5 (D5) a 2 mm deep superficial cut was induced in the main spar and, in damage case 6 (D6), it was expanded to a deep of 4 mm. In Figure 14, the sensor layout and damage locations are detailed in the UAV wing section.

Damage indices for these experiments for both PCA and h-NLPCA are shown in Figure 15. Q indices results for both models showed a similar behavior and the damage detection is performed successfully. In Figure 15c and Figure 15d, T^2 indices are depicted for the PCA and h-NLPCA, respectively, demonstrating high differences with respect to both cases. This is not surprising since h-NLPCA model is expected to be more accurate when nonlinearities are present in the system. Since T^2 index gives an idea of the variation inside the model, it is expected that for a more accurate model there is less variation inside it. This can be evidenced by comparing baseline T^2 indices, h-NLPCA model gave smaller values than the PCA model.

Figure 14. Sensor layout and damage location in the UAV wing section (all dimensions are in mm)

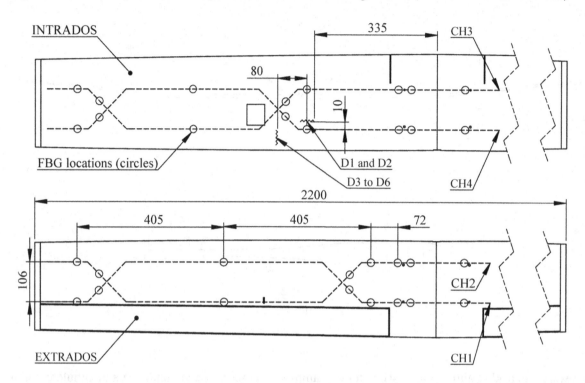

As a summary, considering Q indices for decision-making, the PCA was capable of detecting damages with an accuracy of 98.12% and 99.38% for 95% and 99% confidence level, respectively. In the case of h-NLPCA, the accuracy values are 97.5% and 100%, respectively (Sierra-Pérez & Güemes, 2016).

SMARP UAV

The past case study demonstrated the suitability of the technique for detecting damages in an aircraft structure under simulated operational conditions. This case study aims to demonstrate the methodology under real operational conditions. In order to accomplish this, a UAV, named SMARP, was designed and built using composite materials. The front spar of the aircraft's wing was manufactured integrating 20 FBG sensors in their laminate, a miniature acquisition system was installed on-board as well as a WLAN-based data transmission system for information streaming. The aircraft under operation and its general dimensions are depicted in Figure 17.

The FBGs were distributed equally into four optical fibers located at the upper surface, bottom surface, right-hand lateral surface and left-hand lateral surface of the front spar in order to measure strain from compression, tension and torsion loads, respectively. The distribution of FBG sensors in the beam is depicted in Figure 18, the bottom surface and the right-hand lateral surface are symmetric with upper surface and left-hand lateral surface, respectively, so that the distribution is the same.

A campaign of 16 flights was performed in order to acquire in-flight data. Six flights were performed with the pristine structure and ten after inducing six different artificial damages consisting of bonding 1 mm thick steel plates (covering all the section height) of different lengths to modify structural stiffness resembling a typical damage. The steel plates were located in the right-hand lateral surface (see the dashed rectangle shown in Figure 18). Specifically, for damage cases 5 and 6, an additional steel plate was bonded in the dashed rectangle zone of the bottom surface in order to promote different damage condition.

Damage case 1 (D1) consisted of a 12 mm long plate, damage case 2 (D2) of a 28 mm long plate and damage case 4 (D4) of a 45 mm long plate. In the case of damage case 3 (D3), a 35 mm long plate was bonded over D4. For damage case 5 (D5), in addition to the plates comprising D3, two plates were bonded on the bottom surface (a 62 mm long plate over a 100 mm long plate), finally, for damage case 6 (D6), the 62 mm long was removed.

Figure 15. Damage indices for the UAV wing section. BL corresponds to baseline, D0 to validation data and dashed line and solid line represent damage thresholds associated with 95% and 99% confidence level: (a) Q indices for PCA model, (b) Q indices for h-NLPCA model, (c) T^2 indices for PCA model and c) T^2 indices for h-NLPCA model

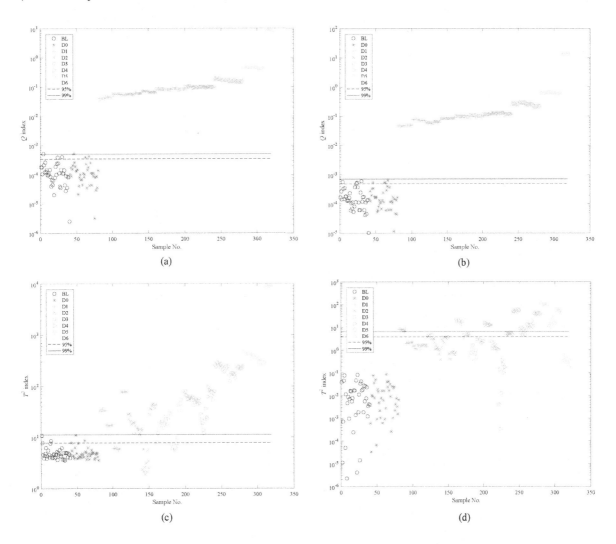

Figure 16. Error analysis for the UAV wing section for a 99% confidence level: (a) PCA model and (b) h-NLPCA model

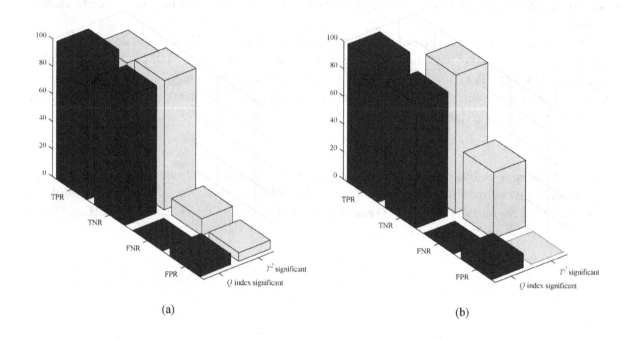

(a)   (b)

Figure 17. SMARP UAV: (a) In-flight photo and (b) drawings with general dimensions (all dimensions are in mm)

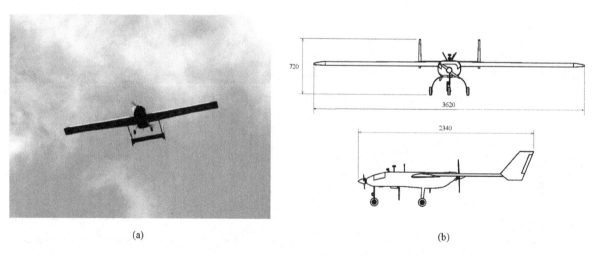

(a)   (b)

Figure 18. Sensor layout for the SMARP UAV and artificial damage locations

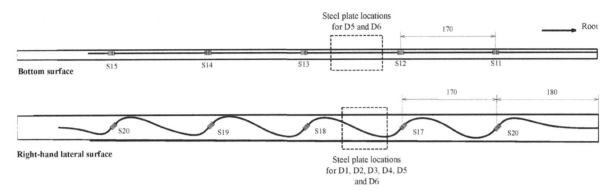

In total 765120 samples (each sample containing 20 strain measurements) were acquired since different operational conditions were expected the OBS procedure based on SOM was used for clustering different operational conditions, although different preprocessing techniques (i.e. standardization and filtering), the best results were obtained with KDE filtering with 95% confidence level and without standardization. This combination produced 127 clusters, that is to say, 127 different operation conditions were found by the DS2L-SOM and an accuracy of 98.5% for 95% confidence level. For more details see (Alvarez-Montoya et al., 2020).

Although the results were adequate, new techniques are being explored in order to either increase accuracy or reduce computational cost. The authors found that by using fuzzy clustering (i.e. fuzzy C-means, Gustason-Kessel, etc.) in the OBS procedure can enhance the methodology by reducing the number of cluster to 20 and with similar accuracies (Alvarez-Montoya & Sierra-Pérez, 2018). On the hand, other feature extraction techniques are promissory, for example, Gaussian process (GP) modeling demonstrated to have a subtly higher performance than when using PCA in the SMARP UAV data (Alvarez-Montoya et al., 2018).

Civil Engineering

SHM is also important in civil engineering since it seeks for maintenance cost reduction and accident prevention. Two applications of strain field pattern recognition will be reviewed, a porticoed structure under dynamic loading and a soil profile prototype for geotechnical monitoring.

Porticoed Concrete Structure

A porticoed concrete structure resembling a scaled building construction was developed and instrumented with 48 FBG sensors bonded in the steel reinforcements. The structure is detailed in Figure 19a, comprising four beams and four columns. The sensors were successfully embedded in the concrete structure by following a previously developed methodology (Correa-Uribe, 2015).

Several experiments were performed by attaching a mechanical shaker capable of exciting the structure in different frequencies. As a proof of concept, the structure was loaded 20 times with a frequency of 15 Hz for 10 seconds of data acquisition; the aim is to simulate the conditions presented during an earthquake. Then, stainless-steel plates were bonded to the structure in order to simulate a damage condition by modifying the structural stiffness. Ten experiments were carried out in the same conditions that the pristine structure.

After processing the acquired data, it is possible to conclude that damage detection by means of strain field pattern recognition is also possible in this kind of structures. Figure 19b shows Q indices for the load cases.

Soil Profile Prototype

A soil movement monitoring system by means of inclinometers was developed using FBG sensors. In order to test the system, a soil profile prototype was built to simulate a representative soil consisting of two layers of different materials. The first layer simulated the compacted soil layer and the second layer simulated the porous and brittle soil layer, which usually slips over the compacted clay area in a landslide event.

The inclinometers comprised two FBG sensors arranged in cylindrical bars. Then, inclinometers can be located in a mesh network to monitoring a determinate ground. In this case, a total of nine inclinometers were installed in the prototype. In order to validate the system, the prototype was divided into 64-equally spaced square elements and the ground was excited in each element with two different load magnitudes. The experimental setup is depicted in Figure 20.

Pattern recognition techniques, namely, support vector machine (SVM), logistic regression, Bayesian classification, multilayer perceptron (MLP), decision trees (DTs) and k-nearest neighbors (k-NN), were then used to detect and locate such excitations in the ground from the strain measurements. The used techniques were compared by means of several performance metrics such as accuracy, Kappa-statistic

Figure 19. Porticoed concrete structure: (a) Experimental setup and (b) Q indices for the PCA model

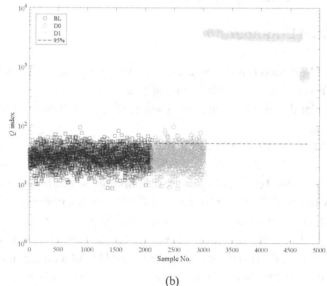

(a)

(b)

Table 1. Pattern recognition comparison for the soil profile prototype

Technique	Correctly Classified Instances [%]	Incorrectly Classified Instances [%]	Kappa Statistic	RMSE
Logistic regression	98.261	1.739	0.982	0.026
MLP	94.783	5.217	0.947	0.046
SVM	21.739	78.261	0.214	0.123
k-NN	98.261	1.739	1	0.011
Bayesian	87.826	12.174	0.876	0.055
DTs	81.739	18.261	0.814	0.07

Figure 20. Soil profile prototype setup

and mean squared error, giving that k-NN is capable of locating the excitation source with the highest accuracy (98.26%). This work aimed to prove the concept of automatic early detection and location of earth mass movement by means of strain field pattern recognition and FBG sensors.

ACKNOWLEDGMENT

This book chapter contains research supported by the Ministerio de Ciencia, Innovación y Universidad of Spain [grant numbers DPI2011-28033-C03-02 and DPI2011-28033-C03-03]; the European Commission [grant number 284562]; and Universidad Pontificia Bolivariana [grant number 636B-06/16-57 and 815B-06/17-23]. The authors thank Centro Nacional de Energías Renovables (CENER), INDRA, Instituto Nacional de la Técnica Aeroespacial (INTA), EADS Astrium, Boeing, Universidad Politécnica de Madrid and Universidad EIA for their support during the experimental stages.

REFERENCES

Alvarez-Montoya, J., Carvajal-Castrillón, A., & Sierra-Pérez, J. (2020). In-flight and wireless damage detection in a UAV composite wing using fiber optic sensors and strain field pattern recognition. *Mechanical Systems and Signal Processing*, 136.

Alvarez-Montoya, J., & Sierra-Pérez, J. (2018). Fuzzy unsupervised-learning techniques for diagnosis in a composite UAV wing by using fiber optic sensors. *Proceedings of the 7th Asia-Pacific Workshop on Structural Health Monitoring, APWSHM 2018*, 682–690.

Alvarez-Montoya, J., Torres-Arredondo, M., & Sierra-Pérez, J. (2018). Gaussian process modeling for damage detection in composite aerospace structures by using discrete strain measurements. *Proceedings of the 7th Asia-Pacific Workshop on Structural Health Monitoring, APWSHM 2018*, 710–718.

Boller, C. (2008). Structural Health Monitoring-An Introduction and Definitions. In Encyclopedia of Structural Health Monitoring. doi:10.1002/9780470061626.shm204

Boresi, A., & Schmidt, R. (2003). *Advanced mechanics of materials*. John Wiley & Sons, Inc.

Correa-Uribe, A. (2015). *Esquema para la implementación de medición de deformaicones en edificaciones de hormigón*. Universidad EIA.

Farrar, C. R., & Worden, K. (2007). An introduction to structural health monitoring. *Philosophical Transactions of the Royal Society A: Mathematical, Physical and Engineering Sciences, 365*(1851), 303–315.

Farrar, C. R., & Worden, K. (2012). Structural Health Monitoring. In Structural Health Monitoring: A Machine Learning Perspective.

Fernández-Lopez, A., Menendez, J. M., & Güemes, A. (2007). Damage Detection in a Stiffened Curved Plate By Measuring Differential Strains. *16th International Conference on Composite Materials*.

Figueiredo, E., & Santos, A. (2018). Machine Learning Algorithms for Damage Detection. In A. S. Nobari & M. H. F. Aliabadi (Eds.), *Vibration-Based Techniques for Damage Detection and Localization in Engineering Structures* (pp. 1–39). Academic Press.

Figueiredo, E., & Santos, A. (2018). Machine Learning Algorithms for Damage Detection. In Computational and Experimental Methods in Structures (Vol. 10, pp. 1–39). doi:10.1142/9781786344977_0001

Giurgiutiu, V. (2016). Structural Health Monitoring of Aerospace Composites. In Polymer Composites in the Aerospace Industry (Vol. 16). Academic Press.

Güemes, A., Fernández-Lopez, A., & Díaz-Maroto, P. (2016). A permanent inspection system for damage detection at composite laminates, based on distributed fiber optics sensing. *Proceedings of the 8th International Symposium on NDT in Aerospace*.

Güemes, A., Fernández-López, A., Díaz-Maroto, P., Lozano, A., & Sierra-Perez, J. (2018). Structural Health Monitoring in Composite Structures by Fiber-Optic Sensors. *Sensors (Basel), 18*(4), 1094. doi:10.339018041094 PMID:29617345

Güemes, A., & Menendez, J. M. (2010). Fiber-Optic Sensors. In Structural Health Monitoring (pp. 225–285). Academic Press.

Guo, H., Xiao, G., Mrad, N., & Yao, J. (2011). Fiber optic sensors for structural health monitoring of air platforms. *Sensors (Basel)*, *11*(4), 3687–3705. doi:10.3390110403687 PMID:22163816

Haroon, M. (2009). Free and Forced Vibration Models. In C. Boller, F.-K. Chang, & Y. Fujino (Eds.), Encyclopedia of Structural Health Monitoring (pp. 1–28). Academic Press.

Inaudi, D. (2010). Overview of fiber optic sensing technologies for structural health monitoring. *Informacije MIDEM*, *40*(4), 263–272.

Johnson, R., & Wichern, D. (2008). *Applied Multivariate Statistical Analysis (6th ed.)*. Pearson.

Kerschen, G., & Golinval, J. (2009). Dimensionality Reduction Using Linear and Nonlinear Transformation. In C. Boller, F.-K. Chang, & Y. Fujino (Eds.), *Encyclopedia of Structural Health Monitoring* (pp. 1–13). John Wiley & Sons, Ltd.

Kramer, M. A. (1991). Nonlinear principal component analysis using autoassociative neural networks. *AIChE Journal. American Institute of Chemical Engineers*, *37*(2), 233–243. doi:10.1002/aic.690370209

Li, H., Beck, F., Dupouy, O., Herszberg, I., Stoddart, P. R., Davis, C. E., & Mouritz, A. P. (2006a). Strain-based health assessment of bonded composite repairs. *Composite Structures*, *76*(3), 234–242. doi:10.1016/j.compstruct.2006.06.032

Li, H., Herszberg, I., Davis, C. E., Mouritz, A. P., & Galea, S. C. (2006b). Health monitoring of marine composite structural joints using fibre optic sensors. *Composite Structures*, *75*(1), 321–327. doi:10.1016/j.compstruct.2006.04.054

Lopez, I., & Sarigul-Klijn, N. (2010). A review of uncertainty in flight vehicle structural damage monitoring, diagnosis and control: Challenges and opportunities. *Progress in Aerospace Sciences*, *46*(7), 247–273. doi:10.1016/j.paerosci.2010.03.003

Menendez, J. M., & Güemes, A. (2006). SHM Using Fiber Sensors in Aerospace Applications. *Optical Fiber Sensors*, MF1.

Mujica, L., Tibaduiza-Burgos, D. A., & Rodellar, J. (2010). Data-driven multiactuator piezoelectric system for structural damage localization. *Fifth World Conference on Structural Control and Monitoring (5WCSCM)*.

Mujica, L. E., Rodellar, J., Fernández, A., & Güemes, A. (2011). Q-statistic and T2-statistic PCA-based measures for damage assessment in structures. *Structural Health Monitoring: An International Journal*, *10*(5), 539–553. doi:10.1177/1475921710388972

Nomikos, P., & MacGregor, J. F. (1994). Monitoring batch processes using multiway principal component analysis. *AIChE Journal. American Institute of Chemical Engineers*, *40*(8), 1361–1375. doi:10.1002/aic.690400809

Ostachowicz, W., Soman, R., & Malinowski, P. (2019). Optimization of sensor placement for structural health monitoring: A review. *Structural Health Monitoring*, *18*(3), 963–988. doi:10.1177/1475921719825601

Perafan, J. (2018). *An unsupervised clustering methodology by means of an improved dbscan algorithm for operational conditions classification in a structure.* Universidad Pontificia Bolivariana.

Popov, E. (1998). *Engineering Mechanics of Solids (2ⁿᵈ ed.).* Pearson.

Qing, X., Li, W., Wang, Y., & Sun, H. (2019). Piezoelectric transducer-based structural health monitoring for aircraft applications. Sensors, 19(3).

Röbler, C. (2008). Design of a fuel cell powered UAV wing and V-tail building instruction. Munich: Academic Press.

Rocha, B., Silva, C., Keulen, C., Yildiz, M., & Suleman, A. (2013). Structural Health Monitoring of Aircraft Structures. In W. Ostachowicz & A. Güemes (Eds.), *New Trends in Structural Health Monitoring.* Udine: Springer. doi:10.1007/978-3-7091-1390-5_2

Scholz, M., Fraunholz, M., & Selbig, J. (2008). Nonlinear Principal Component Analysis: Neural Network Models and Applications. In A. N. Gorban, B. Kégl, D. C. Wunsch, & A. Y. Zinovyev (Eds.), *Principal Manifolds for Data Visualization and Dimension Reduction* (pp. 44–67). Berlin: Springer Berlin Heidelberg. doi:10.1007/978-3-540-73750-6_2

Sierra, J., Frövel, M., Del Olmo, E., Pintado, J. M., & Güemes, A. (2014). A robust procedure for damage identification in a lattice spacecraft structural element by mean of Strain field pattern recognition techniques. *16th European Conference on Composite Materials, ECCM 2014.*

Sierra-Pérez, J. (2014). *Smart aeronautical structures: development and experimental validation of a structural health monitoring system for damage detection.* Universidad Politécnica de Madrid.

Sierra-Pérez, J., & Alvarez-Montoya, J. (2018). Damage detection in composite aerostructures from strain and telemetry data fusion by means of pattern recognition techniques. *9th European Workshop on Structural Health Monitoring, EWSHM 2018,* 1–12.

Sierra-Pérez, J., & Güemes, A. (2016). Damage detection in aerostructures from strain measurements. *Aircraft Engineering and Aerospace Technology, 88*(3), 441–451. doi:10.1108/AEAT-11-2013-0210

Sierra-Pérez, J., Güemes, A., Mujica, L. E., & Ruiz, M. (2015). Damage detection in composite materials structures under variable loads conditions by using fiber Bragg gratings and principal component analysis, involving new unfolding and scaling methods. *Journal of Intelligent Material Systems and Structures, 26*(11), 1346–1359. doi:10.1177/1045389X14541493

Sierra-Pérez, J., Torres-Arredondo, M.-A., & Alvarez-Montoya, J. (2018). Damage detection methodology under variable load conditions based on strain field pattern recognition using FBGs, nonlinear principal component analysis, and clustering techniques. *Smart Materials and Structures, 27*(1), 015002. doi:10.1088/1361-665X/aa9797

Sierra-Pérez, J., Torres-Arredondo, M. A., & Güemes, A. (2016). Damage and nonlinearities detection in wind turbine blades based on strain field pattern recognition. FBGs, OBR and strain gauges comparison. *Composite Structures, 135,* 156–166. doi:10.1016/j.compstruct.2015.08.137

Sohn, H., & Oh, C. K. (2009). Statistical Pattern Recognition. In C. Boller, F.-K. Chang, & Y. Fujino (Eds.), Encyclopedia of Structural Health Monitoring (pp. 1–18). Academic Press.

Tibaduiza, D. A., Mujica, L. E., & Rodellar, J. (2011). Comparison of several methods for damage localization using indices and contributions based on PCA. *Journal of Physics: Conference Series*, *305*, 12013. doi:10.1088/1742-6596/305/1/012013

Wang, X., Kruger, U., Irwin, G. W., McCullough, G., & McDowell, N. (2008). Nonlinear PCA With the Local Approach for Diesel Engine Fault Detection and Diagnosis. *IEEE Transactions on Control Systems Technology*, *16*(1), 122–129. doi:10.1109/TCST.2007.899744

Wold, S., Kettaneh, N., Fridén, H., & Holmberg, A. (1998). Modelling and diagnostics of batch processes and analogous kinetic experiments. *Chemometrics and Intelligent Laboratory Systems*, *44*(1), 331–340. doi:10.1016/S0169-7439(98)00162-2

Yan, R., Chen, X., & Mukhopadhyay, S. C. C. (2017). Advanced Signal Processing for Structural Health Monitoring. In *Smart Sensors* (Vol. 26). Measurement and Instrumentation.

ADDITIONAL READING

Boller, C., & Staszewski, W. J. (2004). Aircraft Structural Health and Usage Monitoring. In Health Monitoring of Aerospace Structures.

Downey, A., Ubertini, F., & Laflamme, S. (2017). Algorithm for damage detection in wind turbine blades using a hybrid dense sensor network with feature level data fusion. *Journal of Wind Engineering and Industrial Aerodynamics*, *168*, 288–296. doi:10.1016/j.jweia.2017.06.016

Holnicki-Szulc, J., Motylewski, J., & Koakowski, P. (*n.d.*) (pp. 1–9). Introduction to Smart Technologies. In Smart Technologies for Safety Engineering.

Karbhari, V. M., & Ansari, F. (2009). *Structural health monitoring of civil infrastructure systems*. Woodhead Pub. doi:10.1533/9781845696825

Kessler, S. S., Brotherton, T., & Gordon, G. A. (2011). Certifying Vehicle Health Monitoring Systems. In System Health Management: With Aerospace Applications (pp. 185–195). doi:10.1002/9781119994053.ch11

Kinet, D., Mégret, P., Goossen, K., Qiu, L., Heider, D., & Caucheteur, C. (2014). Fiber Bragg Grating Sensors toward Structural Health Monitoring in Composite Materials: Challenges and Solutions. *Sensors (Basel)*, *14*(4), 7394–7419. doi:10.3390140407394 PMID:24763215

Li, D., Ho, S.-C. M., Song, G., Ren, L., & Li, H. (2015). A review of damage detection methods for wind turbine blades. *Smart Materials and Structures*, *24*(3), 033001. doi:10.1088/0964-1726/24/3/033001

Martinez-Luengo, M., Kolios, A., & Wang, L. (2016). Structural health monitoring of offshore wind turbines: A review through the Statistical Pattern Recognition Paradigm. *Renewable & Sustainable Energy Reviews*, *64*, 91–105. doi:10.1016/j.rser.2016.05.085

Myung, H., Jeon, H., Bang, Y.-S., & Wang, Y. (2014). Sensor Technologies for Civil Infrastructures. In Sensor Technologies for Civil Infrastructures.

Ogai, H., & Bhattacharya, B. (2018). Smart Sensors for Structural Health Monitoring. In Intelligent Systems, Control and Automation: Science and Engineering (Vol. 89, pp. 153–184). doi:10.1007/978-81-322-3751-8_8

KEY TERMS AND DEFINITIONS

Composite Material: As used in this chapter, a material composed of stiff fibers (continuous or discontinuous) surrounded by a weaker matrix.

Damage: A change in the physical or geometrical properties of the material causing effect in the performance of the system or structure.

Distributed Sensing: Optical fiber-based sensing technology allowing to gather continuous measurements along an optical fiber so than the whole fiber can be seen as a continuous strain or temperature sensor with a high-spatial accuracy.

Fiber Bragg Grating (FBG): A type of fiber optic sensor capable of measuring strain and temperature with considerable advantages over traditional strain gauges. It consists of a periodic modulation of the refractive index fiber's core in a point.

Principal Component Analysis (PCA): A dimensionality reduction technique based on linear transformation for representing high-dimensionality, redundant, noisy data in a lower dimension while retaining the most variability.

Self-Organizing Maps: Classification technique based on unsupervised-learning artificial neural networks allowing to group data into clusters.

Smart Structure: A structure capable of sense disturbances, processing the information and evoking the reaction at the actuators. Structures implementing self-sensing capabilities by means of sensors and pattern recognition techniques lie into the smart definition.

Strain Field: Distribution of strains through a region of a body. For a more formal point of view, the strain field refers to a tensor field assigning a particular strain tensor to each point of the region.

Structural Health Monitoring: The integration of sensors and actuators in a material or structure in order to perform load monitoring, damage detection, damage diagnosis and damage prognosis so that nondestructive testing becomes online and in situ.

Chapter 2
Detection of Network Attacks With Artificial Immune System

Feyzan Saruhan-Ozdag
Istanbul University-Cerrahpasa, Turkey

Derya Yiltas-Kaplan
iD https://orcid.org/0000-0001-8370-8941
Istanbul University-Cerrahpasa, Turkey

Tolga Ensari
Istanbul University-Cerrahpasa, Turkey

ABSTRACT

Intrusion detection systems are one of the most important tools used against the threats to network security in ever-evolving network structures. Along with evolving technology, it has become a necessity to design powerful intrusion detection systems and integrate them into network systems. The main purpose of this research is to develop a new method by using different techniques together to increase the attack detection rates. Negative selection algorithm, a type of artificial immune system algorithms, is used and improved at the stage of detector generation. In phase of the preparation of the data, information gain is used as feature selection and principal component analysis is used as dimensionality reduction method. The first method is the random detector generation and the other one is the method developed by combining the information gain, principal component analysis, and genetic algorithm. The methods were tested using the KDD CUP 99 data set. Different performance values are measured, and the results are compared with different machine learning algorithms.

INTRODUCTION

Network security became more important because of increasing usage of internet and network technologies. Diversity of user profiles and increase of users have a big effect on network attacks. Various security policies are developed in the internal structures of institutions to protect the accessibility, integrity, and confidentiality of data against these security threats. These policies are supported by security applications

DOI: 10.4018/978-1-7998-1839-7.ch002

that are important for the network system and can be software or hardware such as firewall and intrusion detection systems (IDS). IDS need a learning process to decide whether the activities occurring on the network are going to be attacked or not. Machine learning is a popular research area that aims to enable machines to make decisions on their own by adopting human learning tactics. There are many studies that apply different machine learning methodologies to detect network attacks in the literature. Nguyen and Choi (2008) used J48 for detection, Koc et al. (2012) used Hidden Naive Bayes Multiclass Classifier, Hosseinpour et al. (2014) benefited from artificial immune system (AIS), Dasgupta and González (2002) used niching technique with genetic algorithm (GA) for generate detector in AIS, and Gupta and Shrivastava (2015) implemented support vector machine (SVM) and bee colony together for anomaly detection. This study is based on AIS algorithms to detect the attacks. AIS has been inspired by the human immune system. This system has been investigated especially for being an effective methodology to detect virus, fraud and fault control (Dasgupta, 1998). Negative selection algorithm (NSA) is effective on categorizing attacks, therefore in this paper the focus will be on this point. NSA uses antigen and anticors' structures which exist within natural immune system. The attacks in a network are assumed as antigen and the self-cells are recognized in the learning process as anticors.

In the literature, the IDS studies are based on two different detection methods: Anomaly detection and the misuse detection. In misuse detection, events on the system are evaluated according to signatures based on the weak points of the system or security policies. Every attack defined on the system has a signature. Behaviors that are not within these signatures are called normal. There should be defined signature for new attacks and it should be introduced to these systems. The main problem with this method is that the new attack type will not be recognized. Whereas anomaly detection determines the behavior on the system as normal or abnormal. In general, the traffic on the network is monitored and an assessment is made according to the thresholds. The system is trained through normal and abnormal situations, then it is expected to make an evaluation for each new data.

There are limited numbers of papers involving IDS and AIS together. A concurrent study was designed with Particle Swarm Optimization (Tabatabaefar et al., 2017). The experimental results were given as the detection and performance measurement rates. There has not been a comparison with any machine learning algorithm. Another IDS study is based solely on regular AIS (Suliman et al., 2018). DoS and Probing attack classes were detected and the results for correctly predicted attacks were calculated. Another study in the literature (Igbe et al., 2017) presents a method to detect DoS/DDoS attacks by using the Dendritic Cell Algorithm involved in AIS algorithms. In this method, the network traffic features were gathered as a vector and then used to retrieve antigen and signals. The signals were processed to be distinguished into dangerous and safe categories. The output signal was determined as anomaly or normal according to the computational values of the antigens.

This paper suggests an approach for IDS using AIS and for detector generation using both classical methods and GA. In order to observe the effect of the system on the decision making ability, feature selection and dimensionality reduction are used together in the pre-process step. KDD CUP 99 data set is used in the experiments. This data set, which is a different version of DARPA 1988 and DARPA 1999, has been prepared in the DARPA 1988 MIT Lincoln laboratory. Sections of this paper are prepared as following: Section 2 demonstrates a short background on IDS. Section 3 contains information about artificial and natural immune systems, GA, feature selection, and dimensionality reduction. Section 4 gives information about the applied approach. Section 5 illustrates experimental results. Finally, Section 6 gives the discussion on the study and Section 7 is about the conclusion and future works.

BACKGROUND

The IDS history started 40 years ago with increasing of enterprise networks. The problem in these networks can be explained as controlling the access levels of systems and following user activiy in information systems.

The first IDS concept was explained by James P. Anderson in 1980. He published a study with the purpose of improving computer security auditing and surveillance of customer systems (Anderson, 1980). The main idea behind this study was detecting unauthorized access to help administrators review audit trails by using user access logs, file access logs, and system event logs.

Between 1984 and 1986, Dorothy E. Denning and Peter G. Neumann designed an IDS model generating the basis for many systems still used today. This model was named Intrusion Detection Expert System (IDES). IDES had two main aproaches with a rule-based Expert System to detect intrusions and a statistical anomaly detection system based on profiles of users, host systems, and target systems. This model has been developed to a new form called Next-Generation IDES.

In 1987, Fred Cohen demonstrated that the valid systems could not detect all attacks in every case.

In the 1990s, IDS technologies were developed to detect complex and numerous network attacks. The development of anomaly detection and commercial applications of IDS was started during these years.

IDSs can be classified in two main categories as:

- Network-based IDS are the most commonly used IDS. They capture and analyze packets within network traffic. Because of monitoring the entire network, they can detect any attack occurring on any computer on the network. Network-based IDS consist of single-purpose sensors or servers positioned separately on the network.
- Host-based IDS were the first systems developed and implemented. They are usually built on critical computers and hosts. These systems are installed separately on each computer or host and follow only the device on which they are installed.

ARTIFICIAL IMMUNE SYSTEM

The immune system is a complex mechanism that fights against various diseases in human body. By analyzing each cell entering or contacting the body, AIS decides whether it is a healthy or unhealthy cell and tries to protect body against disease by fighting with unhealthy cells. In this mechanism, self and non-self cells are distinguished, pathogen and tumors are destroyed to protect the body. The immune system has a multilayered protection structure. The layers are skin, biochemical barriers, innate immune system, and adaptive immune system. The substances entering the body pass through these layers in order to protect the body against diseases.

AIS is inspired by the theories and models of the human immune system and applied for the solution of complex problems. There are different AIS algorithms in the literature to solve different problems. One of these algorithms is NSA. This algorithm was firstly used by Forrest et al. (1994) to detect computer viruses. The system developed in this study is based on immunity which is more secure and has more sophisticated concept than the existing operating systems. Virus threatening increases in parallel with growing network structures and increasing number of users. There have been many researches on the question of how to protect computers from these attacks. Resolutions produced by AIS are very effective. The problem for protecting computers from harmful viruses is a sample of problem separating self-components from others and NSA studies according to this approach. This method is used when an unexpected situation occurs in a system.

NSA has been designed on the ground of T cell maturation in thymus. Firstly, self-cells are recognized and then the detector generation phase is started. The detectors that do not match with self-cells are transferred to the detector set as in Figure 1. For matching process, some distance (similarity) calculation methodologies like Euclidean, Manhattan, and Minkowski are used for real value of data (Dasgupta, 2012). Any one of these computations can be used, because they have analogous formulas.

Feature Selection and Dimensionality Reduction

Feature selection becomes an important issue for machine learning with increasing feature numbers in the data sets. To decide related data and running some process on them give advantages for performance and productivity of produced results. Feature selection aims to eliminate unrelated, irrelevant, and redundant features on the data set to get more correct results and higher accuracy rates. There are many methods to create the feature subsets. Information gain is a filter-based feature selection method. It is a statistical method that decides whether a feature will remain or not remain in the data set by assigning a score to each feature. Information gain is measured for each feature in the data set. Entropy that provides the measure of uncertainty level for the data set is calculated for each feature (Yazıcı et al., 2015). It is considered that the features having higher entropies have more information about the data. The entropy formula can be seen in Equation 1.

$$\text{Entropy(D)} = -\sum_{i=1}^{m} p_i \log_2^{p_i} \qquad (1)$$

Here, Entropy(D) represents the entropy value of the data set D. p_i is the probability value of the class i in D and m is the number of all sub-classes. After the entropy calculation, information gain (represented with Gain) is calculated in Equation 2 for each feature and the features which have more gain

Figure 1. Detector generation process in NSA

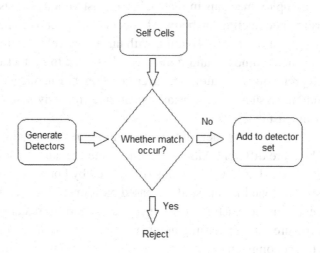

are selected to remain within the data set (Jin et al., 2009). Information gain calculates the importance of a feature in classifying the data.

$$\text{Gain } (D) = D - \Sigma p \ (Di)E(Di) \tag{2}$$

Dimensionality reduction creates new features from the existing features in a data set while feature selection creates new subsets by adding/subtracting features to/from the existing data set. Machine learning starts with a proper data design and for getting efficient results these methods can be performed. There are two main purposes for dimensionality reduction. The first one is to ensure that the data is restructured in the best possible way. The second one is to make the most efficient prediction. One of the dimensionality reduction methods is Principal Component Analysis (PCA) (Jolliffe, 2011). PCA was developed by Harold Hotelling as mentioned by Abdi and Williams (2010). As stated by Vasan and Surendiran (2016), PCA is a feature extraction technique that produces linear combination from the data set. It aims to determine values of correlation within data. Its goal is to produce uncorrelated features from correlative features by providing linearity between the data. These features are called as principal components (PCs). PC explains linear equality of features within a data set. Number of these features can be equal to or less than the original features. The transformation made during the PC finding process such as the first PC has the greatest variance. The next PC has less variance than the previous one and this notation continues to the last PC. The dimensions at which the data most associated with each other are found.

A *dxd* covariance matrix is created for the data set with *d* dimensions, *d* eigenvalues and eigenvectors. PCA provides a representation of the data as a vector and emphasizes the similarities and differences (statistical properties) of the data for this representation. The main purpose is to be able to work with minimum loss by keeping the variance high (Vasan & Surendiran, 2016).

Genetic Algorithm

GA is a search algorithm based on the natural selection in which biological process and evolution of genetic are included. GA aims to generate a set with the fittest solutions within search space. GA works with "survival of the fittest" principle. It tries to simulate remaining the fittest individuals which have been generated in the population during a problem solution. Every individual represents a possible solution in the search space and the individuals within the population are evolved in generating a better solution. GA works based on the following rules:

1. Individuals within a population compete for resources.
2. After each competition, the most successful individuals contribute to the new generation much more than the weak ones.
3. The genes coming from the fittest individuals are transferred to generate more fit individuals.
4. Each new generation produces much better solution set.

The individuals within a population are coded to define the length and get the defined values. Individuals can be assumed as chromosome where their values are as gene (Hitesh & Kumari, 2018). There are different ways in coding. Each way is specific to the problem. In this paper, value coding is used. In the value coding, genes can become real numbers.

The individuals can remain in a population according to their fitness values. The individuals which have optimal or nearly optimal fitness values are added to the solution set. A fitness function is calculated specifically according to the problem. GA aims to generate the best solutions with the concatenation of the genes which come from the chromosomes by following the rules. Parents are selected from a population based on their fitness values. These parents generate new individuals according to some rules and this process is repeated until the fittest individuals are generated. After random determination of the beginning population, three basic genetic operators are used in GA (Sujitha et al., 2012; Pandey, 2016). These operators are:

- **Selection**: It is based on the selection of the fittest individual. It gives high priority to the fittest individuals and selects them as parents without any changes. The main purpose is transferring the best genes to the next generations.
- **Crossover**: This is the most important feature that separates genetic algorithm amongst the optimization techniques. Two individuals are selected within a population by the selection operator. A crossover point is randomly selected (Haldurai et al., 2016). The front part before that point of the first parent and the back part after that point of the second parent are merged. If the individual that is generated after crossover has high fitness value, it is added to the solution set.
- **Mutation**: Biological mutation is referenced for this operator. Mutation is an operation that does small changes on the genes to generate new solutions. Every individual does not have to be mutated. Using mutation operator is bounded to a probability value and this value is inversely proportional to the mutation probability.

PROPOSED APPROACH

In this study NSA, which is the most suitable algorithm for intrusion detection, is used with AIS algorithms. The aim of the system is to decide the behavior of network as normal or attack. In this study, the PCA method is used for AIS. The implementation of the proposed IDS has three main phases, called Pre Process, Train, and Test, as shown in Figure 2.

The data set has three string columns, namely protocol type, service information, and flag information. So, the digitization is made for these columns. Relevant columns of the data set and their numerical values matched are given in Table 1, 2, and 3.

It is observed that the gain values of the properties 9, 15, 20, and 21 are 0. Here, the calculations are made by using Equation 1 and Equation 2. The features with 0 gain are removed from the data set before proceeding to the next step. While implementing PCA, the goals are to improve the factors such as accuracy rate and decision-making speed at the same time with the reducing number of features train

Table 1. Numerical matching of protocol type column of the data set

No	Protocol Name
1	tcp
2	udp
3	icmp

Figure 2. The phases of the proposed IDS

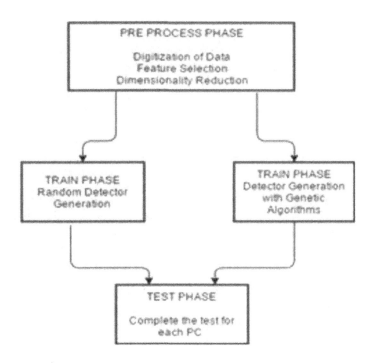

Table 2. Numerical matching of service column of the data set

No	Service	No	Service	No	Service	No	Service
1	http	18	link	35	printer	52	netbios_ssn
2	smtp	19	remote_job	36	efs	53	netbios_dgm
3	domain_u	20	gopher	37	courier	54	sql_net
4	finger	21	ssh	38	uucp	55	vmnet
5	auth	22	name	39	klogin	56	bgp
6	telnet	23	whois	40	kshell	57	Z39_50
7	ntp_u	24	domain	41	echo	58	ldap
8	ftp	25	login	42	discard	59	netstat
9	ecr_i	26	imap4	43	systat	60	urh_i
10	eco_i	27	daytime	44	supdup	61	X11
11	other	28	ctf	45	iso_tsap	62	urp_i
12	private	29	nntp	46	hostnames	63	pm_dump
13	pop_3	30	shell	47	csnet_ns	64	tftp_u
14	ftp_data	31	IRC	48	pop_2	65	tim_i
15	rje	32	nnsp	49	sunrpc	66	red_i
16	time	33	http_443	50	uucp_path	-	-
17	mtp	34	exec	51	netbios_ns	-	-

Table 3. Numerical matching of flag column of the data set

No	Flag	No	Flag
1	SF	7	RSTO
2	S0	8	RSTR
3	S1	9	RSTOS0
4	REJ	10	OTH
5	S2	11	SH
6	S3	-	-

system easily. The PCA implementation has 5 steps. Firstly, the input data set is prepared as a matrix. Then the mean value of the data is computed and these values are subtracted from the data vector. After then the covariance matrix is evaluated. From this covariance matrix eigenvector and eigenvalue are computed. Finally, eigenvectors that have largest eigenvalues are chosen (Jain & Singh, 2018).

In the proposed method, 21 PCs are produced according to the determined eigenvalue and eigenvectors. Within the scope of AIS algorithms, the training phase is the detector generation part. In this study, two different methods are used for detector generation. Euclidean distance formula is used to calculate the similarity between the detectors and the normal samples.

The classical method is known as random generation. For the random detector generation, the minimum and maximum values of each column are calculated and new column values are produced between these values. This procedure is repeated for each column to obtain a vector which is compared to the normal samples in the data set. Based on the comparison made after the calculated Euclidean value, it is decided whether the generated vector is a detector or not. Random detector generation algorithm is shown in Figure 3.

On the other hand, GA uses both normal and attack samples during the detector generation phase. The reason for using GA is to rescue the system from randomness and to see the best solution within GA procedures and its effect on the decision of the system. Detector generation with GA is shown in Figure 4.

Figure 3. Random detector generation algorithm for AIS

```
1: Calculate min euclid distance for nonself samples as minEuclid

2: Calculate max euclid distance for self samples as maxEuclid

3: Generate a vector randomly

4: for The number of self samples do

5:      Calculate euclid distance between vector and self samples as

6:      if min> minEuclid && max > maxEuclid

7:         Accept vector as detector

8:      end if

9: end for
```

Figure 4. Detector generation with GA for AIS

```
1: Initialize population by selecting individuals

   from the space S and A.

2: Calculate the fitness function for each individuals

3: for The number of generation do

4: for The number of population do

5:       Select two individuals one from S and one from A as parents

         by Roulette wheel Selection

6:       Apply crossover

7:       Apply mutation

8:       Calculate the fitness function for child as d2

9:       for The number of detectors do

10:          if d2 < di

11:              Replace child with detector(i)

12:          end if

13:      end for

14:  end for

15: end for
```

Euclidean distance is used to calculate the fitness values of the individuals. The efficiency of AIS algorithms is mostly based on the detector generation. Each PC that is produced by PCA can be considered as a gene. Therefore, when these PCs come together, a chromosome is obtained. In the roulette wheel selection, the operation is based on the creation of a base for the next generations by stochastic selections of the parents (Pandey, 2016). The crossover point is randomly chosen and the genes are combined according to this point. Individuals which have the fittest value have more chance to be selected than the weak ones. It does not mean that the weak ones do not have chance to be selected. They may have useful genes to transfer to the next generations as in nature. The steps in the application are as below:

1. Sum of all fitness values within the population is calculated.
2. Assignment of a random value to x.
3. Calculation of sum starts with the first fitness value and continues until the sum becomes larger than x, the individual that makes the sum larger than x is selected.

In the detection phase, samples are compared with each detector as in Figure 5. If any match occurs, the related sample is obtained as an attack or non-self. Otherwise, the sample is normal and categorized as self.

Figure 5. Anomaly detection process in NSA

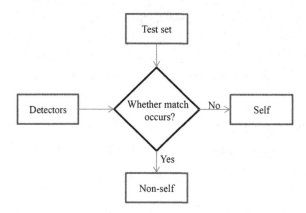

EXPERIMENTAL RESULTS

Data Set

The testing step as well as the design phase is important for IDS. Reliable and comprehensive data sets are needed to test the system. One of the most frequently used data set for IDS is KDD CUP 99 that is also used in this study. In KDD CUP 99 data set the data is labelled according to whether it is attack or not. The data set has 41 features and 22 different attack types. The information about the types and features of this data set can be seen from Table 4 and Table 5.

Performance Parameters

There are different methods for evaluating the models created as a result of machine learning techniques. A confusion matrix is a table that contains the performance metrics for the actual and predicted values. A confusion matrix of an example with two classes (Positive or Negative) can be seen in Figure 6.

The confusion matrix on Figure 6 is detailed as follows:

- **True Positive**: The algorithm returns the same result as the current situation. (The algorithm correctly predicts the attack record.).
- **False Positive**: The algorithm returns different result than the current situation. (The normal record is predicted as attack by the algorithm.).
- **False Negative**: The algorithm returns different result than the current situation. (The attack record is predicted as normal by the algorithm.).
- **True Negative**: The algorithm returns the same result as the current situation (The algorithm correctly predicts the normal record.).

Accuracy and detection rate values are calculated for the developed AIS algorithms as in Equation 3 and Equation 4 with benefiting from Figure 6.

$$Accuracy = (TP + TN) / (TP + FP + TN + FN) \tag{3}$$

Table 4. Attack types

No	Attack Type	No	Attack Type
1	back	12	ipsweep
2	buffer_overflow	13	imap
3	ftp_write	14	loadmodule
4	guess_passwd	15	nmap
5	neptune	16	land
6	phf	17	perl
7	pod	18	warezmaster
8	warezclient	19	teardrop
9	spy	20	multihop
10	satan	21	smurf
11	portsweep	22	rootkit

Table 5. Features list

No	Feature Name	No	Feature Name
1	Duration	22	is_guest_login
2	protocol_type	23	Count
3	Service	24	srv_count
4	Flag	25	serror_rate
5	src_bytes	26	srv_serror_rate
6	dst_bytes	27	rerror_rate
7	Land	28	srv_error_rate
8	wrong_fragment	29	same_srv_rate
9	Urgent	30	diff_srv_rate
10	Hot	31	srv_diff_host_rate
11	num_failed_logins	32	dst_host_count
12	logged_in	33	dst_host_srv_count
13	num_compromised	34	dst_host_same_srv_rate
14	root_shell	35	dst_host_diff_srv_rate _
15	su_attempted	36	dst_host_same_src_port_rate
16	num_root	37	dst_host_srv_diff_host_rate
17	num_file_creations	38	dst_host_serror_rate
18	num_shells	39	dst_host_srv_serror_rate
19	num_access_files	40	dst_host_rerror_rate
20	num_outbound_cmds	41	dst_host_srv_rerror_rate
21	is_host_login	-	-

Figure 6. Confusion matrix for performance measurements

		PREDICTED		TOTAL
		TRUE	FALSE	
ACTUAL	TRUE	TP True Positive	FN False Negative	Actual Positive
	FALSE	FP False Positive	TN True Negative	Actual Negative
TOTAL		Predicted Positive	Predicted Negative	Total Samples

Detection Rate $= TN/(TN + FP)$ (4)

Within the scope of the study, three different IDS applications are developed.

- **AIS1**: Feature selection and dimensionality reduction are not applied. Random detector generation method is used.
- **AIS2**: Feature selection and dimensionality reduction are applied. Random detector generation method is used.
- **AIS3**: Feature selection and dimensionality reduction are applied. Detector generation with GA method is used.

AIS1 is developed for evaluating the results of the hybrid methods of AIS2 and AIS3. With this method, 41 features in the data set are also used. It is observed that the distributions of the numerical values of these features can be very different. The normalization process is preferred in order to determine the threshold value to be used in the AIS and the training period to be short. AIS algorithms are tested on the KDD CUP 99 data set. Table 6 shows the results for AIS1.

AIS2 and AIS3 algorithms are tested according to two different scenarios. The first scenario is based on the PC numbers. The number of detectors and population size is assumed as 100. Table 7 shows the results for AIS2 when the detectors are generated randomly.

AIS2 gives the best results when the PC number is 4. By handling this value, AIS2 is compared with the other machine learning algorithms in Figure 7. Algorithms used to compare are mentioned below:

- **J48**: The J48 algorithm is a classification algorithm which is also known as C4.5 in the literature and is an extension of ID3. J48 is used as a Decision tree algorithm in machine learning.
- **Naive Bayes**: Naive Bayes algorithm is based on the Bayes theorem. The algorithm calculates the probability of each element for all cases and classifies them according to the highest probability value (Belouch et al., 2018).
- **Sequential Minimal Optimization**: Sequential Minimal Optimization (SMO) Algorithm works with the SVM-based approach. SMO algorithm solves any optimization problem of SVM during the training phase occurrence.
- **Artificial neural networks**: Artificial neural networks (ANN) cover mathematical model inspired by the structure and function of biological neural networks. ANN have connected units that work together to process information. According to these units the meaningful results occur.
- **BayesNet**: It is used to explain data modeling and state transitions. Statistical networks and the edges that are switching between nodes are selected according to statistical decisions.

Table 6. AIS1 results

Detector	Accuracy	Detection Rate
50	92.75	91.29
100	96.2	90.9
150	93.28	91.64
200	90.13	94.3
250	90.11	87.19

Table 7. AIS2 results

Principal Component	Accuracy	Detection Rate
1	92.53	95.11
2	95.17	92.75
3	94.95	93.19
4	98.5	97.18
5	96.05	93.02
6	88.13	72.34
7	90.61	76.73
8	91.18	87.89
9	86.93	75.81
10	92.74	89.81
11	88.86	80.9
12	90.22	87.36
13	88.54	76.4
14	86.09	79.21
15	90.68	85.1
16	89.68	87.27
17	89.16	90.23
18	88.44	90.96
19	90.83	89.47
20	81.84	88.96
21	88.21	88.2

It is seen on Figure 7 that AIS2 has better accuracy rates than almost all the others.

Table 8 shows the results for AIS3 when the detectors are generated randomly. AIS3 gives the best results when the PC number is 12. Due to the operational nature of the GA operators, better results are obtained with a greater number of the PCs.

AIS3 is also compared with other machine learning algorithms Naïve Bayes, J48, SMO, and ANN. It is represented in Figure 8 that AIS3 has better accuracy rates than almost all the others like AIS2.

Figure 7. Accuracy values of AIS2 and other algorithms

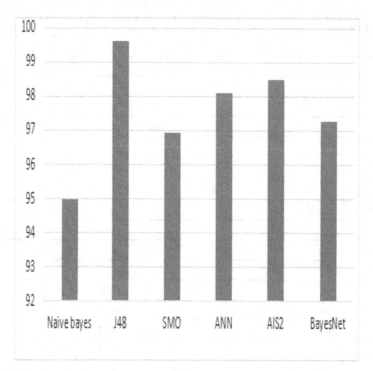

In scenario 1 the PC number that gives the best results is found. In scenario 2, changing the detector numbers with the PCs shows the effect of this experiment. Table 9 and Table 10 illustrate the results for AIS2 and AIS3 respectively when the detectors are generated with GA.

Figure 8. Accuracy values of AIS3 and other algorithms

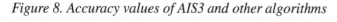

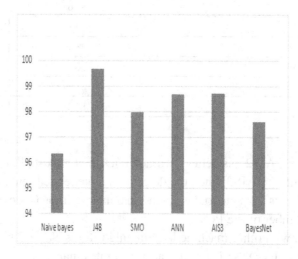

Table 8. AIS3 results

Principal Component	Accuracy	Detection Rate
1	91.21	95.66
2	96.01	90.38
3	96.09	93.87
4	96.41	93.23
5	96.08	94.5
6	96.59	96.09
7	97.24	95.89
8	96.53	95.88
9	96.74	96.26
10	96.71	96.3
11	97.79	96.91
12	98.73	98.01
13	97.2	96.18
14	97.31	96.06
15	96.84	96.09
16	97.08	96.41
17	97.12	96.09
18	96.61	96.53
19	97.53	95.92
20	97.73	95.75
21	95.87	95.29

Table 9. AIS2 results depending on various detector numbers

Detector	Accuracy	Detection Rate
50	94.08	90.96
100	98.5	97.18
150	96.28	91.64
200	96.02	90.93
250	96.12	90.41

Table 10. AIS3 results depending on various detector numbers

Detector	Accuracy	Detection Rate
50	97.1	95.92
100	98.73	98.01
150	96.4	96.29
200	97	95.39
250	97.52	96.12

DISCUSSION

All three models proposed in this paper, namely AIS1, AIS2, and AIS3, have produced quite close results in comparison with each other. The highest accuracy of AIS1 is 96.3% with 100 detectors. Also the highest value is obtained when the number of phase detectors is 100 in both AIS2 and AIS3. As a result of the comparisons made with other algorithms for both AIS2 and AIS3, they outperformed many algorithms, AIS2 with an accuracy rate of 98.5% and AIS3 with 98.7%. The fact that the decision trees are simpler than other algorithms, J48 achieved higher accuracy. The lowest accuracy values were obtained in Naive Bayes algorithm as shown in Figure 7 and Figure 8. Naive Bayes can produce poor results for multidimensional data sets. Hosseinpour et al. (2014) have accuracy values 77.1% and 60.7% for density-based spatial clustering of applications with noise (DBSCAN) and k-means methods, respectively. The proposed methods have better rates as shown in Figure 8, Table 9 and Table 10. The algorithm of Panda and Patra (2007) has 95% detection rate, but AIS3 has better rates for different number of detectors as shown in Table 10. In the study of Tavallaee et al. (2009), ANN, Naïve Bayes, and J48 algorithms have 92.26%, 81.66% and 93.82% rates, respectively. But, AIS3 has higher rates as illustrated in Figure 8. Finally, when we analyze all results of the algorithms, as stated at the beginning of the study, the novel AIS algorithm is very suitable for IDS because it detects the negative situations.

CONCLUSION AND FUTURE SCOPE

In this paper, three experiments and approaches based on AIS were developed. In order to get better results in pre-processing phase, both feature selection and PC analysis were used in AIS2 and AIS3. A positive effect on the results was observed by using these methods together. When the results of AIS1 and other algorithms were compared, it was observed that the aim of increasing the accuracy rates at the beginning of the study was realized. In the stage of detector generation due to the principle of producing the best solution provided by GA, there were better results than the random detector generation. Especially combining of the feature selection, PC analysis, and GA for AIS is much more effective. This part was the main contribution of the proposed method.

The KDD CUP 99 data set was used for the training and testing phases. The data set contains 4 categories for attacks as follows: DoS (denial of service), Probe, u2r (User to Root), r2l (Remote to Local) and totally has 22 attack types. The proposed approach is effective for detecting the attacks and can also be used for different types of attacks. It is a future work to predict the attack types by using various data and features. The approach was also compared with the previous published results and better rates were obtained.

REFERENCES

Abdi, H., & Williams, L. J. (2010). Principal Component Analysis. *Wiley Interdisciplinary Reviews: Computational Statistics, 2*(4), 433–459. doi:10.1002/wics.101

Belouch, M., Hadaj, S. E., & Idhammad, M. (2018). Performance evaluation of intrusion detection based on machine learning using Apache Spark. *Procedia Computer Science, 127*, 1–6. doi:10.1016/j.procs.2018.01.091

Dasgupta, D. (1998). An Overview of Artificial Immune Systems and Their Applications, *Artificial Immune Systems and Their Applications*. In D. Dasgupta (Ed.), (pp. 3–21). Berlin, Germany: Springer-Verlag Berlin Heidelberg.

Dasgupta, D. (2012). Immunity-Based Intrusion Detection System: A General Framework. In *Proceedings of Seventh Int. Conf. on Bio-Inspired Computing: Theories and Applications* (pp. 417-428). Springer.

Dasgupta, D., & González, F. (2002). An Immunity-based Technique to Characterize Intrusions in Computer Networks. *IEEE Transactions on Evolutionary Computation, 6*(3), 281–291. doi:10.1109/TEVC.2002.1011541

Forrest, S., Perelson, A. S., Allen, L., & Cherukuri, R. (1994). Self-Nonself Discrimination in a Computer. In *Proc. of IEEE Symposium on Research in Security and Privacy* (pp. 202- 212). Academic Press.

Gupta, M., & Shrivastava, S. K. (2015). Intrusion Detection System based on SVM and Bee Colony. *International Journal of Computers and Applications, 111*(10), 27–32. doi:10.5120/19576-1377

Haldurai, Madhubala, Rajalakshmi, & Anderson. (1980). Computer Security Threat Monitoring and Surveillance. Technical Report, Fort Washington.

Haldurai, L., Madhubala, T., & Rajalakshmi, R. (2016). A Study on Genetic Algorithm and its Applications. *International Journal on Computer Science and Engineering, 4*(10), 139–143.

Hitesh, K. A. C. (2018). Feature Selection Optimization in SPL using Genetic Algorithm. *Procedia Computer Science, 132*, 1477-1486.

Hosseinpour, F., Amoli, P. V., Farahnakian, F., Plosila, J., & Hämäläinen, T. (2014). Artificial Immune System Based Intrusion Detection: Innate Immunity using an Unsupervised Learning Approach. *Int. J. Digit. Content Technol. and its Appl., 8*(5), 1-12.

Igbe, O., Ajayi, O., & Saadawi, T. (2017). Denial of Service Attack Detection using Dendritic Cell Algorithm. In *2017 IEEE 8th Annual Ubiquitous Computing, Electronics and Mobile Communication Conference (UEMCON)* (pp. 294-299). New York, NY: IEEE. 10.1109/UEMCON.2017.8249054

Jain, D., & Singh, V. (2018). An Efficient Hybrid Feature Selection model for Dimensionality Reduction. *Procedia Computer Science, 132*, 333-341.

Jin, C., Lin, L., & Xiang, M. (2009). An Improved ID3 Decision Tree Algorithm. In *Proceedings of International Conference on Computer Science and Education* (pp. 127-130). Academic Press.

Jolliffe, I. T. (2011). Principal Component Analysis. In International Encyclopedia of Statistical Science. Springer.

Koc, L., Mazzuchi, T. A., & Sarkani, S. (2012). A Network Intrusion Detection System Based on a Hidden Naive Bayes Multiclass Classifier. *Expert Systems with Applications, 39*(18), 13492–13500. doi:10.1016/j.eswa.2012.07.009

Nguyen, H. A., & Choi, D. (2008). Application of Data Mining to Network Intrusion Detection: Classifier Selection Model. In *Asia-Pacific Network Operations and Management Symposium*. Springer-Verlag.

Panda, M., & Patra, M. R. (2007). Network Intrusion Detection Using Naïve Bayes. *Int. Journal of Computer Science and Network Security, 7*(12), 258–263.

Pandey, H. M. (2016). Performance Evaluation of Selection Methods of Genetic Algorithm and Network Security Concerns. In *Int. Conf. on Information Security and Privacy* (no. 78, pp. 13-18). 10.1016/j.procs.2016.02.004

Sujitha, B. B., & Ramani, R. R., & Parameswari. (2012). Intrusion Detection System using Fuzzy Genetic Approach. *Int. J. Adv. Res. Comp. Commun. Eng., 1*(10), 827–831.

Suliman, S. I., Abd Shukor, M. S., Kassim, M., Mohamad, R., & Shahbudin, S. (2018). Network Intrusion Detection System Using Artificial Immune System (AIS). In *Proc. of 2018 3rd International Conference on Computer and Communication Systems (ICCCS)* (pp. 178-182). Nagoya, Japan: Academic Press. 10.1109/CCOMS.2018.8463274

Tabatabaefar, M., Miriestahbanati, M., & Grégoire, J.-C. (2017). Network Intrusion Detection through Artificial Immune System. In *2017 Annual IEEE International Systems Conference (SysCon)*. Montreal, QC, Canada: IEEE. 10.1109/SYSCON.2017.7934751

Tavallaee, M., Bagheri, E., Lu, W., & Ghorbani, A. A. (2009). A Detailed Analysis of the KDD Cup 99 Data Set. *IEEE Int. Symposium on Computational Intelligence for Security and Defense Applications.* 10.1109/CISDA.2009.5356528

Vasan, K. K., & Surendiran, B. (2016). Dimensionality reduction using principal component analysis for network intrusion detection. *Perspectives on Science, 8,* 510–512. doi:10.1016/j.pisc.2016.05.010

Yazıcı, B., Yaslı, F., Gürleyik, H. Y., Turgut, U. O., Aktas, M. S., & Kalıpsız, O. (2015). Veri Madenciliğinde Özellik Seçim Tekniklerinin Bankacılık Verisine Uygulanması Üzerine Araştırma ve Karşılaştırmalı Uygulama, Ulusal Yazılım Mühendisliği Sempozyumu (a national conference). İzmir.

Chapter 3

Fog Computing and Edge Computing for the Strengthening of Structural Monitoring Systems in Health and Early Warning Score Based on Internet of Things

Leonardo Juan Ramirez Lopez

https://orcid.org/0000-0002-6473-5685

Universidad Militar Nueva Granada, Colombia

Gabriel Alberto Puerta Aponte

https://orcid.org/0000-0003-1730-170X

Universidad Militar Nueva Granada, Colombia

ABSTRACT

Currently, with the implementation of IoT, it is expected that medicine and health obtain a great benefit derived from the development of portable devices and connected sensors, which allow acquiring and communicating data on symptoms, vital signs, medicines, and activities of daily life that can affect health. Despite the possible benefits of health services assisted by IoT, there are barriers such as the storage of data in the cloud for analysis by physicians, the security and privacy of the data that are communicated, the cost of communication of the data that is collected, and the manipulation and maintenance of the sensors. This chapter intends to deploy and develop the context of the IoT platforms in the field of health and medicine by means of the transformation of edge and fog computing, as intermediate layers that provide interfaces between heterogeneous networks, networks inherited infrastructure, and servers in the cloud for the ease of data analysis and connectivity in order to implement a structural health monitoring based on IoT for application of early warning score.

DOI: 10.4018/978-1-7998-1839-7.ch003

INTRODUCTION

The new platforms, being a great collaborator and promoter of the new tools, models, instruments and appearances of the health sector (Marolla, 2018). The most implemented applications through IoT is the monitoring, monitoring and management of sector contents (Hsieh, Lee, & Chen, 2018). These applications show great problems according to the sensitive characteristic of the information that is managed. Some of the most representative issues of IoT implementation, in the e-health electronic health management systems, are due to the need for information privacy, secure communication in the media, authentication, protocols control, transport and service orchestration (Aghili, Mala, Shojafar, & Peris-Lopez, 2019). In addition to this the growth of these applications and users, can cause the growth of large volumes of data, which will require great resources or for their transport and analysis (Annamalai, Bapat, & Das, 2019).

IoT can define as a global infrastructure for the information society, which through available services enables the interconnection of physical and virtual objects based on interoperable business-to-business communication technologies (Networks, 2012).The IETF defines IoT as the Internet that considers simultaneous TCP / IP and non-TCP / IP sets and devices or items as objects "created" by single directives (Valdivieso, Peral, Barona & García, 2014). IEEE IoT in its Special Report on "Internet of Things" as defined: A device that connects devices with detection storage (Minerva, Biru, & Rotondi, 2015). Now redefine as a red that connects uniquely identifiable virtual and physical devices using new or existing communication protocols. Verify that dynamically configured devices or devices and user interfaces are accessible from a distance from the Internet (Mirón Rubio et al., 2018).

INTERNET OF THINGS

Currently, science debates about the Internet of Things (IoT) paradig, given the technological advance and the increase of application fields (Lee & Lee, 2015). Even more so when technological trends and innovation at the global level, the results for the diversification of services for society and industry in general (Vögler Matrikelnummer, 2016a). Thus, applications based on the use of content, the possibilities in the services of smart and connected homes, industries 4.0, the orange economy, specialized medical care, health self-care, environmental monitoring, adaptive logistics, the national defense, the automatic transport and the cybersecurity (O. Salman, Elhajj, Chehab, & Kayssi, 2018) .

For the implementation of IoT there are different platforms, hardware and architectures, they are used for different providers, they are given the opportunity to have heterogeneous networks of inherited technology and a variety of them, or the means of communication, the formats of the data and the means of transmission (Cha et al., 2016; S. Lee, Bae, & Kim, 2017). On the other hand, interoperability becomes the priority of the integration of communication systems, the interconnected media, not only at a physical level, but at the level of applications and services (Jabbar, Ullah, Khalid, Khan, & Han, 2017; Bhattarai & Wang, 2018). As an alternative, IoT technology designers use protocols such as: ZigBee, Z-wave, LTE, Wi-Fi, Ethernet, X10, Bluetooth, among others; However, in the standard IoT application network (Minerva et al., 2015) it can be used as an intermediate manager between the legacy data networks and the final sensors.

The International Telecommunications Union (ITU), expected before 2020, the IoT platforms are around 50,000 million network devices (ITU Corporation, 2015). Forbes publishes that the prediction of the IoT markets for the year 2021 will be approximately 521 billion dollars, more than double the year 2017 (Forbes, 2018). This is a sample that in the future. This trend shows that in the near future there will be more connected devices than people, which will have an impact on almost every aspect of our lives (Bramsen, 2017). The communication capacity will not be limited simply to mobile devices, on the contrary this capacity will expand to the things we live with every day (Atzori, Iera, & Morabito, 2017). This means that not only objects or things can communicate, but contents and other media also enter into this new scenario, which can change the paradigm of legacy technologies and networks (O. Salman, Elhajj, Chehab, & Kayssi, 2018).

IoT can be defined as a global infrastructure for the information society, which allows through advanced services the interconnection of physical and virtual objects, based on interoperable communication technologies between existing and evolving (Mehner, e.d.). The IETF defines IoT as the Internet that considers sets of TCP / IP and not TCP / IP simulate us and devices or things as "objects" identified by unique addresses. IEEE IoT in its special report on "Internet of Things", defined it as: A network that connects devices with detection capabilities (J. Wang & Li, 2018). Now it redefines it, as a network that connects identifiable virtual and physical devices in a unique way, using new or existing communication protocols. Where things or devices can be dynamically configured and have interfaces that must be accessible remotely via the Internet (Atzori, Iera & Morabito, 2017).

The concept of IoT, is related to other concepts and tools, which in turn can complement these platforms, for this, some of the connections that were considered the most relevant are described by next figure.

Figure 1. Concepts related to internet of things

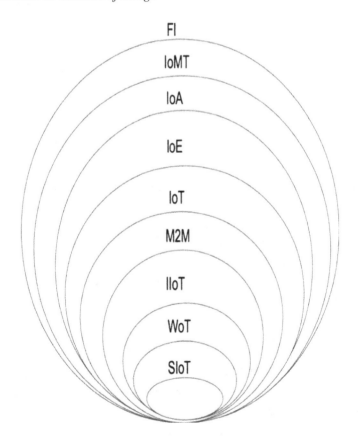

- **SIoT**: Social Internet of Things (SIoT), is considered as the ability of objects as instruments for the establishment of social connectivity and autonomous relationship with users, for the discovery of network services and resources (Panda & Bhatnagar, 2020b). The social network of objects can find novel resources for the best domain in the implementation of user services. In social networks, physical or virtual objects are interconnected to the network to transmit the virtual dimension in decision making to the physical world (Roopa et al., 2019). The SIoT, allows objects to connect to social networks, can publish autonomous information in the owners' social loops, this for the realization of actions and activities related to the users of an automatic machine (Takiddeen & Zualkernan, 2019). The realization of activities automatically and the interaction with the social devices of the users and devices, implies that SIoT establishes and configures new rules, uses and benefits (Panda & Bhatnagar, 2020a).

- **WoT:** The Web of Things (WoT), describes the integration of everyday objects with the web, for it requires integrated computer systems that allow communication and interoperability in the platforms and domains of the applications, that is, WoT generates mechanisms that give the possibility that the devices and services communicate with each other, and allows access and control of IoT resources and applications using conventional web technologies. Unlike IoT, WoT tries to integrate the network of objects, things, people, systems and applications, but still has several challenges of scalability and security (Médini et al., 2017; Chauhan & Babar, 2017).

- **IIoT:** Industrial Internet of Things (IIoT), is how IoT platforms are considered, which provide interconnection and intelligence to industrial systems through the implementation of detection devices, sensors and actuators capable of belonging to ubiquitous and heterogeneous networks (Li & Wang, 2019). The IIoT platforms are positioning themselves as an effective and impacting industry 4.0, a concept of industrial development trends globally (Kabugo, Jämsä-Jounela, Schiemann, & Binder, 2020).

- **M2M:** Machine to Machine (M2M), refers to communications and interactions between machines and devices, these interactions can occur through one or more communication networks or through a cloud computing infrastructure (Shala, Trick, Lehmann, Ghita, & Shiaeles, 2019). M2M offers the means to manage devices interactively, while collecting data from machines or sensors, facilitating communications with mechanical automation (Azad, Bag, Hao, & Salah, 2018). M2M applications cover different areas such as intelligent networks, transport systems, industrial automation, environmental monitoring and e-healthcare (R. H. Ahmad & Pathan, 2016).

- **IoE:** The Internet of Everything (IoE), is a global dynamic information network that links any object and people on the Internet, this is how the internet of things involves, but also expands to include people, processes and data in a broader scope, that allows to make connections generating the capacity to collect large amounts of data and the analysis of them (Iannacci, 2018). The exchange of information and communications is one of the most relevant aspects, which enables the development of other applications such as monitoring biometric data in real time (Galitsky & Parnis, 2019).

- **IoA:** Internet of Agents (IoA), can be defined as an evolutionary IoT process, where the participation of the final devices is considered as a fundamental factor for the adaptability to the behavior of the networks within the IoT platforms (Pico-Valencia, Holgado-Terriza, Herrera-Sánchez, & Sampietro, 2018). This adaptation of network resources is obtained by customizing specific agents at runtimes of the services in IoT (R. Wang et al., 2020). Other authors identify IoA, as an intelligent ecosystem of agents that can manage the resources associated with IoT objects. Resources are management can achieve a level of cognition through software agents that employ semantic techniques that are embedded within the infrastructure (R. Wang et al., 2020). In this way agents have the ability to manage context, loops or social circles, services and resources of IoT (Mostafa, Gunasekaran, Mustapha, Mohammed, & Abduallah, 2020).

- **IoMT:** The Internet of Medical products or the Internet of Medical Things (IoMT), can be defined as the connectivity of a medical device to a health care system through an online network, such as a cloud, which often involves machine communication machine. This refers to the devices based on medical sensors that are incorporated with IoT and whose data collected with these devices are combined with the electronic medical records systems to achieve the transformation of health systems, making them more efficient and efficient. Thus, with the growing development of IoT health sensors, the collection and analysis of medical data is achieved, to reduce the annual costs of managing chronic diseases by about one third (Basatneh, Najafi, & Armstrong, 2018).
- **FI:** Future Internet (FI), is a global network that will encompass all the above-mentioned networks (You, 2016). The FI has six fundamental principles that respond to the requirements of innovation and operation; The six basic principles are the following: Connectivity, Context, Collaboration, Cognition, Cloud and Content (Salamatian, 2011). This network will allow the connection of any type of device, which will generate large volumes of data; this data that will be handled by technologies based on cloud computing, which in turn will contribute to the analysis and transformation of behavior and the context of collaboration between devices, users, contents and things (You, 2016).

Challenges of Internet of Things

For the implementation of IoT there are different platforms and architectures, designed by different providers, which causes heterogeneous networks of inherited technology and a diverse range of physical or virtual sensor devices, communication protocols, data formats and transmission media (Farris, Taleb, Khettab, & Song, 2018). Moreover, interoperability is appropriate in the priority of integrating existing communication systems with new ones to interconnect, on the ground at the physical level, bell at the level of applications and services (Jabbar Ullah, Khalid, Khan & Han, 2017; Farris Taleb, Khettab & Song, 2018). Alternatively, IoT technology designers use protocols such as: ZigBee, Z-wave, LTE, Wi-Fi, Ethernet, X10, Bluetooth, among others; It is not possible to use some standardization protocol for IoT applications which acts as an intermediate manager between inherited data networks and final sensors (Vögler Matrikelnummer, 2016).

The Internet of things presents major challenges, due to its growth and dynamism (Bhattarai & Wang, 2018); in addition to its lack of standards, despite the spur of the moment and business and government initiatives for the development of open source code and hardware. For this reason, researchers have confronted solutions from different fields (Tayyaba, Shah, Khan, & Ahmed, 2017). Some authors define IoT architecture in ways such as internet-centric or home-centric (Almusaylim & Zaman, 2018).

- **Connectivity:** The action of connecting thousands of millions of devices in red is not easy (Atzori Iera & Morabito, 2017). Summed up by the number of devices, it is the variety and variety of different categories and types of devices that constitute the heterogeneity of these platforms, it seems to be one of the most difficult but to achieve connectivity in these conditions (You, 2016). These major differences can break communication and expected performance metrics (Son, 2018).

- **Energy Consumption:** The basic input for the operation of any electronic device is the energy. This input can be supplied by an electrical network, batteries or other forms of energy collection (Liu & Ansari, 2019). In any case, as a result of the scalability and size of the IoT platforms, the devices must be designed and built on the basis of low energy consumption (Ramirez Lopez, Puerta Aponte, & Rodriguez Garcia, 2019).

- **Interoperability and Integration**: IoT implementations are based on many devices from different providers, which sometimes use different technologies (Grigoryan, Njilla, Kamhoua, & Kwiat, 2017). Successful solo integration is possible with IoT systems operating over open standards (El-Mougy, Al-Shiab, & Ibnkahla, 2019). Of any way there may be a number of standards for the same scope of application, but should be established and guaranteed interoperability between them (Soursos, Zarko, Zwickl, Gojmerac, Bianchi & Carrozzo, 2016).

- **Computing and Storage Capacity:** The devices that can compose an IoT system generate large data chants (van Oorschot & Smith, 2019). These data can be presented in continuous form of ragages, can be structured or unstructured. For further analysis, data are transported, stored, and serviced by a large number of hardware and connectivity features (Dolui & Datta, 2017).

- **Security, Reliability and Privacy of Information**: Establishing and penetrating IoT figures on a day-to-day basis, there is a need for generation and implementation of tailor-made secure solutions (Dolui & Datta, 2017). It should also be foreseen that the number of devices and the tendency to increase the number of devices in the design of secure IoT systems (Krishnan, Najeem, & Achuthan, 2018). For this reason, security solutions adapted to IoT must be portable, multi-device and reliable (Bull, Austin, Popov, Sharma, & Watson, 2016).

Different authors and organizations have initiated the response to the challenges of IoT implementation, from several trends, some focused on software and others focused on hardware (Atzori Iera & Morabito, 2017). From the conception of hardware and architecture, some researchers have come up with new solutions to the challenges of implementing IoT platforms, from the perspective of computing paradigms, cloud computing, fog computing and edge computing can be considered as another way to meet Iot's challenges in search of solutions (Li & Wang, 2019).

- **Cloud Computing:** Cloud computing can be described as a paradigm that influences the form of design, development, implementation, prevention, and maintenance of applications across the network (Bull o.fl., 2016). Cloud computing is built as a utility model for the deployment of storage resources and computing capacity (Andreev, Petrov, Huang, Lema, & Dohler, 2019). This utility model is supported by network resources, which support the ability to access remote services (Dolui & Datta, 2017).

From the infrastructure in the cloud it is possible to access the sensors layer (things), where you can get, manage, store, analyze and view the data and services of users (Mourad, Nassehi, Schaefer, & Newman, 2020). Because of its storage and process capability, cloud computing allows for an important number of storage and processing service requests (K. Ahmad, Mohammad, Atieh, & Ramadan, 2019). Cloud computing is a tool for enhancing IoT platforms, where you can meet IoT implementation challenges, as long as you take care of the connectivity and network resources required over the latencies and connection times (Rayes & Salam, 2016). The Figure shows the principal features that must have a node to belong to the cloud computing paradigm.

Figure 2. Features of cloud computing
Source Author

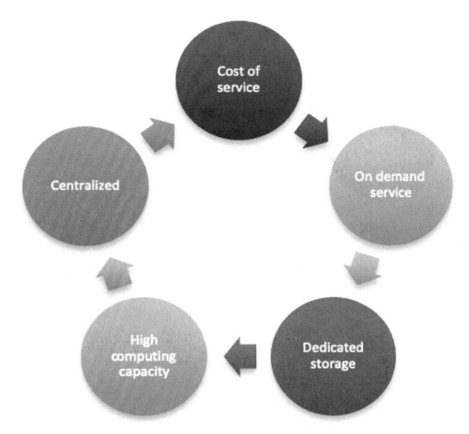

In some IoT applications, they are considered critical to bandwidth latency and another's requirements and response times; In these circumstances, it is not practically dependent on cloud computing due to the need for small response times (Avasalcai & Dustdar, 2020). In these scenarios, where critical bandwidth resources and latencies are critical, where the fog computing paradigm nevertheless promises to address the challenges and contribute to local and distributed solutions (Andreev, Petrov, Huang, Lema & Dohler, 2019; Yang et al., 2018).

- **Fog Computing:** This paradigm has a main objective, this objective is ensure the inherited networks and infrastructure in the cloud, a large amount of storage, processing and connectivity requirements, performing around the end user (Avasalcai & Dustdar, 2020). The next figure shows the features of Fog computing.

The nearness of computation in the network with devices and users allows us to address problems related to bandwidth and latency requirements on a smaller scale (Casadei et al., 2019). From this location the architecture in the whole, it can improve the distributed nature of the applications, contributing to the increase of agility and efficiency (El-Mougy, Al-Shiab & Ibnkahla, 2019). However, this computing paradigm in some ways contributes to solving certain challenges of the IoT platforms, by itself, computing

Figure 3. Features of fog computing
Source Author

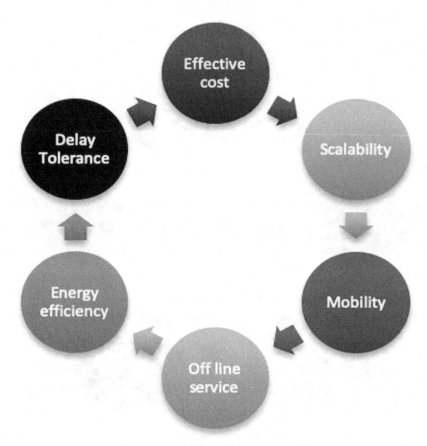

in its own presents its own challenges, which must be addressed so as to provide an acceptable computing framework (Naeem, Bashir, Amjad, Abbas, & Afzal, 2019). Some of the challenges of computing in the whole are given by sensitivity to delay and latencies, limitations of bandwidth, privacy, movement, and dynamism in topologies (Vilela, Rodrigues, Solic, Saleem, & Furtado, 2019).

- **Edge Computing:** There is another computing paradigm that is ubiquitous but close to users (devices and persons), this paradigm presents a distributed computing that implements services and resources in a decentralized way (Janjua, Vecchio, Antonini, & Antonelli, 2019). Decentralization allows labor to be carried out from the domains of data, devices, and applications (Galitsky & Parnis, 2019). The figure represents the main features for a node to fall into the computing paradigm at the edge.

Some researchers discuss the need Fog computing and Edge computing, as the only mechanism for meeting the requirements and overcoming the limitations of mobile devices (Cao, Zhang, & Shi, 2018). These limitations are added when mobile devices, which are immersed in personal area networks (PANs), there is a need to minimize the computing capacity that requires the areas, energy consumption, and

Figure 4. Features of edge computing
Source Author

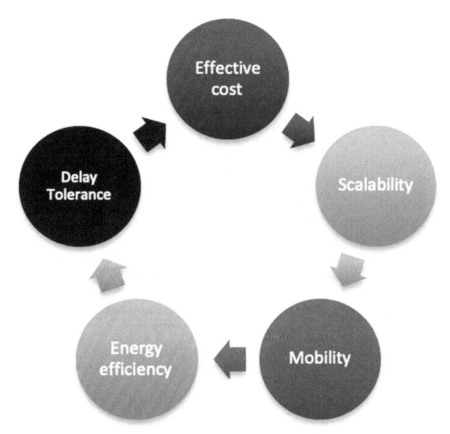

data storage (Dolui & Datta, 2017). The next figure shows the principals and commons features of the three computations paradigms.

In an ideal scenario the constraints on the problem, as all would be without computer limitations on the infrastructure, the bandwidth would always be sufficient and redundant, the latency would be negligible and all the resources would be available for any request (Dolui & Datta, 2017). But for the implementation of applications, services and IoT platforms, in abrasive scenarios, with major restrictions on bandwidth, interoperability and connectivity, where remote discharge can be an unviable scenario; Occurring with the likelihood of what you think, in these scenarios, the computational paradigms in the field and in the border can manage resources efficiently, converting into an important tool in the search for solutions and capabilities in IoT platforms (Cao, Zhang & Shi, 2018).

Internet of Things Infrastructure and Architecture

This infrastructure, coast of an intermediate node architecture, this architecture will be the basis for the design and the way to ensure the deployment of resources and transport and connectivity services, which can enable the functioning of the applications in IoT, under critical conditions, with bandwidth finder restrictions and connectivity to the public cloud infrastructure. The solution must comply with

Figure 5. Features of cloud computing
Source Author

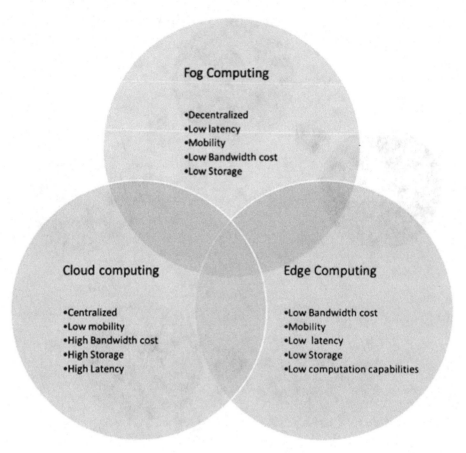

the Service Level Agreement (SLA) parameters to achieve compliance with the maximum levels except for delays and latencies . The figure shows the general structure of IoT, with each part of the structure relating some resources and example services.

It is difficult to find a universal consensus on IoT architecture (Sethi & Sarangi, 2017a). Different architectures have been proposed from different fields between them and the field of research (Yaqoob et al., 2017). If I consider the different models of architecture proposed, the models of three and five layers (Madakam, Ramaswamy, & Tripathi, 2015). The Figure shows three and five layer architectural models for IoT. In these models we can see the architectural approaches, the first three-layer and resource-oriented, the second five-layer, service-oriented model.

Three-layer Internet of Things Architecture

This architecture defines the main idea of the IoT platforms, but is not sufficiently embraced for IoT deepening and research (Madakam, Ramaswamy & Tripathi, 2015). For this reason, some authors propose more architectures with more layers; but this three-layer architecture can be considered as the starting point for the implementation of other more complex and number of layers (O. Salman, Elhajj, Chehab & Kayssi, 2018).

Figure 6. IoT Infrastructure
Source Author

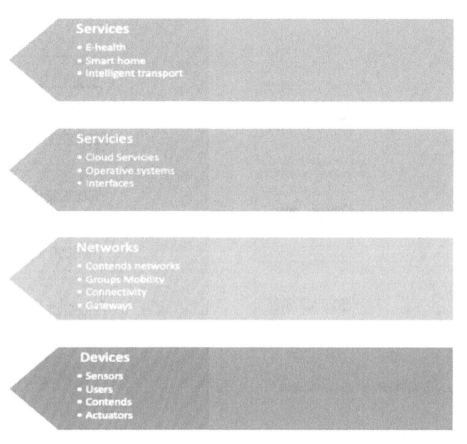

- **Perception Layer:** This layer is the physical layer of the architecture, in this layer if the sensors, personas the generators or collectors of information and contained (Santos, Wauters, Volckaert, & de Turck, 2018) (Poongodi, Krishnamurthi, Indrakumari, Suresh, & Balusamy, 2020). This layer detects physical parameters, recognizes their surroundings, detects objects or people around them (Tao, Zuo, Xu, & Zhang, 2014).
- **Network Layer:** This layer has the responsibility to connect with other devices, users, network nodes and servers (Kulkarni & Bakal, 2019). The main features of this layer are the ability to transmit and process sensor actuator data (Hsieh, Lee & Chen, 2018).
- **Application Layer:** The responsibility for this layer is for the purpose of delivering specific services for managed applications (Karagiannis, Chatzimisios, Vazquez-Gallego, & Alonso-Zarate, 2015) (Shang, Yu, Zhang, & Droms, 2016). If fields and applications on which IoT platforms are being implemented, some of the most popular applications are given in the fields of smart homes, smart transport, smart cities, smart farms, e-health, self-care systems and IoTM., among others (Chen et al., 2019)

Figure 7. IoT architecture
Source Author

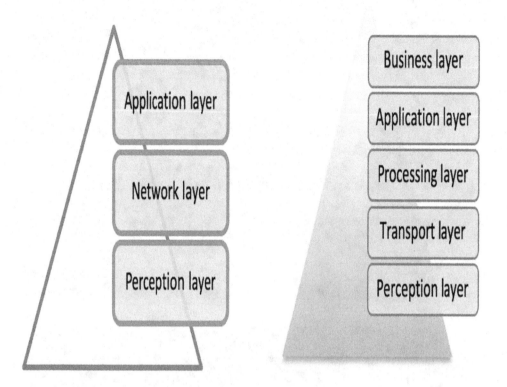

Five-layer IoT Architecture

The architecture of five layers, is based on the architecture of three layers, adding the layers but called the next way; the processing layer and business layer (Kumar & Mallick, 2018) (Kumar & Mallick, 2018). Adding the layers, the model drops as follows; perception, transportation, processing, application and business. The role of the layers of perception and application varies with respect to their exposed in the model of three layers, as it is in the object of clarification in this section.

- **Transport Layer:** This layer is charged with the transfer of data from the perception layer devices to the sense processing layer, through the use of inherited network services (Kumar & Mallick, 2018). Some examples of services and protocols networks: Zigbee, Z-wave, wifi, LTE, 3g, Bluetooth, among others.
- **Processor Layer:** The processing layer is also called middleware. This layer analyzes and processes large data volumes, which are generated in the perception layer and transport layer (Kumar & Mallick, 2018). This layer is able to provide services to lower layers in accordance with requests and available resources (Mathew, Atif, & El-Barachi, 2017). Here you can find the full range of technologies such as data bases, processing modules and big data (Mathew, Atif, & El-Barachi, 2017).

- **Business Layer:** Business layer is the task of managing the entire IoT platform, including all profit and commercial applications (Mathew, Atif, & El-Barachi, 2017). This layer includes efforts to protect the privacy of the user (Ikävalko, Turkama, & Smedlund, 2018a). This layer focuses on services and micro services that can be generalized from applications to users and users to IoT platforms (Ikävalko, Turkama, & Smedlund, 2018b).

Internet of Things Protocol Stack Architecture

IoT has been developed on the composition between networks, a heterogeneous network, an inherited network and a dedicated infrastructure (Soursos, Zarko, Zwickl, Gojmerac, Bianchi & Carrozzo, 2016). For this reason the operation of the IoT platforms is supported in some cases by protocols belonging to the inherited networks (Sethi & Sarangi, 2017b). The development of proprietary protocols for operation and support on IoT platforms is given the premise of meeting the requirements for devices with low memory and process capability, but also limited bandwidth capability and high latency. (T. Salman & Jain, 2017). The next figure shows the protocol stack for IoT.

Figure 8. IoT protocols stack
Source (C.P, 2016)

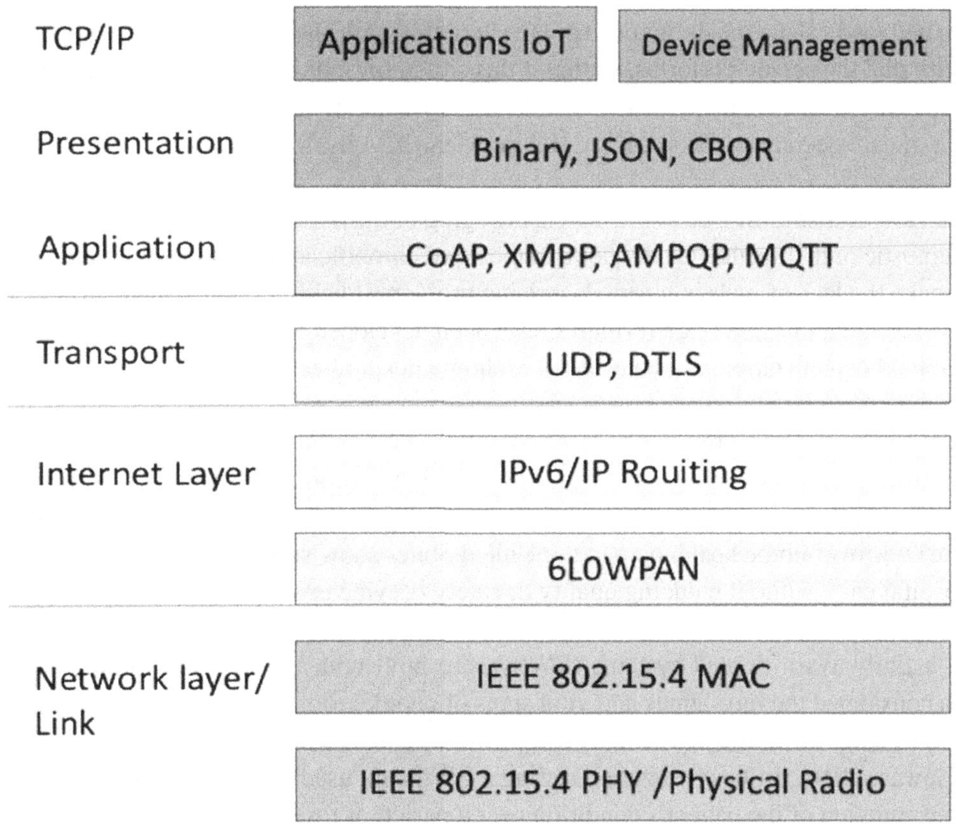

INTERNET OF THINGS HEALTH APPLICATIONS

One of the great applications and uses of IoT in health is remote monitoring systems, this systems can become an effective tool for health services, since these systems could provide several benefits in different contexts (McDaniel, Novicoff, Gunnell, & Cattell Gordon, 2018). It could provide feasible access for older people living independently in the home or people living in rural areas. In particular, it minimizes stress for health care users, such as non-critical patients at home instead of staying in hospitals (McCurdy, 2012). The remote health monitoring system provides better control so that patients maintain their health at all times. The remote monitoring medical care system minimizes the cost and overcrowding of hospital beds (Bozorgchami, Member, Sodagari, & Member, e.d.).

- **E-health:** The application of IoT technologies enables the application of electronic health, an assisted health modality that allows remote monitoring of patients in real time, creating a continuous record of record of data in the form of texts and graphics in the cloud (Journal, Engineering, & Kumar, 2019). On the other hand, the detection of variations in the signals, can be informed in real time to whom it corresponds (Kulkarni & Bakal, 2019).

The data monitored by e-health correspond to vital signs such as heart rate, which are complemented with the data record of the environment in which the individual is located, such as humidity and temperature, factors that can affect the health of people and whose knowledge allows to improve the monitoring via e-health and medical diagnosis, for example the variation of the heart rate can be presented according to the activity that the person performs, so that if this situation is known, false alerts will be avoided. In this way, the health service can be managed continuously, making the quality of medical care improve and the cost of care decreases (Zikria, Kim, Hahm, Afzal, & Aalsalem, 2019).

- **Home Hospitalization:** Due to the increasing aging of the population, as well as the advancement in diagnostic or therapeutic techniques and the rapid chronification of noncommunicable diseases, they make the lack of beds in hospitals a rising problem (Puchi-Gómez, Paravic-Klijn, & Salazar, 2018). To weigh this, new care formulas have been developed for patient care, as an alternative to traditional hospitalization, as is the case with home hospitalization (Commonwealth, 2017).

Home hospitalization must respond to three basic premises: improve the quality of life of the patient and their families; reduce the rate and severity of infections, avoiding the onset of nosocomial infections; and reduce transfer and entry costs (Brody et al., 2019). The findings of the first randomized pilot study on the control of home health care in the United States show significantly lower costs, compared to usual hospital care, without reducing quality or safety (Levine et al., 2018).

- **EWSs:** Early Warn Scored systems (EWSs), are built under the composition of scales, these scales considered the bio signals and vital signs of people, some examples of these are; heart rate, blood pressure, respiratory rate, body temperature, position among others (Albur, Hamilton, & MacGowan, 2016). In hospitalization settings, EWSs are used to assess the status, improvement or determination of the patient's condition over a period of time (Gerry et al., 2017).

Proposed Architecture for E-health and EWS Systems

In this architecture three intermediate layers are considered where the hardware associated with each one of the three computing paradigms is available; cloud, fog and edge computing.

This architecture have a four layers, three layers of them are the computer paradigms for IoT and the fourth layer are associated with devices and users (Naeem, Bashir, Amjad, Abbas, & Afzal, 2019). The layer of devices and users, works in principle as the layer of perception of the previous architectures, for this reason, it will not enter it in this section.

- **Cloud Layer:** The cloud layer, as a rule, is burdened by the tasks that do not have to proceed in real time, these tasks are provided by data analysis, data storage and device management services. These services are preceded by non-access to network resources. Transmission time should not be a critical requirement for the delivery of required services, as the latency and delay time in this case is likely to be high (R. Wang et al., 2020; Cavalcante et al., 2016).
- **Fog Layer:** The fog layer usually handles tasks, which requires a moderately high computational cost, but low latency times (Cavalcante et al., 2016). This layer is responsible for distributing tasks throughout its neighbors and storing a reduced volume of data (O. Salman, Elhajj, Chehab, & Kayssi, 2017). The hardware used in this capacity allows the process of micro analytics, analytics, storage and limited processing (Hsieh, Lee & Chen, 2018).
- **Edge Layer:** This layer is oriented to cover particular requirements, such as basic network resources, connectivity between devices, real-time micro analytics and notification (Yu et al., 2017). This layer is the one that is closest to the devices, so it can provide cooperation between nodes and intermediation to upper and lower layers (Nunna & Ganesan, 2017).
- **Devices Layer:** In this layer are the devices, users and things, it is in this layer where the perception or the execution of final tasks is carried out. This layer represents the heterogeneity of dispositive networks and platform protocols (O. Salman, Elhajj, Chehab, & Kayssi, 2018) .

Proposed E-health and EWS Systems

The general design proposed for the implementation of e-health and EWS systems, based on two possible scenarios; The first scenario takes place with cloud connectivity and the second does not have that type of connectivity. In the case of the first scenario, the sensors are connected to an edge device, which performs data aggregation of all sensors and preprocessing them. After the data are avoided through the internet to the cloud infrastructure and replicated to the fog node. Once the data is in the cloud node, it is processed and stored; for the relationship of alerts, validations, alerts, alarms and the establishment of patterns and rules. The node in the cloud is responsible for verifying the codifications and issuing alarms when necessary.

In the second scenario, the data is collected by the edge device, then sent to the fog node, where they are temporarily stored, until connectivity to the cloud infrastructure is restored. These data are treated locally and micro analytical are performed to determine if it is necessary to measure alerts, alarms or notifications. Alerts and alarms are sent from the fog node to the other available fog nodes, local visualization is also provided for users and health professionals. Once it is determined that the connection to the cloud infrastructure is lost, it notifies the health professional of remote monitoring and the start of local services. The next figure shows de general overview of system proposed.

Figure 9. IoT architecture
Source Author

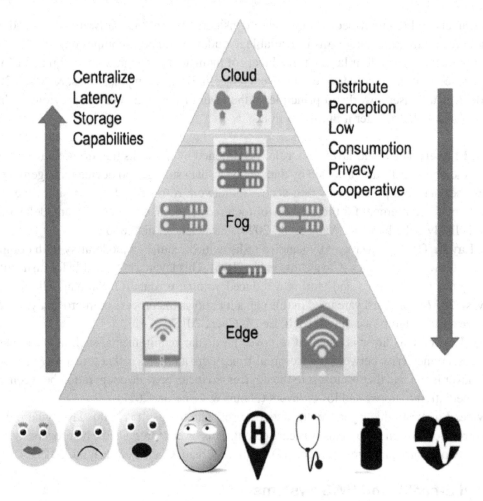

Proposed Data Flow for E-health and EWS Systems

The proposed detailed workflow can be seen in the following figure. The edge and fog nodes provide local services and support for cloud services when connectivity is not available. The fog node provides the service of visualization and local notification to the user to the health professional as the case may be. Once the data is collected, the edge or fog nodes are able to generate a corresponding alert or response for the actuators.

On the edge device the data of the sensors, devices and users are extracted, they are added and processed to be sent to the cloud or to the fog depending on the connectivity conditions. Once the data is sent, it is analyzed and stored; This data can be the object of integration or merger with other funetes of data, for the validation, comparison or verification of alerts. Once the alert is sent and processed the fog or cloud nodes send a response pattern to the edge nodes, so that in this way a preventive action is executed. Since fog devices have limited computational capabilities, data analysis and data storage activities are done in a modular manner, in the absence of cloud connectivity.

Figure 10. General overview E-health and EWSs work flow
Source Author

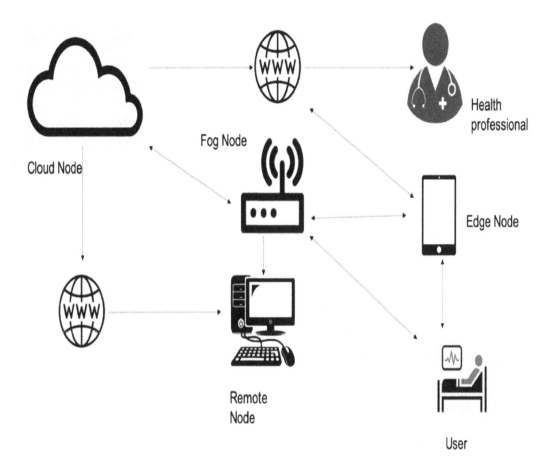

Figure 11. E-health and EWSs work flow
Source Author

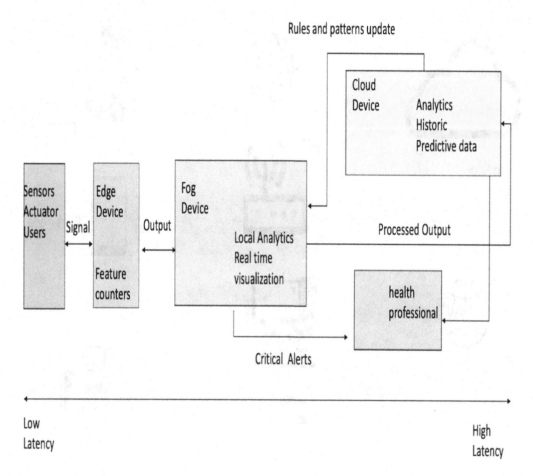

ACKNOWLEDGMENT

The authors are grateful for the support of the Universidad Militar Nueva Granada through its Research Vice-Rectory with project code IMP-ING-2660.

REFERENCES

Aghili, S. F., Mala, H., Shojafar, M., & Peris-Lopez, P. (2019). LACO: Lightweight Three-Factor Authentication, Access Control and Ownership Transfer Scheme for E-Health Systems in IoT. *Future Generation Computer Systems*, *96*, 410–424. doi:10.1016/j.future.2019.02.020

Ahmad, K., Mohammad, O., Atieh, M., & Ramadan, H. (2019). IoT: Architecture, Challenges, and Solutions Using Fog Network and Application Classification. ACIT 2018 - 19th International Arab Conference on Information Technology, 1–7. 10.1109/ACIT.2018.8672696

Ahmad, R. H., & Pathan, A.-S. K. (2016). A study on M2M (machine to machine) system and communication: Its security, threats, and intrusion detection system. Security Solutions and Applied Cryptography in Smart Grid Communications. doi:10.4018/978-1-5225-1829-7.ch010

Albur, M., Hamilton, F., & MacGowan, A. P. (2016). Early warning score: A dynamic marker of severity and prognosis in patients with Gram-negative bacteraemia and sepsis. *Annals of Clinical Microbiology and Antimicrobials*, *15*(1), 1–10. doi:10.1186/s12941-016-0139-z PubMed

Almusaylim, Z. A., & Zaman, N. (2018). A review on smart home present state and challenges: Linked to context-awareness internet of things (IoT). *Wireless Networks*, *5*, 1–12. doi:10.100711276-018-1712-5

Andreev, S., Petrov, V., Huang, K., Lema, M. A., & Dohler, M. (2019). Dense Moving Fog for Intelligent IoT: Key Challenges and Opportunities. *IEEE Communications Magazine*, *57*(5), 34–41. doi:10.1109/MCOM.2019.1800226

Annamalai, P., Bapat, J., & Das, D. (2019). Emerging Access Technologies and Open Challenges in 5G IoT: From Physical Layer Perspective. *2018 IEEE International Conference on Advanced Networks and Telecommunications Systems (ANTS)*, 1–6. 10.1109/ants.2018.8710133

Atzori, L., Iera, A., & Morabito, G. (2017). Understanding the Internet of Things: Definition, potentials, and societal role of a fast evolving paradigm. *Ad Hoc Networks*, *56*, 122–140. doi:10.1016/j.adhoc.2016.12.004

Avasalcai, C., & Dustdar, S. (2020). Latency-aware distributed resource provisioning for deploying IoT applications at the edge of the network. Lecture Notes in Networks and Systems, 69, 377–391. doi:10.1007/978-3-030-12388-8_27

Azad, M. A., Bag, S., Hao, F., & Salah, K. (2018). M2M-REP: Reputation system for machines in the internet of things. *Computers & Security*, *79*, 1–16. doi:10.1016/j.cose.2018.07.014

Basatneh, R., Najafi, B., & Armstrong, D. G. (2018). Health Sensors, Smart Home Devices, and the Internet of Medical Things: An Opportunity for Dramatic Improvement in Care for the Lower Extremity Complications of Diabetes. *Journal of Diabetes Science and Technology*, *12*(3), 577–586. doi:10.1177/1932296818768618 PubMed

Bhattarai, S., & Wang, Y. (2018). Internet of Things Security and Challenges. IEEE Computer, 78, 544–546. Retrieved from https://www.sciencedirect.com/science/article/pii/S0167739X17316667

Bozorgchami, B., Member, S., Sodagari, S., & Member, S. (n.d.). Spectrally Efficient Telemedicine and In-Hospital Patient Data Transfer. Academic Press.

Bramsen, P. (2017). Exploring a New IoT Infrastructure. University of California at Berkley. Retrieved from http://www2.eecs.berkeley.edu/Pubs/TechRpts/2017/EECS-2017-56.html

Brody, A. A., Arbaje, A. I., DeCherrie, L. V., Federman, A. D., Leff, B., & Siu, A. L. (2019). Starting Up a Hospital at Home Program: Facilitators and Barriers to Implementation. *Journal of the American Geriatrics Society*, *67*(3), 588–595. doi:10.1111/jgs.15782 PubMed

Bull, P., Austin, R., Popov, E., Sharma, M., & Watson, R. (2016). Flow based security for IoT devices using an SDN gateway. Proceedings - 2016 IEEE 4th International Conference on Future Internet of Things and Cloud, FiCloud 2016, 157–163. doi:10.1109/FiCloud.2016.30

Cao, J., Zhang, Q., & Shi, W. (2018). Challenges and opportunities in edge computing. Í. SpringerBriefs in Computer Science. doi:10.1007/978-3-030-02083-5_5

Casadei, R., Fortino, G., Pianini, D., Russo, W., Savaglio, C., & Viroli, M. (2019). A development approach for collective opportunistic Edge-of-Things services. *Information Sciences, 498*, 154–169. doi:10.1016/j.ins.2019.05.058

Cavalcante, E., Pereira, J., Alves, M. P., Maia, P., Moura, R., & Batista, T. … Pires, P. F. (2016). On the interplay of Internet of Things and Cloud Computing: A systematic mapping study. *Computer Communications*. doi:10.1016/j.comcom.2016.03.012

Cha, S., Ruiz, M. P., Wachowicz, M., Tran, L. H., Cao, H., & Maduako, I. (2016). The role of an IoT platform in the design of real-time recommender systems. 2016 IEEE 3rd World Forum on Internet of Things (WF-IoT), 448–453. 10.1109/WF-IoT.2016.7845469

Chauhan, M. A., & Babar, M. A. (2017). Using Reference Architectures for Design and Evaluation of Web of Things Systems: A Case of Smart Homes Domain. Managing the Web of Things: Linking the Real World to the Web. doi:10.1016/B978-0-12-809764-9.00009-3

Chen, W., Zhang, Z., Hong, Z., Chen, C., Wu, J., & Maharjan, S. (2019). *Zhang, Y*. Cooperative and Distributed Computation Offloading for Blockchain-Empowered Industrial Internet of Things. IEEE Internet of Things Journal; doi:10.1109/jiot.2019.2918296

Commonwealth. (2017). Hospital at Home" Programs Improve Outcomes, Lower Costs But Face Resistance from Providers and Payers. Author.

Corporation, I. T. U. (2015). Internet of Things Global Standards Initiative. Internet of Things Global Standards Initiative. Retrieved from http://www.itu.int/en/ITU-T/gsi/iot/Pages/default.aspx

C.P., V. (2016). Security improvement in IoT based on Software Defined Networking (SDN). *International Journal of Science Engineering and Technology Research, 5*(1), 291–295.

Dolui, K., & Datta, S. K. (2017). Comparison of edge computing implementations: Fog computing, cloudlet and mobile edge computing. GIoTS 2017 - Global Internet of Things Summit Proceedings. doi:10.1109/GIOTS.2017.8016213

El-Mougy, A., Al-Shiab, I., & Ibnkahla, M. (2019). Scalable Personalized IoT Networks. *Proceedings of the IEEE*. 10.1109/JPROC.2019.2894515

Farris, I., Taleb, T., Khettab, Y., & Song, J. S. (2018). A survey on emerging SDN and NFV security mechanisms for IoT systems. *IEEE Communications Surveys and Tutorials*, (c): 1–26. doi:10.1109/COMST.2018.2862350

Forbes. (2018). IoT Market Predicted To Double By 2021, Reaching $520B. Author.

Galitsky, B., & Parnis, A. (2019). Accessing Validity of Argumentation of Agents of the Internet of Everything. Artificial Intelligence for the Internet of Everything. doi:10.1016/B978-0-12-817636-8.00011-9

Gerry, S., Birks, J., Bonnici, T., Watkinson, P. J., Kirtley, S., & Collins, G. S. (2017). Early warning scores for detecting deterioration in adult hospital patients: A systematic review protocol. *BMJ Open*, *7*(12), 1–5. doi:10.1136/bmjopen-2017-019268 PubMed

Grigoryan, G., Njilla, L., Kamhoua, C., & Kwiat, K. (2017). *Enabling Cooperative IoT Security via Software Defined Networks*. SDN; doi:10.1109/ICC.2018.8423017

Hsieh, H.-C., Lee, C.-S., & Chen, J.-L. (2018). Mobile Edge Computing Platform with Container-Based Virtualization Technology for IoT Applications. *Wireless Personal Communications*, *1*. doi:10.100711277-018-5856-5

Iannacci, J. (2018). Internet of things (IoT); internet of everything (IoE); tactile internet; 5G – A (not so evanescent) unifying vision empowered by EH-MEMS (energy harvesting MEMS) and RF-MEMS (radio frequency MEMS). *Sensors and Actuators. A, Physical*, *272*, 187–198. doi:10.1016/j.sna.2018.01.038

Ikävalko, H., Turkama, P., & Smedlund, A. (2018a). Enabling the Mapping of Internet of Things Ecosystem Business Models Through Roles and Activities in Value Co-creation. Proceedings of the 51st Hawaii International Conference on System Sciences. doi:10.24251/HICSS.2018.620

Ikävalko, H., Turkama, P., & Smedlund, A. (2018b). *Value Creation in the Internet of Things: Mapping Business Models and Ecosystem Roles*. Technology Innovation Management Review; doi:10.22215/timreview/1142

Jabbar, S., Ullah, F., Khalid, S., Khan, M., & Han, K. (2017). Semantic {Interoperability} in {Heterogeneous} {IoT} {Infrastructure} for {Healthcare}. *Wireless Communications and Mobile Computing*, *2017*, e9731806. doi:10.1155/2017/9731806

Janjua, Z. H., Vecchio, M., Antonini, M., & Antonelli, F. (2019). IRESE: An intelligent rare-event detection system using unsupervised learning on the IoT edge. *Engineering Applications of Artificial Intelligence*, *84*(May), 41–50. doi:10.1016/j.engappai.2019.05.011

Journal, I., Engineering, A., & Kumar, S. (2019). An Extensive Review on Sensing as a Service Paradigm in IoT : Architecture, Research Challenges. *Lessons Learned and Future Directions*, *14*(6), 1220–1243.

Kabugo, J. C., Jämsä-Jounela, S.-L., Schiemann, R., & Binder, C. (2020). Industry 4.0 based process data analytics platform: A waste-to-energy plant case study. *International Journal of Electrical Power & Energy Systems*, *115*. doi:10.1016/j.ijepes.2019.105508

Karagiannis, V., Chatzimisios, P., Vazquez-Gallego, F., & Alonso-Zarate, J. (2015). *A Survey on Application Layer Protocols for the Internet of Things*. Transaction on IoT and Cloud Computing; doi:10.5281/ZENODO.51613

Krishnan, P., Najeem, J. S., & Achuthan, K. (2018). SDN framework for securing IoT networks. Lecture Notes of the Institute for Computer Sciences. *Social-Informatics and Telecommunications Engineering, LNICST*, *218*, 116–129. doi:10.1007/978-3-319-73423-1_11

Kulkarni, N. J., & Bakal, J. W. (2019). E-Health: IoT Based System and Correlation of Vital Stats in Identification of Mass Disaster Event. Proceedings - 2018 4th International Conference on Computing, Communication Control and Automation, ICCUBEA 2018, 1–6. 10.1109/ICCUBEA.2018.8697529

Kumar, N. M., & Mallick, P. K. (2018). The Internet of Things: Insights into the building blocks, component interactions, and architecture layers. Í. *Procedia Computer Science*. doi:10.1016/j.procs.2018.05.170

Lee, I., & Lee, K. (2015). The Internet of Things (IoT): Applications, investments, and challenges for enterprises. *Business Horizons*, *58*(4), 431–440. doi:10.1016/j.bushor.2015.03.008

Lee, S., Bae, M., & Kim, H. (2017). Future of IoT Networks: A Survey. Applied Sciences, 7(10), 1072. doi:10.3390/app7101072

Levine, D. M., Ouchi, K., Blanchfield, B., Diamond, K., Licurse, A., Pu, C. T., & Schnipper, J. L. (2018). Hospital-Level Care at Home for Acutely Ill Adults: A Pilot Randomized Controlled Trial. Journal of General Internal Medicine, 1–8. doi:10.100711606-018-4307-z

Li, W., & Wang, P. (2019). Two-factor authentication in industrial Internet-of-Things: Attacks, evaluation and new construction. *Future Generation Computer Systems*, *101*, 694–708. doi:10.1016/j.future.2019.06.020

Liu, X., & Ansari, N. (2019). Toward Green IoT: Energy Solutions and Key Challenges. *IEEE Communications Magazine*, *57*(3), 104–110. doi:10.1109/MCOM.2019.1800175

Madakam, S., Ramaswamy, R., & Tripathi, S. (2015). Internet of Things (IoT): A Literature Review. Journal of Computer and Communications. doi:10.4236/jcc.2015.35021

Marolla, C. (2018). Information and Communication Technology for Sustainable Development. Information and Communication Technology for Sustainable Development. Springer Singapore. doi:10.1201/9781351045230

Mathew, S. S., Atif, Y., & El-Barachi, M. (2017). From the Internet of Things to the web of things-enabling by sensing as-A service. *Proceedings of the 2016 12th International Conference on Innovations in Information Technology, IIT 2016*. 10.1109/INNOVATIONS.2016.7880055

McCurdy, B. R. (2012). hospital-at-home programs for patients with acute exacerbations of chronic obstructive pulmonary disease (COPD): An evidence-based analysis. Ontario Health Technology Assessment Series.

McDaniel, N. L., Novicoff, W., Gunnell, B., & Cattell Gordon, D. (2018). Comparison of a Novel Hand-held Telehealth Device with Stand-Alone Examination Tools in a Clinic Setting. Telemedicine Journal and e-Health. doi:10.1089/tmj.2018.0214

Médini, L., Mrissa, M., Khalfi, E. M., Terdjimi, M., Le Sommer, N., Capdepuy, P., ... Touseau, L. (2017). Building a Web of Things with Avatars: A comprehensive approach for concern management in WoT applications. Managing the Web of Things: Linking the Real World to the Web. doi:10.1016/B978-0-12-809764-9.00007-X

Mehner, S. (n.d.). Secure and Flexible Internet of Things using Software Defined Networking. Academic Press.

Minerva, R., Biru, A., & Rotondi, D. (2015). Towards a definition of the Internet of Things (IoT). Retrieved from https://iot.ieee.org/definition.html

Mirón Rubio, M., Ceballos Fernández, R., Parras Pastor, I., Palomo Iloro, A., Fernández Félix, B. M., Medina Miralles, J., ... Alonso-Viteri, S. (2018). Telemonitoring and home hospitalization in patients with chronic obstructive pulmonary disease: Study TELEPOC. *Expert Review of Respiratory Medicine*, *12*(4), 335–343. doi:10.1080/17476348.2018.1442214 PubMed

Mostafa, S. A., Gunasekaran, S. S., Mustapha, A., Mohammed, M. A., & Abduallah, W. M. (2020). Modelling an adjustable autonomous multi-agent internet of things system for elderly smart home. Advances in Intelligent Systems and Computing, 953, 301–311. doi:10.1007/978-3-030-20473-0_29

Mourad, M. H., Nassehi, A., Schaefer, D., & Newman, S. T. (2020). Assessment of interoperability in cloud manufacturing. *Robotics and Computer-integrated Manufacturing*, *61*(June), 101832. doi:10.1016/j. rcim.2019.101832

Naeem, R. Z., Bashir, S., Amjad, M. F., Abbas, H., & Afzal, H. (2019). Fog computing in internet of things: Practical applications and future directions. *Peer-to-Peer Networking and Applications*, *12*(5), 1236–1262. doi:10.1007/s12083-019-00728-0

Nunna, S., & Ganesan, K. (2017). Mobile edge computing. Health 4.0: How Virtualization and Big Data are Revolutionizing Healthcare. doi:10.1007/978-3-319-47617-9_9

Panda, C. K., & Bhatnagar, R. (2020a). Social Internet of Things in Agriculture: An Overview and Future Scope. doi:10.1007/978-3-030-24513-9_18

Panda, C. K., & Bhatnagar, R. (2020b). Toward Social Internet of Things (SIoT): Enabling Technologies, Architectures and Applications (B. 846). doi:10.1007/978-3-030-24513-9

Pico-Valencia, P., Holgado-Terriza, J. A., Herrera-Sánchez, D., & Sampietro, J. (2018). Towards the internet of agents: An analysis of the internet of things from the intelligence and autonomy perspective. *Ingenieria e Investigacion*, *38*(1), 121–129. doi:10.15446/ing.investig.v38n1.65638

Poongodi, T., Krishnamurthi, R., Indrakumari, R., Suresh, P., & Balusamy, B. (2020). Wearable Devices and IoT. doi:10.1007/978-3-030-23983-1_10

Puchi-Gómez, C., Paravic-Klijn, T., & Salazar, A. (2018). Indicators of the quality of health care in home hospitalization: An integrative review. *Aquichan*, *18*(2), 186–197. doi:10.5294/aqui.2018.18.2.6

Ramirez Lopez, L. J., Puerta Aponte, G., & Rodriguez Garcia, A. (2019). Internet of Things Applied in Healthcare Based on Open Hardware with Low-Energy Consumption. *Healthcare Informatics Research*, *25*(3), 230. doi:10.4258/hir.2019.25.3.230 PubMed

Rayes, A., & Salam, S. (2016). Internet of things-from hype to reality: The road to digitization. doi:10.1007/978-3-319-44860-2

Roopa, M. S., Pattar, S., Buyya, R., Venugopal, K. R., Iyengar, S. S., & Patnaik, L. M. (2019). Social Internet of Things (SIoT): Foundations, thrust areas, systematic review and future directions. *Computer Communications*, *139*, 32–57. doi:10.1016/j.comcom.2019.03.009

Salamatian, K. (2011). Toward a polymorphic future internet: A networking science approach. *IEEE Communications Magazine, 49*(10), 174–178. doi:10.1109/MCOM.2011.6035832

Salman, O., Elhajj, I., Chehab, A., & Kayssi, A. (2017). Software Defined IoT security framework. 2017 Fourth International Conference on Software Defined Systems (SDS), 75–80. doi:10.1109/SDS.2017.7939144

Salman, O., Elhajj, I., Chehab, A., & Kayssi, A. (2018). IoT survey: An SDN and fog computing perspective. *Computer Networks, 143*, 221–246. doi:10.1016/j.comnet.2018.07.020

Salman, T., & Jain, R. (2017). Networking protocols and standards for internet of things. Internet of Things and Data Analytics Handbook. doi:10.1002/9781119173601.ch13

Santos, J., Wauters, T., Volckaert, B., & de Turck, F. (2018). Fog computing: Enabling the management and orchestration of smart city applications in 5G networks. *Entropy (Basel, Switzerland), 20*(1). doi:10.3390/e20010004

Sethi, P., & Sarangi, S. R. (2017a). Internet of Things: Architectures, Protocols, and Applications. *Journal of Electrical and Computer Engineering, 2017*, 1–25. doi:10.1155/2017/9324035

Sethi, P., & Sarangi, S. R. (2017b). *Review Article Internet of Things : Architectures, Protocols, and Applications*. Academic Press.

Shala, B., Trick, U., Lehmann, A., Ghita, B., & Shiaeles, S. (2019). Novel Trust Consensus Protocol and Blockchain-based Trust Evaluation System for M2M Application Services. Internet of Things. doi:10.1016/j.iot.2019.100058

Shang, W., Yu, Y., Zhang, L., & Droms, R. (2016). Challenges in IoT Networking via TCP/IP Architecture. NDN Project, Tech. Rep. NDN-0038.

Son, J. (2018). Integrated Provisioning of Compute and Network Resources in Software-Defined Cloud Data Centers. Retrieved from https://minerva-access.unimelb.edu.au/bitstream/handle/11343/212287/thesis.pdf?sequence=1&isAllowed=y

Soursos, S., Zarko, I. P., Zwickl, P., Gojmerac, I., Bianchi, G., & Carrozzo, G. (2016). Towards the cross-domain interoperability of IoT platforms. EUCNC 2016 - European Conference on Networks and Communications, 398–402. 10.1109/EuCNC.2016.7561070

Takiddeen, N., & Zualkernan, I. (2019). Smartwatches as IoT Edge Devices: A Framework and Survey. 2019 Fourth International Conference on Fog and Mobile Edge Computing (FMEC), 216–222. doi:10.1109/FMEC.2019.8795338

Tao, F., Zuo, Y., Da Xu, L., & Zhang, L. (2014). IoT-Based intelligent perception and access of manufacturing resource toward cloud manufacturing. *IEEE Transactions on Industrial Informatics*. doi:10.1109/TII.2014.2306397

Tayyaba, S. K., Shah, M. A., Khan, O. A., & Ahmed, A. W. (2017). Software defined network (SDN) based internet of things (IoT): A road ahead. ACM International Conference Proceeding Series, Part F1305. doi:10.1145/3102304.3102319

Valdivieso, A., Peral, A., Barona, L., & García, L. (2014). Evolution and Opportunities in the Development IoT Applications. Retrieved from Http://Journals.Sagepub.Com/Doi/Full/10.1155/2014/735142

van Oorschot, P. C., & Smith, S. W. (2019). The Internet of Things: Security Challenges. *IEEE Security and Privacy*, *17*(5), 7–9. doi:10.1109/MSEC.2019.2925918

Vilela, P. H., Rodrigues, J. J. P. C., Solic, P., Saleem, K., & Furtado, V. (2019). Performance evaluation of a Fog-assisted IoT solution for e-Health applications. *Future Generation Computer Systems*, *97*, 379–386. doi:10.1016/j.future.2019.02.055

Vögler Matrikelnummer, M. (2016a). Efficient IoT Application Delivery and Management in Smart City Environments. Universität Wien. Retrieved from http://www.infosys.tuwien.ac.at/Staff/sd/papers/Diss_Voegler_Michael.pdf

Vögler Matrikelnummer, M. (2016b). *Efficient IoT Application Delivery and Management in Smart City Environments*. Academic Press.

Wang, J., & Li, D. (2018). Adaptive computing optimization in software-defined network-based industrial internet of things with fog computing. Sensors (Switzerland), 18(8), 2509. doi:10.339018082509

Wang, R., Li, M., Peng, L., Hu, Y., Hassan, M. M., & Alelaiwi, A. (2020). Cognitive multi-agent empowering mobile edge computing for resource caching and collaboration. *Future Generation Computer Systems*, *102*, 66–74. doi:10.1016/j.future.2019.08.001

Yang, R., Wen, Z., Mckee, D., Lin, T., Xu, J., & Garraghan, P. (2018). *Chapter #: Fog Orchestration and Simulation for IoT Services*. Academic Press.

Yaqoob, I., Ahmed, E., Hashem, I. A. T., Ahmed, A. I. A., Gani, A., Imran, M., & Guizani, M. (2017). Internet of Things Architecture: Recent Advances, Taxonomy, Requirements, and Open Challenges. *IEEE Wireless Communications*, *24*(3), 10–16. doi:10.1109/MWC.2017.1600421

You, T. (2016). *Toward the future of Internet architecture for IoE*. Academic Press.

Yu, W., Liang, F., He, X., Hatcher, W. G., Lu, C., Lin, J., & Yang, X. (2017). A Survey on the Edge Computing for the Internet of Things. *IEEE Access : Practical Innovations, Open Solutions*. doi:10.1109/ACCESS.2017.2778504

Zikria, Y., Kim, S. W., Hahm, O., Afzal, M. K., & Aalsalem, M. Y. (2019). Internet of Things (IoT) Operating Systems Management: Opportunities, Challenges, and Solution. *Sensors (Basel)*, *19*(8), 1793. doi:10.3390/s19081793 PubMed

Chapter 4
Signal Processing and Pattern Recognition in Electronic Tongues:
A Review

Jersson X. Leon-Medina
Universidad Nacional de Colombia, Colombia

Maribel Anaya Vejar
Universidad Santo Tomás, Colombia

Diego A. Tibaduiza
Universidad Nacional de Colombia, Colombia

ABSTRACT

This chapter reviews the development of solutions related to the practical implementation of electronic tongue sensor arrays. Some of these solutions are associated with the use of data from different instrumentation and acquisition systems, which may vary depending on the type of data collected, the use and development of data pre-processing strategies, and their subsequent analysis through the development of pattern recognition methodologies. Most of the time, these methodologies for signal processing are composed of stages for feature selection, feature extraction, and finally, classification or regression through a machine learning algorithm.

INTRODUCTION

Rapid progress has been made in the advancement of several key areas of science and technology, such as artificial intelligence, design of digital electronic sensors, materials science, microcircuit design, software innovations and electronic systems integration. This has stimulated the development of electronic sensors and intelligent systems applicable to various areas of human activity (Wilson & Baietto, 2011).

DOI: 10.4018/978-1-7998-1839-7.ch004

Specifically, the development of systems allows magnifying or at least improving some of the senses in human beings. Although human senses are essential for our common activities, in many cases they are insufficient given their limitations. For this reason, systems based on artificial vision, electronic noses, electronic tongues, among others, have gained popularity and have been applied to various industrial processes where, for example, an operator carried out quality control processes.

Recently, a new concept of sensor application has arisen, which includes the use of a matrix of non-selective sensors along with a mathematical data processing unit using pattern recognition methods. This concept, which imitates human perception, has been applied in the development of analytical tools such as "the electronic nose" (Gardner & Bartlett, 1994) and "electronic tongue". The latter was introduced in 1995 as a result of a Russian-Italian joint research (Vlasov et al., 2000). The electronic tongue is an analytical system applied to the analysis of liquids and is formed by a set of sensors that generate multidimensional information, plus a signal processing tool to extract meaning from these complex data (Del Valle, 2010). Figure 1 illustrates an electronic tongue sensor array composed of several electrodes.

Electronic tongues are used in different fields such as: environmental monitoring, water quality, food industry: adulteration, quality of liquid food, determination of origin, safety to detect hazardous substances, in biomedical topics like the identification of components in biological fluids or in vitro/in vivo analysis, and finally in pharmaceutical industry. Figure 2 illustrates the different applications of electronic tongues.

Some of the advantages of using electronic tongues for liquid variable monitoring processes include:

- Low cost compared to the use of an expert panel.
- Possibility of validation of results through standard methods used in the food industry.
- Use of instrumentation and precise acquisition systems.
- Portability of the equipment.
- Wide variety of sensors for the analysis of different samples.
- Possibility of process development as an automated system, which allows its constant monitoring and the possibility of online consultation and analysis of the data.
- Possibility of using sensor networks in large processes that require it.

Figure 1. Electronic tongue sensor array
Source: The authors

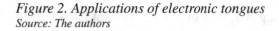

Figure 2. Applications of electronic tongues
Source: The authors

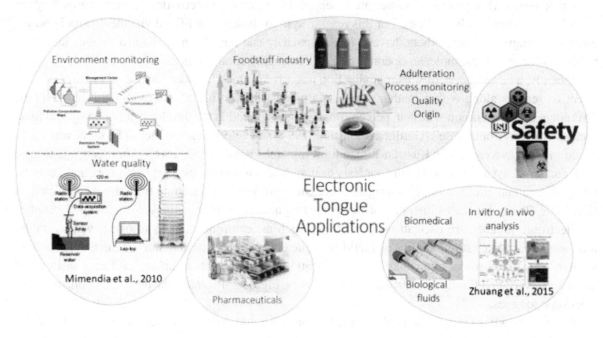

Although there are many advantages in the use of systems of sensor arrangements such as noses or electronic tongues, there are also some drawbacks related to their use and implementation, including:

- Most of these devices are still in the development stage, which means that their use in practical applications differs from applications in controlled laboratory environments.
- The sensors can suffer degradation processes as a result of their interaction with the environment in which they are submerged, which affects the variability of the data obtained over time.
- Some sensors require periodic calibration processes.
- The use of specialized data acquisition systems, sometimes at a high cost, is required.
- Some issues related with data acquisition such as measurement errors, sensor failures, offset, data variability and generation of outliers. This causes that the analysis requires the use of many sensor samples to increase its accuracy.
- Use of large capacity memories for data storage.
- Development of robust algorithms for the analysis of data.
- Appropriate handling of uncertainties.
- High computational cost according to the algorithm and the data analysis that will be implemented.

As a contribution, this work presents a review of the different approaches of signal processing from electronic tongues sensor arrays with the purpose of taste recognition as a classification problem in machine learning. The remainder of this chapter is organized as follows. The background section describes different types of sensors used in electronic tongues, electrochemical cells and theoretical concepts like selectivity of sensors, non-specific low-selectivity sensors arrays and electrochemical transduction techniques used by electronic tongue sensors like potentiometry and voltammetry, lastly an important topic on signal conditioning of the acquired signals is also discussed. Then, the signal processing and pattern recognition methods used in electronic tongues section is devoted to present the different stages in the pattern recognition methodologies like data preprocessing, exploratory analysis, dimensionality reduction, feature extraction, feature selection, classification and class modeling, finally different validation strategies are defined as shown in Figure 3. The last sections of the chapter are the future research directions and conclusion where the main findings of the analysis performed in this review are shown.

BACKGROUND

Types of Sensors in Electronic Tongues

At the implementation level, there are several types of sensors used in liquids, such as optical and gravimetric ones, among others. However, the most commonly used sensors in electronic tongues are electrochemical (potentiometric or voltammetric) (Beebe et al., 1998). The review work of Salles and Paixão (2014) proposes a classification of publications related to electronic tongues, distinguishing the type of electrochemical sensor used and the type of liquid analyzed. Figure 4 shows a classification of the different types of sensors used in electronic tongues.

Figure 3. Main steps of pattern recognition methodologies in electronic tongues
Source: The authors

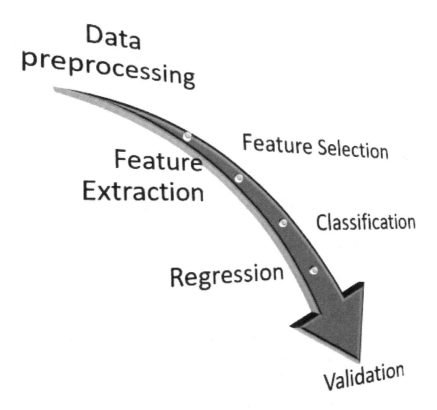

Figure 4. Different types of sensors used in electronic tongues
Adapted from (Banerjee et al., 2016)

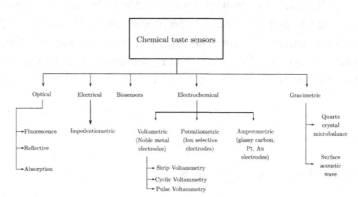

Electrochemical Cells and Theoretical Concepts

An electro-chemical cell is a device capable of obtaining electrical energy from chemical reactions or, producing chemical reactions through the introduction of electrical energy, when the cell is being charged. Electroanalytical techniques study the electrical properties of a solution in the so-called "electrochemical cell". These techniques are characterized by high sensitivity, high selectivity and high precision. An important distinction must be made between analysis, determination and measurement. An analysis provides chemical or physical information about a sample. The component of interest in the sample is called analyte, and the rest of the sample is the matrix. An analysis can determine the identity, concentration or properties of an analyte, measuring one or more of the physical or chemical properties (Harvey, 2010).

The definition of electroanalysis according to the Royal Academy of Engineering of Spain is: "Method that relates some of the electrical magnitudes that characterize the electrochemical circuit with the concentration of a substance that reacts or can react in the electrodes and is used for the determination of the amount of composed of electrical magnitudes such as coulometry, potentiometry, conductimetry, amperometry and polarography "(Raing, 2019) .

Selectivity of Sensors

The selectivity of a sensor refers to the ability to correctly identify the analyte(s) of interest in the presence of expected chemical/physical interferences. The selectivity can be evaluated by testing the appropriate matrix blanks and the enriched matrix blanks. The blanks in the matrix should contain the chemical species, apart from the analyte(s) of interest that are reasonably expected to be present in a real sample (Mishalanie et al., 2005).

Selectivity is a performance characteristic that demonstrates the ability of the method to generate useful data for analytes, their levels and matrices defined within the scope of the method. Selectivity is especially important at different levels of concentration and must be demonstrated in the presence of known or predicted species to pose analytical challenges to the method. In general terms, selectivity is demonstrated by providing information that justifies the identity of the analytes in the presence of the expected matrix components (Mishalanie et al., 2005). Selectivity is a precondition for accuracy; therefore, an accurate method is automatically selective (Wilken et al., 2016).

Non-Specific Low-Selectivity Sensors Arrays

Many times, it is sought that an electrochemical measurement device has high selectivity. This is achieved with the development of highly selective sensors for a specific analyte. However, electronic tongue-type sensor arrays use a series of non-specific sensors that respond to primary and interfering ions, as the taste buds of human beings do in nature. As electronic tongues use a set of sensors, the information obtained has high dimensionality and many variables to analyze. These analyses are usually done by multivariate chemometric methods to reach final conclusions, such as identifying a type of flavor, similarly to how the human brain does it (Pravdová et al., 2002).

An electronic tongue "comprises a series of non-specific, non-selective chemical sensors with partial specificity (cross-sensitivity) to different compounds in a solution, and a chemometric tool suitable for data processing" (Holmberg et al., 2004). The property of cross-sensitivity has great importance in electronic tongues, since cross-sensitivity is a characteristic of the organization of the animal senses. This is the case of the smell, where the receptors of the nasal organ are grouped and slightly differ one from another, thus giving different responses to different smells. Thus, the sense of smell manages a set of responses formed by individual responses, that is, cross sensitivity. Subsequently, the brain of the animal is responsible for data processing (Del Valle, 2010). The cross-sensitivity property allows electronic tongues to simultaneously determine different analytes in a mixture (Del Valle, 2007). In addition, with the same type of electronic tongue, different liquids such as milk, wine, coffee, etc. can be classified. As electronic tongues use non-selective sensors, they can indirectly determine specific analyte concentrations in the presence of interfering agents, through the correlation of the analyte with the measurements made in the solution (Del Valle, 2010).

Electrochemical Transduction Techniques Used by Electronic Tongue Sensors

There are mainly two types of electronic tongue systems that distinguish the nature of the most commonly used electrochemical sensors. First, the systems are potentiometric when using a matrix of ion-selective electrodes (ISE). Secondly, the systems are voltammetric when using a number of metal electrodes of different nature or a quantity of modified electrodes. When both types of sensors form an electronic tongue, the system is called "hybrid electronic tongue" (Del Valle, 2017). For more information on potentiometric electronic tongues, authors suggest further reading of the detailed review by Bratov et al. (2010) and the work of Wei et al. (2018) for voltammetric electronic tongues.

Potentiometry

The most used potentiometric sensors in electronic tongues are ion selective electrodes (ISE). The principle of operation of the ion-selective electrodes is based on the measurement of their potential changes against a reference electrode under conditions of zero current (Gomes, 2009). The potential of the ion-selective electrodes is a function of the activity of the ionic species in a sample solution and is formed in the ion-sensitive membrane, where the selective complexation (ion recognition) of the analyte molecules occurs (Kimmel et al., 2011). The main disadvantages of potentiometric measurements are the temperature dependence, the influence of the solution changes and the adsorption of the components of the solution that affect the nature of the charge transfer. However, the effect of these factors can be minimized by the temperature control. Figure 5 illustrates different signals obtained by a potentiometric electronic tongue.

Figure 5. Example of response signals in a potentiometric electronic tongue
Taken from (Cuartero et al., 2017)

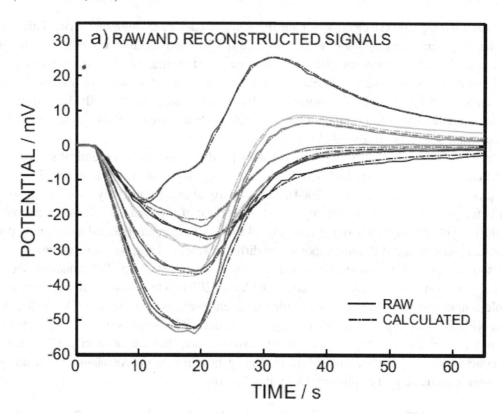

Cyclic Voltammetry

Voltammetric measurements are taken when equilibrium is not reached, and the signal obtained is the current-potential relationship. The simplest measurement configuration uses three electrodes: reference, working and auxiliary electrodes (Gomes, 2009). It is assumed that the potential of the reference electrode is constant, and the current flows between the working and auxiliary electrodes. The electrolysis reaction occurs at the working electrode and is responsible for the current generation. The current is a function of the rate of electrolysis, which in turn is governed by the transport of electroactive species present in the sample (i.e. diffusion coefficients and concentrations of electroactive species) (Harvey, 2010).

In cyclic voltammetry, the potential of the working electrode is measured against a reference electrode that maintains a constant potential, and the resulting applied potential produces an excitation signal like the one in Figure 6(a). In the forward scanning of the Figure 6(a), the first potential scan is negative, starting from a larger potential (a) and ending in a lower one (d). The potential at the end (d) is called switching potential, and it is the point where the voltage is sufficient to cause the oxidation or reduction of an analyte (Harvey, 2010). The inverse scan occurs from (d) to (g), where the potential positively scans. Figure 6(a) shows a typical reduction that occurs from (a) to (d) and an oxidation that occurs from (d) to (g). It is important to keep in mind that some analytes are oxidized first and as a result the potential would be scanned positively. This cycle may be repeated, and the scanning speed may vary. Figures 6(b) and (c) show cyclic voltammograms obtained as responses of electronic tongues.

Figure 6. Example of signals in an electronic tongue of cyclic voltammetry: (a) excitation signal taken from (Harvey, 2010) ; (b) voltammetric responses of an Au electrode submerged in five brands of pasteurized milk during its first day of storage, taken from (Bougrini et al., 2014) ; and (c) 3D view of different voltammograms obtained in an electronic tongue measure, taken from (Cetó et al., 2016)

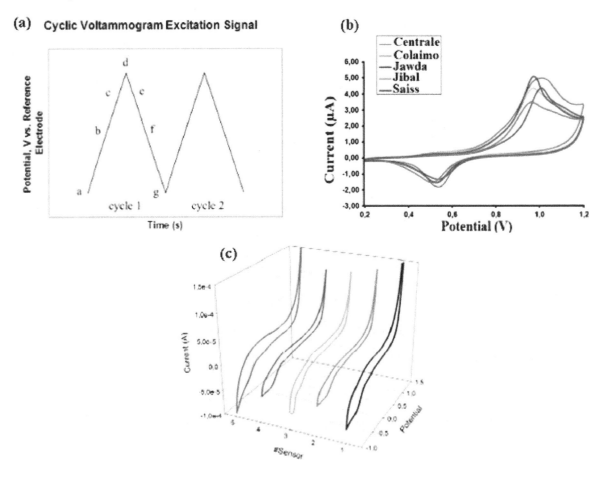

Multifrequency Large Amplitude Pulse Signal Voltammetry (MLAPV)

One of the voltammetry techniques used to analyze signals from electronic tongue-type sensor arrays is the Large Amplitude Pulse Voltammetry (LAPV). Pulse voltammetry is of special interest due to its greater sensitivity and resolution (Winquist et al., 1997). Subsequently, the Multifrequency Large Amplitude Pulse Signal Voltammetry -MLAPV technique was used, which is composed of a series of individual LAPV waveforms with different step lengths (Tian et al., 2007). Each cycle of the potential pulse step starts from the transient state of the reaction process of the last step of the cycle, so that the MLAPV obtains additional information (Wei et al., 2018) . The MLAPV family includes Multifrequency Pulse Voltammetry (Multifrequency Hackle Pulse Voltammetry MHPV), Multifrequency Rectangle Pulse Voltammetry (Multifrequency Rectangle Pulse Voltammetry MRPV) and Multifrequency Staircase Pulse Voltage (MSPV). The MSPV technique has proven to be the best waveform for classifying different yogurts (Wei et al., 2013) .

Figure 7. Example of response signals in an electronic tongue of pulse voltammetry: (a1) Hackle's Volta-metry, (a2) response curve of the VE tongue based on Hackle's Voltametry; (b1) rectangle voltammetry, (b2) response curve of the VE tongue based on the rectangle voltammetry; (c1) staircase voltammetry, (c2) response curve of the VE tongue based on staircase voltammetry. Taken from (Wei et al., 2013)

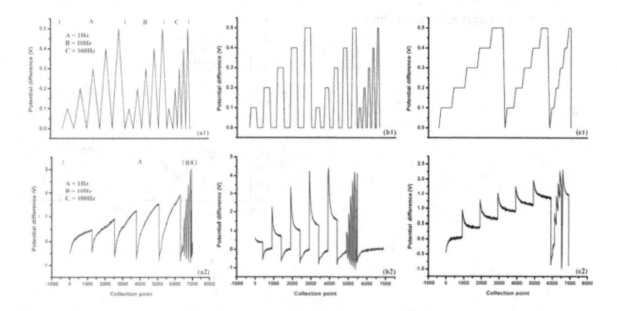

Signal Conditioning

The conditioning of the signal establishes the connection between the two fundamental modules of an electronic tongue, the sensors and the pattern recognition unit. Several electronic circuits are involved in the integration of pattern analysis algorithms with the mechanisms of chemical transduction (Banerjee et al., 2016) . First, the responses of the taste sensors must be measured and converted into an electrical signal. This operation is carried out using transducers. Second, the electrical signal is subjected to an analog conditioning to improve its information content. This implies for example amplification operations and filtering, among others. Third, the analog signal is sampled, digitized and stored in the memory of a computer or a dedicated processing system. Finally, the sampled signal is pre-processed in order to correct errors inherent in the acquisition of data and in order to correctly compare the signals from different sensors in complex sensor networks. The electrical signals generated by the sensor interface circuits are often unsuitable for acquisition in a computer and may need to be processed by several signal conditioning circuits.

SIGNAL PROCESSING AND PATTERN RECOGNITION METHODS USED IN ELECTRONIC TONGUES

Despite the type of sensor used, the pattern recognition methodologies used to process the signals obtained by the sensors vary in the different stages, including data preprocessing, feature extraction and selection and use of the classifier and regression algorithms. Figure 8 shows the differents stages of the process and in the remainder of this section, an extensive review of these methodologies is made.

Figure 8. Process performed by an electronic tongue: Sample, sensor network, signal processing, pattern recognition, classification, regression
Source: Authors

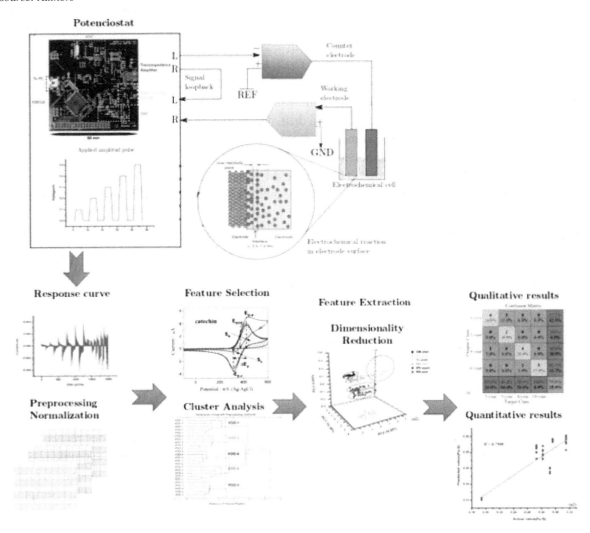

Data Preprocessing

As an array of sensors is used in the electronic tongue, there are very large amounts of multivariable data, which produce the need to transform the data obtained to understand them. The signals obtained with the electronic tongue sensors are complex and to treat them several pre-processing tools can be used for three main purposes. Firstly, for the elimination or reduction of random noise; secondly, for the elimination or reduction of unwanted systematic variations; and finally, for the reduction or compression of data (Oliveri et al., 2010)

Unwanted systematic variations may be due to instrumental obstacles, experimental conditions or physical characteristics of the samples. When improving the experimental configuration, unwanted signal variations should be avoided. However, the experimental approach is not always feasible because it is necessary to save time and cost, therefore, it would imply complex pre-treatments of the sample, precise control of the temperature, etc. For these reasons, when possible, it may be preferred to eliminate signal variations later, using pre-processing data (Marco & Gutierrez-Galvez, 2012). Zhao et al. (2008) classify data preprocessing techniques into five categories: i) Noise Removal, ii) Baseline Removal, iii) Signal Alignment, iv) Outlier Detection, and v) Data Normalization.

Palit et al. (2010) studied the influence of different data preprocessing techniques on the accuracy of classification for LAPV electronic tongues. In particular, 10 different data preprocessing techniques are studied, yielding the best result when combining the autoscaling method with the baseline subtraction method, since the latter eliminates the local drift. The signal processing methodology began with the application of a data normalization method, then, the Wavelet transform method and, finally, data classification using a supervised neural network.

Banerjee et al., (2016) carried out a study on the different techniques of data preprocessing in the fusion of electronic nose and tongue. In this work, the authors concluded that the normalization method range scale 1 was the one that best results of classification accuracy, with 98.45%. For the fused data of tongue and nose, the best technique of normalization was relative scale 4, with a classification accuracy of 98.99%

Exploratory Analysis, Dimensionality Reduction

Following, the stages belonging to the dimensionality reduction process in electronic tongue type sensor arrays are described, among which are the feature extraction and the feature selection.

Feature Extraction

The purpose of feature extraction is to obtain informative features from mathematical transformations of the multivariate time responses of the chemical sensor arrays. These transformations reduce the dimensionality of the input space and aim to keep the information related to the objective problem. Some of the different approaches for features extraction that can be found in the literature are classified into five categories: (1) subspace projection methods, (2) curve fitting methods, (3) waveform descriptors, (4) projection of non-linear subspaces and (5) grouping in the feature space, etc. It is also common to concatenate several feature extraction methods. For example, a typical implementation could extract a feature vector based on waveform descriptors, then use PCA to reduce dimensionality and, finally, avoid overfitting with LDA (Marco & Gutierrez-Galvez, 2012).

Plastria et al., (2008) studied the effects of dimensionality reduction using different techniques and different dimensions in six data sets of two classes, with numerical attributes as preprocessing for two classification algorithms. They concluded that a good selection of technique and number of dimensions can have a great impact on the power of classification, generating excellent classifiers. The dimensionality reduction should be used not only for visualization or as preprocessing in very high dimensional data, but also as a general preprocessing technique in numerical data to increase the classification power. However, the difficult selection of both, the dimensionality reduction technique and the reduced dimension, should be based directly on the effects on the classification power, with performance metrics such as accuracy (Plastria et al., 2008).

The most popular method for data compression in chemometrics is Principal Component Analysis (PCA). For example, for cyclic voltammetry, when voltammograms are compressed by PCA, there are some theoretical limitations. PCA is a linear projection method that does not preserve the structure of a non-linear data set. If there is any non-linearity in the voltammograms, it can appear as a small perturbation in a linear solution and will not be described by the first principal components, as in a linear case (Despagne & Massart, 1998).

Many of the works related to chemical sensor arrays are related to electronic noses. Zhang and Tian (2014) showed that SVM seems to be the best selection in a binary classification. However, the complexity of the SVM algorithm for a general problem of various classes can hinder an online real application in the development of a sensor network. Data preprocessing methods, such as feature extraction, and dimensionality reduction methods, such as PCA, have also been combined with ANN or SVM to improve the accuracy of electronic nose prediction. Both the feature extraction and the dimensionality reduction aim to obtain useful features for classification. Dimensionality reduction can lower redundant information, such as noise, but may lose useful information in the original data. In addition, the classification method also has the ability to automatically depress the useless components in the sampling learning process. A feature extraction method called new discriminant analysis framework (NDA) (Zhang and Tian, 2014), which, through the construction of a Laplacian dispersion matrix between classes and a Laplacian dispersion matrix within the classes, resolves an optimization problem. It makes samples between classes more separable and samples within classes are more compactable, thus improving the process of reducing the dimensionality of the data to in turn enhance the results of the classification algorithm. The KNDA method reached a 94.14% of average accuracy. A running time comparison was performed between several algorithms being the KNDA 18.2 times faster than KSVM.

Feature Selection

Feature selection is a dimensionality reduction approach that consists in selecting certain numerical characteristics from all those available for later use by a learning algorithm (Marco & Gutierrez-Galvez, 2012). Depending on the organization of a search process, feature selection algorithms are generally classified as belonging to filters, wrappers or embedded approaches. Combinations of approaches are also constructed. For example, a filter is first used, then a wrapper, or a wrapper can be used as a filter (Stańczyk, 2015).

The filtering methods are based on the general characteristics of the data to evaluate and select subsets of features without involving the chosen learning algorithm. Wrapper methods use the performance of the chosen learning algorithm to evaluate each subset of candidate features. Wrapper methods look for features that best fit the chosen learning algorithm but can be significantly slower than filter methods if the learning algorithm takes a long time to execute (Kohavi, & John, 1997).

Gualdrón et al. (2006) compared five different feature selection techniques: sequential forward selection (SFS), sequential backward selection (SBS), stepwise selection, Genetic Algorithms (GA) and Simulated Annealing. The techniques were used to analyze data from an electronic nose-type sensor.

In many classification tasks, the total number of possible features that can be used is relatively high. The use of all the features would result in high dimensionality that makes processing difficult, or even impractical. The presence of too many variables is inconvenient for most machine learning algorithms, even when these attributes are relevant to the task, in addition to the redundant variables that other patterns can hide (Stańczyk, 2015) .

In the literature, there are works such as Marx et al. (2017), Gutiérrez et al. (2013) and Prieto et al. (2013), related to the use of feature selection algorithms to improve the performance of electronic tongue-type sensor arrays.

Machine Learning in Electronic Tongue Sensor Arrays

The data analysis is divided into two large groups: qualitative and quantitative analysis (Oliveri et al., 2010). The qualitative analysis allows to differentiate liquid samples from each other, for example, to determine a type of milk, coffee, wine among other liquid substances (Zhang et al., 2018). In contrast, quantitative analysis, as the name implies, determines quantities of components in substances, such as pH and content of soluble solids in cherry tomato juice (Hong & Wang, 2014) or bitterness in traditional medicines (Lin et al., 2016). In machine learning problems, the qualitative analysis is considered as a classification problem, while the quantitative analysis is classified as a regression analysis.

Classification and Class Modeling

With the steps of data preprocessing and dimensionality reduction, the next step in data analysis consists in an automatic learning process. In terms of learning strategies, learning processes can be divided into three categories: supervised, semi-supervised and unsupervised learning (Zhao et al., 2008). To carry out the previous analyses, different multivariate techniques have been developed and can be divided into two large groups. The first one contains the chemometric statistical techniques (Gorji-Chakespari et al., 2016) and the second one comprises the different artificial neural networks (ANN). Several comparative studies have been conducted on the best technique used to perform pattern recognition and multivariate analysis in electronic tongues, for example the works of Liu et al., (2013) and that of Ceto et al. (2013).

Kotsiantis et al. (2007) shows a review of supervised machine learning techniques is performed. The learning algorithms compared were: Decision Trees, Neural Networks, Naïve Bayes, kNN, SVM and Rule learners. The performance results are shown on a scale of 1 to 4 stars (**** stars represent the best and * star the worst performance). The key question when it comes to classification in Machine Learning (ML) is not whether a learning algorithm is superior to others, but under what conditions a particular method can significantly outperform others in a given application problem.

The classification of chemical sensor data is, in many cases, fraught with problems. It is quite common to find highly correlated features, dispersion due to concentration or drift, multimodality and distribution of non-Gaussian data per class. In addition, the number of training data is limited due to the high costs of calibration of the acquisition equipment. For all these reasons, it is important to select robust classifiers to tackle the previous problems (Marco & Gutierrez-Galvez, 2012).

Taste Recognition Methodologies in Electronic Tongues

Since its development, electronic tongues have had as a primary component the pattern recognition system to classify with certain accuracy the different liquids analyzed, within said system of pattern recognition and signal processing there are different stages between those found in data normalization, feature extraction and selection, some machine learning classifying algorithms and finally a cross-validation method to find the classification accuracy. Within the literature there are different methodologies that have combined different methods of the above-mentioned stages, some of the most prominent are collected below due to their high classification rate:

Ciosek et al. (2005) presented a new data analysis strategy for artificial taste and odor systems. A direct processing comparison was made (raw data processed by the artificial neural network (ANN), raw data processed by partial-discriminant least squares analysis (PLS-DA)) and two-step processing first (principal component analysis) (PCA) and second, outputs processed by ANN, PLS-DA. It was shown that there was a considerable increase in the classification capacity in the case of the new method proposed by the authors reaching 100% classification with the combination PLS-DA + ANN.

Ciosek and Wróblewski (2006) showed a sensor array composed of partially selective potentiometric electrodes for milk recognition. The task of the system is to distinguish between five brands of milk. For this purpose, five pattern recognition procedures are used: K-nearest neighbors, partial least squares-PLS, soft independent modeling of class analogy-SIMCA, back propagation neural networks -BPNN and learning vector quantization-LVQ. The LVQ networks showed the best performance reaching an accuracy of 99.8%. Their additional advantages, such as fast training and robustness, make them the suggested pattern classifiers for the sensor matrix data.

Bhondekar et al. (2011) Optimize an impedance electronic tongue using the Dynamic Social Impact Theory based optimize method-SITO, which searches for the best hyperparameters of the SVM classifier algorithm, the best number of main components (PCs) that describe the information and the best excitation frequencies EF's. In the optimization stage for calculating the fitness function fuzzy inference (FIS) is used. The performance of the optimization algorithm was compared with genetic algorithms GA and particle swarm optimization PSO. The best classification achieved with this feature selection wrapper approach was 99.77%.

Szöllősi et al. (2012) used several classification performance indicators that include the classification accuracy value (ACC), Cohen's kappa (KAPPA) and the area under the ROC curve (AUC). The objective of this work was to find the best classification method to classify four samples of soft drinks. The results showed that the ACC value and the KAPPA values give similar results. The three best models according to ACC, KAPPA and AUC were " K- nearest neighbors", "random forest" and " Linear discriminant analysis". The results showed that the best classification model was K-nearest neighbors with a corrected ACC of 80%.

Gutierrez et al. (2013) developed a hybrid electronic tongue composed of potentiometric and voltametric sensors for the classification of beers. The signal processing procedure was composed of a normalization stage with the Square Max Value method, then stepwise feature selection was used to later identify some representative features of the waveforms of the obtained signals and finally, LDA as a classifier algorithm. The accuracy achieved by this system was 96%.

Liu et al. (2013) used three different classification algorithms to treat problems with unbalanced, multiclass and small sample data. Thay solve classification problems with 4 different data sets. As a result, they found that Random Forest (RF) shows advantages over Backward Propagation Neural Networks (BPNN) and Support Vector Machines (SVM), since RF performed the classification without preprocessing procedures and RF reached a 99.07% of classification rate.

Liu et al. (2013) used an voltammetric electronic tongue to analyze green and black tea, the signal processing is done with principal components analysis -PCA in a first stage, then obtain qualitative qualification results with SVM, compare three different kernels within SVM being the polynomial Kernel the one that better accuracy obtained with a value of 99.7%.

Buratti et al. (2014) performed a study with different instrumental tools for the classification of coffee varieties. One of them a potentiometric electronic tongue in a first stage of signal processing, the SELECT algorithm is used as a feature selection to then apply the linear discriminant analysis-LDA algorithm as a classifier. The results of classification accuracy reached by the electronic tongue were 78.76%.

Dominguez et al. (2014) used a voltammetric electronic tongue to analyze Mexican coffee, for signal processing a data normalization stage was first included with the blank correction method and the classification performance of the Linear Discriminant Analysis LDA algorithm was compared versus Support Vector Machines SVM. Two types of classifications were made, the first Growing conditions (i.e., organic or non-organic practices and altitude of crops) and the second, based on geographical origin of simple. The best classification result was achieved by SVM with 97.5% accuracy.

Marx et al. (2017) developed an electronic tongue for table olives classification according to the presence and intensity of negative defects. The signal processing comprises a stage of feature selection using the simulates annealing algorithm to then move to the LDA classifier algorithm. A 4-fold cross validation approach is used to evaluate the classification model. The highest classification accuracy reached was $93 \pm 1.2\%$.

Shi et al. (2018) used an electronic tongue type: large amplitude pulsed voltammetry to classify a traditional Chinese herbal medicine known as pericarpium citri reticulatae (PCR). The discrete wavelet transforms (DWT) is used for feature extraction. Seven linear and nonlinear classification methods were then compared, namely, principal component analysis (PCA), cluster analysis (CA), linear discriminant analysis (LDA), back-propagation neural network (BPNN), extreme learning machine (ELM), random forest (RF) and support vector machine (SVM). ELM being the one that obtained the best classification accuracy for the testing set with a value of 95%.

Liu et al. (2018) worked with a multifrecuency large amplitude pulse voltammetry-MLAPV electronic tongue to classify 7 different substances. In this research, discrete wavelet transforms (DWT) was used as a feature selection method. the k-nearest neighbor (k-NN), support vector machine (SVM), and random forest (RF) classifiers were used to evaluate the improvement of the refined features. Two types of approaches of the DWT were used Relative Power Ratio-RPR and Active Feature Selection-ASF with the ASF-DWT together with KNN the best classification result obtained with $84.13\% \pm 0.0125$.

Chen et al. (2018) proposed an alternative method of feature extraction for the improvement of the specificity of electronic tongue-type sensor array, which is called feature specificity enhancement (FSE). The proposed FSE method measures the specificity of the sensor in the paired sensor responses and uses the core function for nonlinear projection. The FSE method was compared against three feature extraction methods: raw, principal component analysis (PCA) and discrete wavelet transform (DWT). Subsequently three classifiers were used to determine their performance: the support vector machine (SVM), the random forest (RF) and the kernel extreme learning machine (KELM). For cross-validation the leave one out (LOO) strategy was used. The new method proposed by the authors reaching 95.24% classification with the FSE+KELM combination. A time consumption comparison is performed being the FSE+KELM combination 70.6 times faster than FSE+RF combination and 42.6 times faster than FSE+SVM combination.

Zhang et al. (2018) developed an electronic tongue of multifrecuency large amplitude pulse voltammetry-MLAPV to perform artificial taste recognition of 13 different substances, the data processing comprises the following stages: it starts with Filter & Feature Selection stage via sliding window-based smooth filter, later a stage of Feature Extraction stage comparing the developed subspace learning algorithm: Local Discriminant Preservation Projection- LDPP algorithm, to finally classify with Kernelized

Extreme Learning Machine-KELM. The LDPP algorithm is compared versus other feature extraction methods like PCA, KPCA, LPP, LPDP. Among the KELM classifier the KSVM and ELM algorithm are also compared in the classification task. The process was validated using the 5-fold cross validation technique, achieving an average accuracy of 98.22% for the LDPP + KELM combination. A comparison of computational time expending by the algorithms is done and as result the LDPP + KELM combination is 24.7 times faster than the LDPP+KSVM combination and 44.34 times faster than the LDPP+ELM combination.

Wesoły and Ciosek (2018) conducted a study on pharmaceutical analysis with a potentiometric electronic tongue, the authors affirmed that, to the best of their knowledge, no in-depth study of the chemometric treatment of sensor responses had been presented for taste discrimination studies by means of an electronic tongue. The best classification results were obtained using SVM-DA and PCA + BPNN, while PCR and KNN provided the worst results. The work concludes that it is not feasible to develop a universal methodology for the analysis of data from an electronic tongue, but some general recommendations can be provided based on the obtained results (Wesoły & Ciosek, 2018):

- The optimization of the pattern recognition technique by adjusting the selected parameters is a crucial step in the data analysis process and significantly influences the results of the classification.
- In the case of the study conducted in (Wesoły & Ciosek, 2018), the best classification results were obtained applying the SVM-DA and PCA + BPNN methods. However, neural networks require an experienced user, who must correctly adjust the parameters that are not directly deduced.
- Using the steady-state responses of the sensor array as data extraction is better for simple classifiers such as KNN.
- The most reliable methods were SVM-DA and PLS, considering the lower standard deviation in the results obtained from the training and test data sets.
- The addition of PCA values to the data sets decreases the classification capabilities of the models.
- The SVM-DA models and the BPNN methods exhibited the best generalization skills, that is, the belonging of most of the test samples was correctly assigned.

Zhong et al. (2019) proposed a convolutional neural network-based auto features extraction strategy (CNN-AFE) in an electronic tongue system for tea classification. First, the sensor response of the e-tongue was converted to time-frequency maps by short-time Fourier transform (STFT). Second, features were extracted by convolutional neural network (CNN) with time-frequency maps as input. Finally, the features extraction and classification results were carried out under a general shallow CNN architecture. Several feature extraction methods were compared, among those found raw response, peak-inflection point, Discrete wavelet transform (DWT), Discrete cosine transform (DCT), singular value decomposition (SVD) and DCT fused with SVD (DCT + SVD). For classification different algorithms of machine learning like SVM, KNN and RF was compared versus the CNN-AFE developed approach. The experimental results are based on the five-fold cross-validation. the highest average recognition rate achieved by the CNN-AFE method was 99.9%.

Lu et al. (2019) developed a joint voltammetry technology, which combines different types of voltammetry, to qualitatively and quantitatively analyze four basic flavors: sweetness, salinity, acidity and bitterness with an array of multiple electrodes. Cyclic voltammetry (CV), differential pulse voltammetry (DPV) and square wave voltammetry (SWV) were used. A wavelet transform was used as feature extraction method. Subsequently, the features ares used as input to to a layer recurrent neuronal network

(LRNN) that could effectively identify the types of stimuli. The accuracies of the training set and the joint voltammetry test set were both greater than those of regular voltammetry, which confirms that the neural network could quantitatively predict the unique taste stimulation of the mixture.

Leon_Medina et al. (2019) introduced a taste recognition methodology, which is composed of several steps including data unfolding, data normalization, principal component analysis for compressing the data, and classification through different machine learning models. The methodology was tested using data from an MLAPV electronic tongue with 13 different liquid substances. A five-fold cross-validation was used and a comparative study was made on the use of the number of components at the entrance to the classification process carried out by the machine learning algorithms. When considering the first eight principal components, the classification accuracy made by KNN reached 94.74%. It was found that results are associated with the number of components. In addition, a comparison to evaluate the methodology is made with different classification performance measures that show the behavior of the process in a single number.

Validation

The development of classification models in machine learning always requires a careful validation, in order to provide information with real veracity and usefulness for the problem studied. If it is not validated, a model is not exploitable in practice, since the reliability of its results is completely unknown (Oliveri et al., 2010). Prediction capacity values must have a confidence interval, which depends to a large extent on the number of objects used for validation. The estimation of the predictive capacity in new objects is a fundamental step in any modeling process, and several procedures have been implemented for this purpose. The most common validation strategies divide the available data into two subsets: a training set used to calculate the model and a test set used to evaluate its reliability. One of the most used strategies is Cross-Validation, which can be divided into (1) Cross-Validation by random sub-sampling, (2) Leave-One-Out Cross-Validation and (3) K-Fold Cross-validation. Other techniques used in the validation of models are the Confusion Matrix and the analysis of ROC (Receiver Operating Characteristics) curves. There are also different types of errors, as well as measures of sensitivity, specificity and precision (Ballabio & Todeschini, 2009).

Ballabio and Todeschini (2009) conducted a study of the multivariable classification in chemometrics, which consists in finding mathematical relationships between a set of descriptive variables and a qualitative variable (belonging to the class). There is a large number of applications of classification methods in the literature, in different types of data and with different objectives, even if basically the final objective of a classification model is always the separation of two (or more) object classes and the assignment of new unknown objects to one of the defined classes (or none of the classes when class modeling approaches are applied). However, sometimes classifiers are chosen only on the basis of personal knowledge and user preferences, although the best classification approach should be preferred based on the characteristics of the data and the purpose of the analysis. In the study of Ballabio and Todeschini (2009), a section referring to the evaluation of the classification performance is included, where different performance measures are described and compared for the estimation of the quality of the classification models, both for adjustment and validation purposes.

Table 1 shows a comparison of works related to classification of substances using electronic tongue-type sensor arrays. Table 1 distinguishes the type of electronic tongue, the data processing stages and the best combination of methods for signal processing that obtained the best recognition accuracy.

Table 1. Comparison of data processing methodologies in electronic tongues

#	Ref	Electronic Tongue Type	Data processing Stages	Best Combination of Methods	Best Recognition Accuracy
1	Leon- Medina et al., 2019	MLAPV	Normalization Feature Extraction Classifier	Group Scaling PCA KNN	94.74%
2	Zhong et al., 2019	MLAPV	Normalization Feature Extraction Classifier	Normalization 0-1 STFT CNN-AFE	99.9%
3	Zhang et al., 2018	MLAPV	Filter & Feature Selection Feature Extraction Classifier	sliding window-based smooth filter LDPP KELM	98.22%
4	Liu et al., 2018	MLAPV	Feature Selection Classifier	ASF-DWT KNN	84.13%
5	Wesoly & Ciosek, 2018	Potentiometric	Feature Extraction Classifier	PCA BPNN	98.3±1.2%
6	Shi et al., 2018	LAPV	Feature Selection Classifier	DWT ELM	95%
7	Chen et al., 2018	MLAPV	Feature Extraction Classifier	FSE KELM	95.24%
8	Marx et al., 2017	Potentiometric	Feature Selection Classifier	SA LDA	93 ± 1.2%
9	Buratti et al., 2014	Potentiometric	Feature Selection Classifier	SELECT LDA	78.76%
10	Domínguez et al., 2014	Voltametric	Normalization Classifier	blank correction SVM	97.5%
11	Liu et al., 2013	Potentiometric	Classifier	RF	99.07%
12	Gutierrez et al., 2013	Potentiometric and voltametric	Normalization Feature Selection Feature Extraction Classifier	Square Max Value stepwise representative features LDA	96%
13	Liu et al., 2013	Voltametric	Feature Extraction Classifier	PCA SVM	99.7%
14	Szöllősi et al., 2012	Potentiometric	Normalization Classifier	mean center- unit deviation KNN	80%
15	Bhondekar et al., 2011	Impedance	Feature Selection Feature Extraction Classifier Multiobjective optimization	AFS PCA SVM FIS SITO	99.77%
16	Palit et al., 2010	MLAPV	Normalization Feature Selection Classifier	Baseline substraction +autoscale DWT RBF ANN	98.33%
17	Ciosek & Wróblewski, 2006	Potentiometric	Normalization Feature Extraction Classifier	Autoscale PCA LVQ	99.8 ±0.2%
18	Ciosek et al., 2005	Potentiometric	Feature Extraction Classifier	PLS-DA ANN	100%

Source: Information obtained in the literature and organized by the authors

As can be seen in Table 1, most accuracy recognition is close to 95% and this has been achieved through different pattern recognition methodologies composed of different stages, the computational efficiency and the required time of each of them is not reported in the literature which could be a future study to evaluate, for example, the computational complexity of each methodology and determine its implementation in embedded systems of portable electronic tongue sensor arrays. From the information collected in the Table 1, the following can be inferred: of the 17 works mentioned in Table 1, all use a machine learning classifier algorithm, 6 use Normalization, 8 use feature selection, 9 use feature extraction and only one uses multi-objective optimization.

FUTURE RESEARCH DIRECTIONS

In futures developments the evolution of signal processing will be concentrated in big data challenges that the use of a sensor array with a high number of sensors implies. These big data challenges include volume with large scale data; variety with heterogeneous, high dimensional and nonlinear data; velocity with real time, stream, high speed needs; veracity for treat uncertain and incomplete data and finally the value challenge with low value density and diverse data meaning. The data obtained by electronic tongue type sensor arrays are characterized of unbalanced, multiclass and small sample data, Therefore, the machine learning community has these challenges to face and solve.

CONCLUSION

Electronic tongue type sensor arrays emerge as an economical, portable and viable option for liquid analysis. The classification of different liquids allows to identify a constant flavor, origin, adulterations, and improvements in quality processes in the production of those. As an electronic tongue is composed of multiple sensors, in each analysis a significant size of information is collected that must be processed by different algorithms. The current chapter book approached the investigations carried out from the point of view of development of algorithms for the signal processing obtained by electronic tongue type sensor arrays. It should be noted that each pattern recognition methodology depends on the type of data. Through normalization, feature extraction and feature selection stages, an analysis can be obtained that can help to further compact the data trough features and distinguish each class such that a machine learning classifier algorithm can yield satisfactory results of classification accuracy.

It is desired to have reliability and high accuracy in the results when using the electronic tongue sensor arrays, however, for online or portable monitoring applications of these sensors it is also necessary that the signal processing be fast. Recently, one of the most promising algorithms that have a good balance between accuracy and calculation time is the Kernel Extreme Learning Machine KELM which has reached 98.22% accuracy and for example is 24.7 times faster than the kernel support vector machine KSVM.

Different stages have been used in the process of recognizing data patterns obtained by electronic tongue type sensors arrays, among which are: The normalization stage is used to make the variables obtained by different sensors comparable regardless of their magnitude. The feature extraction stage decreases the computational load at the input of the classification machine and also excludes irrelevant and noise signals from the analysis. Proper use of the available samples leads to a reliable supervised learning strategy and avoids excessive overfitting. In the validation stage for a case where there are enough samples, the data can be divided into three subsets: training, validation and testing. For a case with fairly limited samples, cross-validation of k-fold is a good option for the construction and testing of the model. The extreme case of the k-fold cross validation process is the strategy of leaving one out (LOO).

According to the analysis performed, there are different pattern recognition methodologies that are composed of different stages, the use of more stages would involve more computing time at the expense of high accuracy, this type of study will be carried out in the future to determine the computational complexity of different pattern recognition methodologies and thereby determine the best methodology depending on the type of application of a specific electronic tongue.

ACKNOWLEDGMENT

The authors express their gratitude to the Administrative Department of Science, Technology and Innovation – Colciencias with the grant 779 – "Convocatoria para la Formación de Capital Humano de Alto Nivel para el Departamento de Boyacá 2017" for sponsoring the research presented herein. Jersson X. Leon-Medina is grateful with Colciencias and Gobernación de Boyacá for the PhD fellowship.

REFERENCES

Banerjee, M. B., Chatterjee, T. N., Roy, R. B., Tudu, B., Bandyopadhyay, R., & Bhattacharyya, N. (2016, April). Multivariate preprocessing techniques towards optimising response of fused sensor from electronic nose and electronic tongue. In *2016 International Conference on Computing, Communication and Automation (ICCCA)* (pp. 949-954). IEEE 10.1109/CCAA.2016.7813875

Banerjee, R., Tudu, B., Bandyopadhyay, R., & Bhattacharyya, N. (2016). A review on combined odor and taste sensor systems. *Journal of Food Engineering*, *190*, 10–21. doi:10.1016/j.jfoodeng.2016.06.001

Beebe, K. R., Pell, R. J., & Seasholtz, M. B. (1998). *Chemometrics: a practical guide* (Vol. 4). Wiley-Interscience.

Bhondekar, A. P., Kaur, R., Kumar, R., Vig, R., & Kapur, P. (2011). A novel approach using Dynamic Social Impact Theory for optimization of impedance-Tongue (iTongue). *Chemometrics and Intelligent Laboratory Systems*, *109*(1), 65–76. doi:10.1016/j.chemolab.2011.08.002

Bougrini, M., Tahri, K., Haddi, Z., El Bari, N., Llobet, E., Jaffrezic-Renault, N., & Bouchikhi, B. (2014). Aging time and brand determination of pasteurized milk using a multisensor e-nose combined with a voltammetric e-tongue. *Materials Science and Engineering C*, *45*, 348–358. doi:10.1016/j.msec.2014.09.030 PMID:25491839

Brazdil, P. B., Soares, C., & Da Costa, J. P. (2003). Ranking learning algorithms: Using IBL and meta-learning on accuracy and time results. *Machine Learning*, *50*(3), 251–277. doi:10.1023/A:1021713901879

Buratti, S., Sinelli, N., Bertone, E., Venturello, A., Casiraghi, E., & Geobaldo, F. (2015). Discrimination between washed Arabica, natural Arabica and Robusta coffees by using near infrared spectroscopy, electronic nose and electronic tongue analysis. *Journal of the Science of Food and Agriculture*, *95*(11), 2192–2200. doi:10.1002/jsfa.6933 PMID:25258213

Cetó, X., Céspedes, F., & del Valle, M. (2013). Comparison of methods for the processing of voltammetric electronic tongues data. *Mikrochimica Acta, 180*(5-6), 319–330. doi:10.100700604-012-0938-7

Cetó, X., Voelcker, N. H., & Prieto-Simón, B. (2016). Bioelectronic tongues: New trends and applications in water and food analysis. *Biosensors & Bioelectronics, 79,* 608–626. doi:10.1016/j.bios.2015.12.075 PMID:26761617

Chen, Y., Liu, T., Chen, J., Li, D., & Wu, M. (2018, November). A Novel Feature Specificity Enhancement for Taste Recognition by Electronic Tongue. In *International Conference on Extreme Learning Machine* (pp. 11-16). Springer.

Ciosek, P., Brzózka, Z., Wróblewski, W., Martinelli, E., Di Natale, C., & D'amico, A. (2005). Direct and two-stage data analysis procedures based on PCA, PLS-DA and ANN for ISE-based electronic tongue—Effect of supervised feature extraction. *Talanta, 67*(3), 590–596. doi:10.1016/j.talanta.2005.03.006 PMID:18970211

Ciosek, P., & Wróblewski, W. (2006). The analysis of sensor array data with various pattern recognition techniques. *Sensors and Actuators. B, Chemical, 114*(1), 85–93. doi:10.1016/j.snb.2005.04.008

Costa, C., Taiti, C., Strano, M. C., Morone, G., Antonucci, F., Mancuso, S., ... Menesatti, P. (2016). Multivariate Approaches to Electronic Nose and PTR-TOF-MS Technologies in Agro-Food Products. In *Electronic Noses and Tongues in Food Science* (pp. 73–82). London, UK: Academic Press. doi:10.1016/B978-0-12-800243-8.00008-1

Cuartero, M., Ruiz, A., Oliva, D. J., & Ortuño, J. A. (2017). Multianalyte detection using potentiometric ionophore-based ion-selective electrodes. *Sensors and Actuators. B, Chemical, 243,* 144–151. doi:10.1016/j.snb.2016.11.129

Del Bratov, A., Abramova, N., & Ipatov, A. (2010). Recent trends in potentiometric sensor arrays—A review. *Analytica Chimica Acta, 678*(2), 149–159. doi:10.1016/j.aca.2010.08.035 PMID:20888446

Del Valle, M. (2007). Potentiometric electronic tongues applied in ion multidetermination. *Comprehensive Analytical Chemistry, 49,* 721–753. doi:10.1016/S0166-526X(06)49030-9

Del Valle, M. (2010). Electronic tongues employing electrochemical sensors. *Electroanalysis, 22*(14), 1539–1555.

Del Valle, M. (2017). Materials for electronic tongues: Smart sensor combining different materials and chemometric tools. In *Materials for Chemical Sensing* (pp. 227–265). Cham: Springer. doi:10.1007/978-3-319-47835-7_9

Despagne, F., & Massart, D. L. (1998). Neural networks in multivariate calibration. *Analyst (London), 123*(11), 157R–178R. doi:10.1039/a805562i PMID:10396805

Díaz, Y. Y. R., Acevedo, C. M. D., & Cuenca, M. (2017). Discriminación De Hidromieles A Través De Una Lengua Electrónica. *Revista Colombiana De Tecnologías De Avanzada (RCTA), 1*(23).

Domínguez, R. B., Moreno-Barón, L., Muñoz, R., & Gutiérrez, J. M. (2014). Voltammetric electronic tongue and support vector machines for identification of selected features in Mexican coffee. *Sensors (Basel), 14*(9), 17770–17785. doi:10.3390140917770 PMID:25254303

Gardner, J. W., & Bartlett, P. N. (1994). A brief history of electronic noses. *Sensors and Actuators. B, Chemical, 18*(1-3), 210–211. doi:10.1016/0925-4005(94)87085-3

Gomes, H. L. (2009). Sensor arrays for liquid sensing (electronic tongue systems). Biossensores, Mestrado Integrado em Eng. Electrónica e Telecomunicações (MIEET-2009/00). Universidade do Algarve, FCT, Campus de Gambelas.

Gorji-Chakespari, A., Nikbakht, A. M., Sefidkon, F., Ghasemi-Varnamkhasti, M., Brezmes, J., & Llobet, E. (2016). Performance comparison of Fuzzy ARTMAP and LDA in qualitative classification of Iranian Rosa damascena essential oils by an electronic nose. *Sensors (Basel), 16*(5), 636. doi:10.339016050636 PMID:27153069

Gualdrón, O., Llobet, E., Brezmes, J., Vilanova, X., & Correig, X. (2006). Coupling fast variable selection methods to neural network-based classifiers: Application to multisensor systems. *Sensors and Actuators. B, Chemical, 114*(1), 522–529. doi:10.1016/j.snb.2005.04.046

Gutiérrez, J. M., Haddi, Z., Amari, A., Bouchikhi, B., Mimendia, A., Cetó, X., & del Valle, M. (2013). Hybrid electronic tongue based on multisensor data fusion for discrimination of beers. *Sensors and Actuators. B, Chemical, 177*, 989–996. doi:10.1016/j.snb.2012.11.110

Harvey, D. (2010). *Analytical Chemistry 2.0*. Retrieved from: https://chem.libretexts.org/Bookshelves/Analytical_Chemistry/Book%3A_Analytical_Chemistry_2.0_(Harvey)

Holmberg, M., Eriksson, M., Krantz-Rülcker, C., Artursson, T., Winquist, F., Lloyd-Spetz, A., & Lundström, I. (2004). 2nd workshop of the second network on artificial olfactory sensing (NOSE II). *Sensors and Actuators. B, Chemical, 101*(1-2), 213–223. doi:10.1016/j.snb.2004.02.054

Hong, X., & Wang, J. (2014). Detection of adulteration in cherry tomato juices based on electronic nose and tongue: Comparison of different data fusion approaches. *Journal of Food Engineering, 126*, 89–97. doi:10.1016/j.jfoodeng.2013.11.008

Kimmel, D. W., LeBlanc, G., Meschievitz, M. E., & Cliffel, D. E. (2011). Electrochemical sensors and biosensors. *Analytical Chemistry, 84*(2), 685–707. doi:10.1021/ac202878q PMID:22044045

Kohavi, R., & John, G. H. (1997). Wrappers for feature subset selection. *Artificial Intelligence, 97*(1-2), 273–324. doi:10.1016/S0004-3702(97)00043-X

Kotsiantis, S. B., Zaharakis, I., & Pintelas, P. (2007). Supervised machine learning: A review of classification techniques. *Emerging Artificial Intelligence Applications in Computer Engineering, 160*, 3-24.

Leon-Medina, J. X., Cardenas-Flechas, L. J., & Tibaduiza, D. A. (2019). A data-driven methodology for the classification of different liquids in artificial taste recognition applications with a pulse voltammetric electronic tongue. *International Journal of Distributed Sensor Networks, 15*(10). doi:10.1177/1550147719881601

Lin, Z., Zhang, Q., Liu, R., Gao, X., Zhang, L., Kang, B., ... Li, X. (2016). Evaluation of the bitterness of traditional Chinese medicines using an e-tongue coupled with a robust partial least squares regression method. *Sensors (Basel), 16*(2), 151. doi:10.339016020151 PMID:26821026

Liu, M., Wang, M., Wang, J., & Li, D. (2013). Comparison of random forest, support vector machine and back propagation neural network for electronic tongue data classification: Application to the recognition of orange beverage and Chinese vinegar. *Sensors and Actuators. B, Chemical, 177*, 970–980. doi:10.1016/j.snb.2012.11.071

Liu, N., Liang, Y., Bin, J., Zhang, Z., Huang, J., Shu, R., & Yang, K. (2014). Classification of green and black teas by PCA and SVM analysis of cyclic voltammetric signals from metallic oxide-modified electrode. *Food Analytical Methods, 7*(2), 472–480. doi:10.100712161-013-9649-x

Liu, T., Chen, Y., Li, D., & Wu, M. (2018). An active feature selection strategy for DWT in artificial taste. *Journal of Sensors.*

Lu, L., Hu, X., & Zhu, Z. (2019). Joint Voltammetry Technology with a Multi-electrode Array for Four Basic Tastes. *Current Analytical Chemistry, 15*(1), 75–83. doi:10.2174/1573411014666180522100504

Marco, S., & Gutierrez-Galvez, A. (2012). Signal and data processing for machine olfaction and chemical sensing: A review. *IEEE Sensors Journal, 12*(11), 3189–3214. doi:10.1109/JSEN.2012.2192920

Marx, Í., Rodrigues, N., Dias, L. G., Veloso, A. C., Pereira, J. A., Drunkler, D. A., & Peres, A. M. (2017). Sensory classification of table olives using an electronic tongue: Analysis of aqueous pastes and brines. *Talanta, 162*, 98–106. doi:10.1016/j.talanta.2016.10.028 PMID:27837890

Mimendia, A., Gutiérrez, J. M., Leija, L., Hernández, P. R., Favari, L., Muñoz, R., & del Valle, M. (2010). A review of the use of the potentiometric electronic tongue in the monitoring of environmental systems. *Environmental Modelling & Software, 25*(9), 1023–1030. doi:10.1016/j.envsoft.2009.12.003

Mishalanie, E. A., Lesnik, B., Araki, R., & Segall, R. (2005). *Validation and peer review of US Environmental Protection Agency chemical methods of analysis*. Washington, DC: Environmental Protection Agency.

Oliveri, P., Casolino, M. C., & Forina, M. (2010). Chemometric brains for artificial tongues. *Advances in Food and Nutrition Research, 61*, 57–117. doi:10.1016/B978-0-12-374468-5.00002-7 PMID:21092902

Palit, M., Tudu, B., Bhattacharyya, N., Dutta, A., Dutta, P. K., Jana, A., ... Chatterjee, A. (2010). Comparison of multivariate preprocessing techniques as applied to electronic tongue based pattern classification for black tea. *Analytica Chimica Acta, 675*(1), 8–15. doi:10.1016/j.aca.2010.06.036 PMID:20708109

Plastria, F., De Bruyne, S., & Carrizosa, E. (2008, October). Dimensionality reduction for classification. In *International Conference on Advanced Data Mining and Applications* (pp. 411-418). Springer. 10.1007/978-3-540-88192-6_38

Pravdová, V., Pravda, M., & Guilbault, G. G. (2002). Role of chemometrics for electrochemical sensors. *Analytical Letters, 35*(15), 2389–2419. doi:10.1081/AL-120016533

Prieto, N., Oliveri, P., Leardi, R., Gay, M., Apetrei, C., Rodriguez-Méndez, M. L., & De Saja, J. A. (2013). Application of a GA–PLS strategy for variable reduction of electronic tongue signals. *Sensors and Actuators. B, Chemical, 183*, 52–57. doi:10.1016/j.snb.2013.03.114

Real Academia de Ingeniería-Raing. (2018). *Electroanálisis*. Retrieved from: http://diccionario.raing.es/es/lema/electroan%C3%A1lisis

Salles, M. O., & Paixão, T. R. (2014). Application of Pattern Recognition Techniques in the Development of Electronic Tongues. *Advanced Synthetic Materials in Detection Science*, (3), 197.

Shi, Q., Guo, T., Yin, T., Wang, Z., Li, C., Sun, X., ... Yuan, W. (2018). Classification of Pericarpium Citri Reticulatae of Different Ages by Using a Voltammetric Electronic Tongue System. *International Journal of Electrochemical Science*, *13*(12), 11359–11374. doi:10.20964/2018.12.45

Sliwinska, M., Wisniewska, P., Dymerski, T., Namiesnik, J., & Wardencki, W. (2014). Food analysis using artificial senses. *Journal of Agricultural and Food Chemistry*, *62*(7), 1423–1448. doi:10.1021/jf403215y PMID:24506450

Stańczyk, U. (2015). Feature evaluation by filter, wrapper, and embedded approaches. In *Feature Selection for Data and Pattern Recognition* (pp. 29–44). Berlin: Springer. doi:10.1007/978-3-662-45620-0_3

Szöllősi, D., Dénes, D. L., Firtha, F., Kovács, Z., & Fekete, A. (2012). Comparison of six multiclass classifiers by the use of different classification performance indicators. *Journal of Chemometrics*, *26*(3-4), 76–84. doi:10.1002/cem.2432

Tian, S. Y., Deng, S. P., & Chen, Z. X. (2007). Multifrequency large amplitude pulse voltammetry: A novel electrochemical method for electronic tongue. *Sensors and Actuators. B, Chemical*, *123*(2), 1049–1056. doi:10.1016/j.snb.2006.11.011

Vlasov, Y. G., Legin, A. V., Rudnitskaya, A. M., D'amico, A., & Di Natale, C. (2000). «Electronic tongue»—New analytical tool for liquid analysis on the basis of non-specific sensors and methods of pattern recognition. *Sensors and Actuators. B, Chemical*, *65*(1-3), 235–236. doi:10.1016/S0925-4005(99)00323-8

Wang, L., Niu, Q., Hui, Y., & Jin, H. (2015). Discrimination of rice with different pretreatment methods by using a voltammetric electronic tongue. *Sensors (Basel)*, *15*(7), 17767–17785. doi:10.3390150717767 PMID:26205274

Wei, Z., Wang, J., & Jin, W. (2013). Evaluation of varieties of set yogurts and their physical properties using a voltammetric electronic tongue based on various potential waveforms. *Sensors and Actuators. B, Chemical*, *177*, 684–694. doi:10.1016/j.snb.2012.11.056

Wei, Z., Yang, Y., Wang, J., Zhang, W., & Ren, Q. (2018). The measurement principles, working parameters and configurations of voltammetric electronic tongues and its applications for foodstuff analysis. *Journal of Food Engineering*, *217*, 75–92. doi:10.1016/j.jfoodeng.2017.08.005

Wilken, A., Kraft, V., Girod, S., Winter, M., & Nowak, S. (2016). A fluoride-selective electrode (Fse) for the quantification of fluoride in lithium-ion battery (Lib) electrolytes. *Analytical Methods*, *8*(38), 6932–6940. doi:10.1039/C6AY02264B

Wilson, A. D., & Baietto, M. (2011). Advances in electronic-nose technologies developed for biomedical applications. *Sensors (Basel)*, *11*(1), 1105–1176. doi:10.3390110101105 PMID:22346620

Winquist, F., Wide, P., & Lundström, I. (1997). An electronic tongue based on voltammetry. *Analytica Chimica Acta*, *357*(1-2), 21–31. doi:10.1016/S0003-2670(97)00498-4

Zhang, L., & Tian, F. C. (2014). A new kernel discriminant analysis framework for electronic nose recognition. *Analytica Chimica Acta, 816,* 8–17. doi:10.1016/j.aca.2014.01.049 PMID:24580850

Zhang, L., Wang, X., Huang, G. B., Liu, T., & Tan, X. (2018). Taste recognition in e-tongue using local discriminant preservation projection. *IEEE Transactions on Cybernetics, 49*(3), 947–960. doi:10.1109/TCYB.2018.2789889 PMID:29994190

Zhao, W., Bhushan, A., Santamaria, A. D., Simon, M. G., & Davis, C. E. (2008). Machine learning: A crucial tool for sensor design. *Algorithms, 1*(2), 130–152. doi:10.3390/a1020130 PMID:20191110

Zhao, W., & Davis, C. E. (2009). Swarm intelligence based wavelet coefficient feature selection for mass spectral classification: An application to proteomics data. *Analytica Chimica Acta, 651*(1), 15–23. doi:10.1016/j.aca.2009.08.008 PMID:19733729

Zhong, Y. H., Zhang, S., He, R., Zhang, J., Zhou, Z., Cheng, X., ... Zhang, J. (2019). A Convolutional Neural Network Based Auto Features Extraction Method for Tea Classification with Electronic Tongue. *Applied Sciences, 9*(12), 2518. doi:10.3390/app9122518

Zhuang, L., Guo, T., Cao, D., Ling, L., Su, K., Hu, N., & Wang, P. (2015). Detection and classification natural odors with an in vivo bioelectronic nose. *Biosensors & Bioelectronics, 67,* 694–699. doi:10.1016/j.bios.2014.09.102 PMID:25459058

Chapter 5
Cost–Effective Tabu Search Algorithm for Solving the Controller Placement Problem in SDN

Richard Isaac Abuabara
Universidad Santo Tomás, Colombia

Felipe Díaz-Sánchez
Universidad Santo Tomás, Colombia

Juliana Arevalo Herrera
Universidad Santo Tomás, Colombia

Isabel Amigo
IMT Atlantique, France

ABSTRACT

Software-defined networks (SDN) is an emerging paradigm that has been widely explored by the research community. At the same time, it has attracted a lot of attention from the industry. SDN breaks the integration between control and data plane and creates the concept of a network operating system (controller). The controller should be logically centralized, but it must comply with availability, reliability, and security requirements, which implies that it should be physically distributed in the network. In this context, two questions arise: How many controllers should be included? and Where should they be located? These questions comprise the controller placement problem (CPP). The scope of this study is to solve the CPP using the meta-heuristic Tabu search algorithm to optimize the cost of network operation, considering flow setup latency and inter-controller latency constraints. The network model presented considers both controllers and links as IT resources as a service, which allows focusing on operational cost.

DOI: 10.4018/978-1-7998-1839-7.ch005

INTRODUCTION

A software-defined network (SDN), is a network architecture that breaks the integration between the data and control planes of network equipment, as opposed to traditional networks that are vertically integrated. (Open Networking Foundation, 2016). The intention with this paradigm is to provide well-defined programming interfaces between the forwarding devices (switches) and a centralized entity (SDN controller) to enable software development to control the connectivity, network traffic and network function (Open Networking Foundation, 2013). This approach allows providing a unified "Network Operating System" (i.e. the controller) that allows simplifying the planning, deployment, commissioning, and management of the network.

An SDN network architecture consists of three layers (Figure 1):

- **The Application Layer**: groups network functions such as routing, load balancing, and switching.
- **The Control Layer**: where flow setup decisions are given as instructions to forwarding devices.
- **The Data Layer**: in charge of forwarding packets based on the instructions sent by The Control Layer. These instructions are saved in the Flow Table database.

Within the architecture, the control plane is composed of one or more controllers that act as a centralized entity to govern the operation of the entire network. This paradigm promises the possibility to have a high-level view of the network by providing a well-defined, and efficient communication protocol between the data layer and controller layer (Southbound interface). OpenFlow protocol (Mckeown et al., 2008) was created to that end, and it is the DE-facto standard for the Southbound interface.

Some advantages of SDN paradigm over traditional networks are:

- A centralized vision of the network, to deliver services and dynamism for physical and virtual device provisioning.
- Higher innovation since changes are made on software and not in hardware.
- Easier implementation of the complex function in the network using customized algorithms.
- Lower operative costs due to better server utilization, and better virtualization control.

The implementation of an SDN comes with different challenges, and one of them is the amount and physical location of the controllers., This problem is relevant to meet the needs of the network related to latency, performance, scalability, fault tolerance, security, among others, while maintaining a reasonable cost. To optimize the operation, it is necessary to establish an appropriate number of controllers, as well as the location for each controller within the network. In many cases, the parameters optimization will be a trade-off, so it is important to define thresholds as well as optimization objectives.

The calculation to obtain the number and location for the controllers is called the Controller Placement Problem (CPP) and it was first declared by Heller et al. (Heller, Sherwood, & Mckeown, 2012). The authors narrowed the problem to two questions: (1) How many controllers are needed? (2) Where in the topology should they go?.

Heller intends to address the CPP in a network that uses IT resources (controllers and links) as a service.

In this chapter, the authors propose an approach to find a solution to the CPP in SDN with the Tabu Search heuristic optimization method. The algorithm considers OPEX cost as the optimization objective, as well as flow setup and inter-controller latency, and feasibility constraints. First, a description of the

Figure 1. Software-defined network architecture

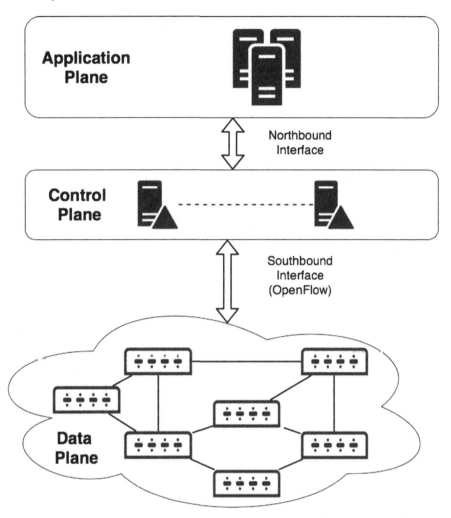

current problem is presented to define a model that optimizes cost, while keeping defined constraints within limits. Then the Tabu Search algorithm is proposed. Finally, it is evaluated using real-life topologies from the Topology Zoo (Knight, Nguyen, Falkner & Roughan, 2011).

BACKGROUND

The CPP is an interesting problem that has attracted attention from the research community (Lu, Zhang, Hu, Yi, & Lan, 2019; Singh & Srivastava, 2018; Wang, Zhao, Huang, Network, & 2017, 2017). A review of the literature shows that there is a growing interest in the CPP since it was introduced in 2012.

Existing approaches to solve the problem include latency minimization (Chen, Chen, Jiang, & Liu, 2018; Heller et al., 2012; Killi & Rao, 2016) reliability maximization (Jiajia Liu et al., 2018; Jiang Liu, Liu, & Xie, 2016) and multi-objective functions (Gong et al., 2016; Ren, Sun, Luo, & Guizani, 2019; Zhang, Wang, & Huang, 2018) as presented in Table 1.

Table 1. Current approaches to solve the CPP

Paper	Optimization Objective	Technique
The Controller Placement Problem (Heller et al., 2012)	Minimize the average latency or worst-case latency	Exhaustive search
Optimal Model for Failure Foresight Capacitated Controller Placement in Software-Defined Networks (Killi et al., 2016)	Minimize the maximum delay from any switch to its Kth reference controller.	Solver based (CPLEX)
Multi-Controller Placement Towards SDN Based on Louvain Heuristic Algorithm (Chen, Chen, et al., 2018)	Minimize the latency between controllers and switches in each control domain under the constraints	Louvain algorithm (heuristic)
Joint Placement of Controllers and Gateways in SDN-Enabled 5G-Satellite Integrated Network (Jiajia Liu et al., 2018)	1. Obtain the minimum average latency. 2. Maximize the average reliability of the integrated network	Exhaustive search and Simulated annealing
Reliability-based controller placement algorithm in software defined networking (Jiang Liu et al., 2016)	Maximize network average reliability (find the shortest path between controller and switches)	Exhaustive search and a greedy algorithm
An efficient and coordinated mapping algorithm in virtualized SDN networks (Gong et al., 2016)	Minimize the cost of embedding the VN for this vSDN request, and minimize the average controller-to-switch delay for this vSDN request.	Co-vsdne (Heuristic)
A Novel Control Plane Optimization Strategy for Important Nodes in SDN-IoT Networks (Ren, Sun, Luo, et al., 2019)	Maximizing the gain caused by the Scontroller, while achieving a good tradeoff between the gain and cost.	Binary particle swarm optimization (Heuristic)
The controller placement problem in software defined networking: a survey (Zhang, et al., 2018)	Minimize packet propagation latency between controllers and switches by partitioning the network into sub-networks	Clustering-Based Network Partition K center
Effective controller placement in controller-based Named Data Networks (Aloulou, et al., 2017)	Minimize the inter-controller and switch-to-controller latency	Clustering k-medoids based Algorithm (Heuristic)
Dynamic controller provisioning in software defined networks (Bari et al., 2013)	Minimize the weighted sum of four different costs: -Statics collection cost -Flow setup cost -Synchronization cost -Switch reassignment cost	Greedy Knapsack Simulated Annealing (Heuristic)
A new framework for reliable control placement in software-defined networks based on multi-criteria clustering approach (Jalili et al., 2019)	Minimize the flow setup time and inter-controller latency	Moth Flame optimization (Heuristic)
Optimal controller placement in Software Defined Networks (SDN) using a non-zero-sum game (Rath et al., 2014)	Minimize the cost by reducing the number of controllers and cost (CAPEX and OPEX) associated with each controller	Non-zero-sum game (Game theoretical approach)
Five Nines of Southbound Reliability in Software-defined Networks (Ros et al., 2014)	Minimize the cost of deploying a controller in a facility and the cost of serving a switch with a controller	Fuzzy logic algorithm.
Expansion Model for the Controller Placement Problem in Software Defined Networks (Sallahi et al., 2017)	Minimize the cost of expansion and upgrade	Solver based (CPLEX)
Dynamic and static controller placement in Software-Defined Satellite Networking (Wu, et al., 2018)	Minimize the cost of the network while considering different expenses: 1) Performance expense (E p), Failure tolerance expense (E f), Load balance expense (E l), Economic expense (E e)	Particle Swarm Optimization
The approach in this study	Minimize the cost of implementation (controller licenses, and link use)	Tabu search (Heuristic)

The solution of the algorithm will depend, not only on the objective but also on the specific metric used in the model. Widespread metrics are average and worst-case switch to controller latency, average and worst-case inter-controller latency, percentage of control path loss in case of failure, and controller imbalance (difference in the number of switches served by different controllers).

Flow setup time is a latency metric also used for some works (Aloulou, Ayari, Zhani, Saidane, & Pujolle, 2017; Bari et al., 2013; Jalili, Keshtgari, & Akbari, 2019). It consists of a round-trip delay for requesting instructions from a switch to a controller to establish a new forwarding path, plus the controller processing delay. Within Openflow, this process uses a "packet-in" message, which sends the initial packet of a new flow from the receiving switch to the controller, so the later can analyze it and establish the new entry on the flow table. The flow setup latency will be the time difference between the moment that the first packet of a flow arrives at the switch until it is forwarded (Khalili, Despotovic, & Hecker, 2018). A graphical representation of the different latencies is presented in Figure 2.

Regarding costs, authors found five studies (Bari et al., 2013; Rath, Revoori, Nadaf, & Simha, 2014; Ros & Ruiz, 2014; Sallahi & St-Hilaire, 2017; Wu, Chen, Yang, Fan, & Zhao, 2018). The studies use different techniques to solve the problem, such as an exact solver (CPLEX), game-theoretical approach or heuristic techniques like simulated annealing or particle swarm optimization.

The study in Bari et al. (2013), is one of the first that used the cost objective approach to solve the CPP. The intention is to adapt the number and location of controllers with changing network conditions, so the problem is called Dynamic CPP. The function seeks to minimize costs for i) switch state collection, (ii) inter-controller synchronization, and (iii) switch to controller reassignment, while ensuring minimal flow setup time and communication overhead. Greedy Knapsack and Simulated Annealing methods are used to solve the problem.

Figure 2. Latencies considered in an SDN

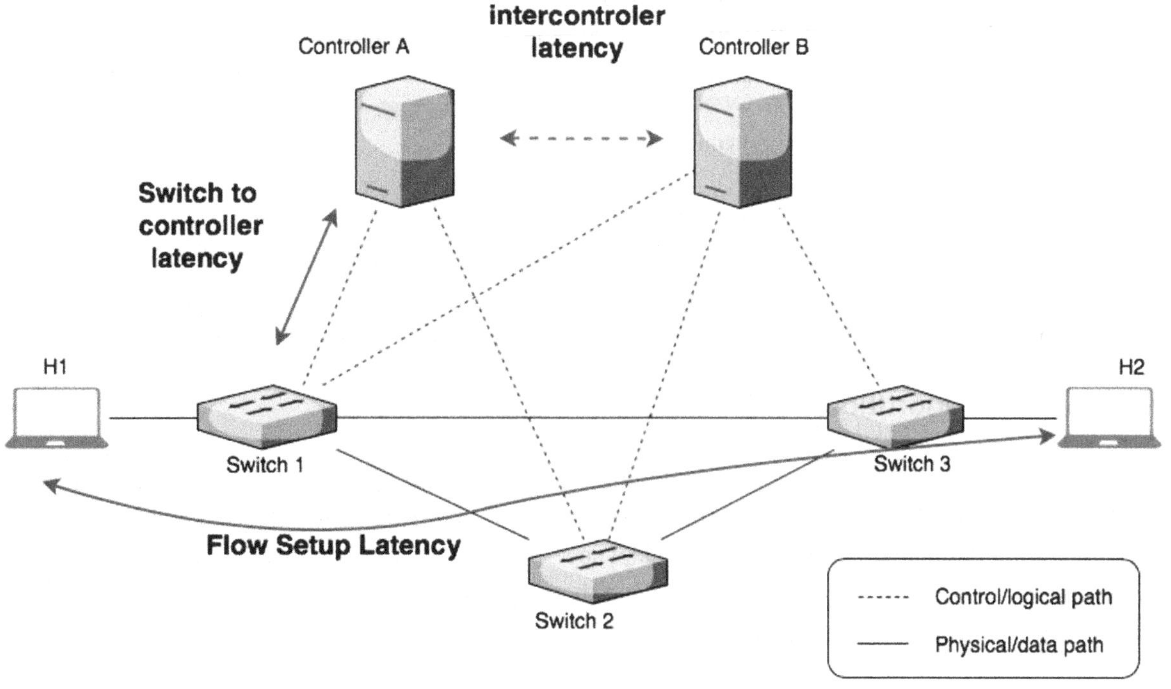

On the other hand, authors in Rath et al. (2014), propose to minimize the cost to operate the controllers using an optimization method based on game theory. The method is presented as low-complexity and with the possibility to run in real-time The objective consider constraints such as (i) load on the controllers is uniform, (ii) maximum utilization is achieved, and (iii) latency is reduced.

Reliability is a constraint proposed in Ros & Ruiz (2014), considered as the possibility that each node has an operational path to any controller. The optimization objective is to reduce the cost of the controller placement using a customized heuristic algorithm.

The planning approach is used in Sallahi & St-Hilaire (2017) to create a new network or expand an existing one. The study presents a modified CPP, as Expansion Model for the Controller Placement Problem (EM-CPP), that considers minimization of cost for three variables: i) installing or removing controllers, ii) adding or removing switch to controller links, and iii) adding or removing inter-controller links. The solving method is an exact one, using CPLEX solver.

Finally, Wu et al. (2018) present two models for controller placement in a satellite network: static (for a single solution in unchanged network conditions) and dynamic (for multiple solutions depending on the changes in network conditions). Authors consider the constraints presented in previous studies to minimize the expense of four elements i) performance, ii) fault tolerance, iii) load balance iv) economic. A heuristic algorithm is used based on the Particle Swarm Optimization (PSO) technique to solve both dynamic and static CPP.

Several other works use heuristic methods different from Tabu search, to solve the CPP (Jalili, Keshtgari, Akbari, & Javidan, 2018; Tanha, Sajjadi, Ruby, & Pan, 2018). Also, they propose objective functions with goals different from cost. Regarding the use of Tabu Search, in (Ahmadi & Khorramizadeh, 2018) an adaptive heuristic algorithm is presented, with the use of a Tabu list; however, that study did not implement the complete algorithm. On the other hand, in (Ramya & Manoharan, 2018) authors present a Tabu Search algorithm to solve the CPP in a multi-objective function, but fails to present test and evaluation of the algorithm, since they only presented a theoretical discussion.

As presented, several heuristic algorithms have been used to solve the CPP, however, none of them experiment with the Tabu search algorithm. Additionally, only five studies use cost as an optimization objective, but none of them considers a model in which controllers and links are provided as IT resources as a service.

The contributions of the presented study are:

- A model in which both the controllers and links are provided as IT resources as a service.
- Tabu search algorithm is used to optimize OPEX cost in the CPP, represented in the controller license (in a period) and the use of network links.
- Evaluation of the proposed algorithm in 19 real-life topologies obtained from the Topology Zoo.

Tabu Search Algorithm

Meta-heuristic methods for optimization problems are preferred due to its efficiency in calculation time for NP-Hard problems, such as the CPP (Heller et al., 2012). These algorithms solve a widespread type of optimization problems without having to adapt specifically to each of them. Their main features are: nature-inspired, using statistical components, and the use of several parameters that must be fixed for the problem (Boussaïd, Lepagnot, & Siarry, 2013).

Tabu Search (Glover, 1989) is a mathematical optimization method for local search. It enhances the performance of a search using two memory structures: 1) a short-term, and 2) long-term memory. The algorithm uses a memory list for the latest solutions or some of its attributes to find the best of them. This list acts as short-term memory.

The process starts with a randomly generated solution and gets a score. The algorithm continues by exploring neighboring solutions that will be stored in the short-term memory to compare the scores. Once it finds the best solution within the local search, it tags it as "TABU", stores it in the long-term memory and then starts a local search in a different solution space. The Tabu solutions are forbidden, so the algorithm searches for alternatives. The use of memory structures enables the algorithm to escape the local minimum.

The length of the tabu list controls the memory of the search process. If the length is small, the search will be held in small areas. On the other hand, a big length forces the process to search larger regions, since it will ban a larger number of solutions.

TABU-SEARCH MODEL FOR CONTROLLER PLACEMENT PROBLEM

This section presents the definitions for the model of the SDN in which the CPP will be solved. In this proposal, an SDN is defined as a WAN with the following network elements: switches, controllers, and links. The data plane is made up of switches and links modeled as a graph where V is the set of switches and E the set of physical links between the switches. Every switch is managed by the controller (control plane). A controller is a software implemented in one or more switch with enough computational power to run it. The switches are capable of hosting a controller, this is known as in-band control (Figure 3). Each controller can have one of two states: active (hosting an active controller), inactive (no hosting a controller). In the model, it has been considered that the number of active controllers in a network will depend on a predefined percentage. This configuration causes the control information to be sent over the same links used for data transport, so it is considered in-band control.

Figure 3. Location of the controllers

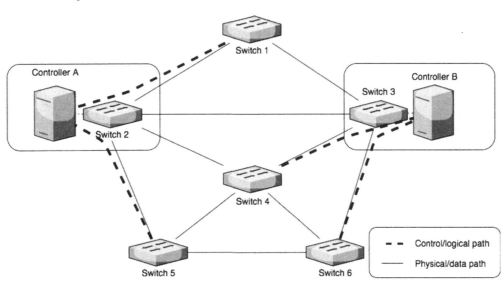

115

When the switches receive a packet that requires a new entry in the flow table, they send a packet-in message to request instructions to the controller. In the context of this study, that request will be called Flow Establishment Request (FER) and they are processed under a policy First-in, First-out, (Figure 4). That means that there is no assigned priority for them, other than its arrival order. The average time to serve the FER is a function of the amount of queued FERs and the processing delay for FER processing. On the one hand, the number of queued FER corresponds to the sum of total FER from the switches managed by the same controller.

The links are paid as a service, so there is an associated periodical cost for their use. Additionally, there are several server providers p that offer the links with a defined latency and $l_{u,v}^{p}$ cost $\lambda_{u,v}^{p} u, v \in V$. Moreover, there can be one or more links between each pair of switches (Figure 5).

This scenario considers the cloud computing paradigm, that allows consuming IT resources as a service (A. F. Diaz-Sanchez, 2014). If the provider offers a public service, the SDN may set up its links at their will as presented in (F. Diaz-Sanchez et al., 2014). The same consideration is made for the controllers as software resources that run under a licensing model per time unit with a price B.

Figure 4. Processing of packet-in messages

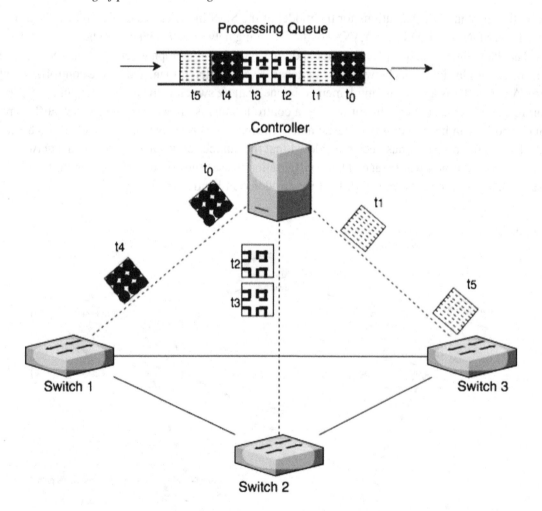

To ensure the quality of the solution, there are two feasibility constraints:

- Any switch can host a controller.
- One controller license in a switch has a fixed cost in time.
- A switch can only be governed by one and only one controller.
- Only one or cero controllers can be instantiated in a switch.

The goal of this approach is to minimize the related costs with network control, as follows:

- Cost of use of controller licenses C_c as the sum of total used licenses in a defined period.
- Cost of the switch to controller communication C_s as the sum of prices of active links between switches and controllers.
- Cost of inter-controller communication C_w as the sum of prices of active links between controllers

The formal definition of the model is presented below:

SDN network is modeled as an undirected and connected graph $G=(V,E)$. With V being the set of switches (*i.e.* nodes) and E the set of links (*i.e.* edges). Each switch in the graph may become a controller ($C \subseteq V$), therefore the control-plane network is not separated from the data-plane network (*i.e.* in-band control). A switch can become a controller by activating a software license. It is assumed that controllers are overloaded, in other words, in scenarios of a high number of flow setup requests, the controller performance is degraded. Particularly, this fact impacts the average processing time for each flow setup request that doubles if the number of request doubles. For this reason, the buffers in switches used to queue flow-setup requests are considered infinite, which implies that a controller deployed into a switch can handle any number of flow-setup requests.

Figure 5. More than one provider per link

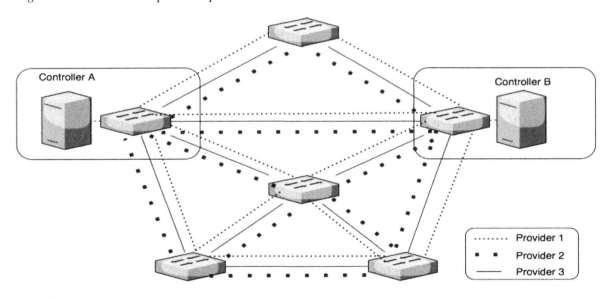

To formulate the mathematical model, we assume the following information is known:

- The maximum delay allowed in the network for the switch to controller communications (flow-setup latency) and inter-controller communications.
- The cost of allocating a license to run a controller on a switch.
- The hardware capacities of switches for running controllers' licenses and handling flow setup requests in terms of the processing delay for flow setup.
- The forwarding latency and the cost of using a link belonging to a given service provider between each pair of nodes.

Based on this information, we formulate the problem as follows:

1. Notation
   - Input sets and parameters:
     - V, the set of nodes.
     - E, the set of links with $E \subseteq V x V$.
     - $G(V,E)$, the network with node set V and link set E.
     - C, the set of possible controller locations with $C \subseteq V$.
     - $L_{u,v}^p$, the forwarding latency (in milliseconds) from node $u \in V$ to node $v \in V$ for service provider p.
     - B, the cost (in $) for running a controller license.
     - α_c, the number of average requests present in a controller.
     - σ_c, the average time (in milliseconds) for handling a given number of flow setup request in a controller.
     - ρ_c, the processing delay (in milliseconds) for flow setup with $\rho_c = \alpha_c \sigma_c, \forall c \in C$.
     - $\lambda_{u,v}^p$, the cost (in $) of using a forwarding path from service provider p between node $u \in V$ to node $v \in V$.
   - Constraints:
     - ϕ, the maximum flow setup latency (in milliseconds).
     - η, the maximum inter-controller latency (in milliseconds).
   - Variables:
   - x_c, a binary variable such that $x_c = 1$ if and only if a controller is deployed into a possible controller location with $c \in C$.
   - $k_{u,c}$, a binary variable such that $k_{u,c} = 1$ if and only if a node $u \in V$ is assigned to controller $c \in C$.
   - $z_{u,v}$, a binary variable such that $z_{u,v} = 1$ if and only if node $u \in C$ and node $v \in C$.
   - δ_u a variable representing the latency between a node $u \in V$ and its controller $c \in C$.
2. Cost Function

The objective is to minimize the costs related with the control of the network given by:

- The cost for running controller licenses.

$$C_c(x) = \sum_{c \in C} B\, x_c \tag{1}$$

- The cost for sending control information between a switch and its assigned controller.

$$C_s(k) = \sum_{u \in V} \sum_{c \in C} \sum_{p \in P} \lambda_{u,v}^p k_{u,c} \tag{2}$$

- The cost for sending control information between controllers.

$$C_w(z) = \sum_{u \in C} \sum_{v \in C} \sum_{p \in P} \lambda_{u,v}^p z_{u,v} \tag{3}$$

3. The Model

The Controller Placement Problem (CPP) has been modeled as follows.
CPP:

$$Minimize\left(C_c(x) + C_s(k) + C_w(z)\right) \tag{4}$$

Subject to the following constraints:

- Assure that only one controller can be placed in a node.

$$\sum_{c \in C} x_c \leq 1, \forall c \in C \tag{5}$$

- Assure that each switch is controlled by only one controller.

$$\sum_{c \in C} k_{u,c} = 1, \forall u \in V \tag{6}$$

- Assures that a switch is assigned to a location in which a controller is placed.

$$\sum_{u \in V} k_{u,c} \leq |V| x_c, \forall c \in C \tag{7}$$

- Assures round-trip flow setup latencies below or equal to the maximum flow setup latency.

$$2L_{u,v}^p + \rho_c \leq \phi, \forall u \in V, \forall p \in P \tag{8}$$

- Assures round-trip inter-controller latency below or equal to the maximum inter-controller latency.

$$2\sum_{u \in C} L_{u,v}^p z_{u,v} \leq \eta, \forall v \in V, \forall p \in P \tag{9}$$

- Assures a control latency $\delta_u = 0$ when the location of the switch and the controller coincide, otherwise $\delta_u = L_{u,v}^p$.

$$\delta_u = \sum_{u \in V} L_{u,v}^p k_{u,c}, \forall c \in C, \forall p \in P \tag{10}$$

Tabu Algorithm to Solve the CPP

To solve the objective function, a heuristic Tabu Search Algorithm is defined (Figure 6). In it, an SDN modeled with the conditions mentioned above is presented as the solution. The set of all possible solutions is defined as Y. The quality of the solution is measured with its cost (the lower the cost, the better the solution).

To apply the algorithm to the CPP, the authors considered the following elements:

- Tabu search is a local search strategy with a flexible memory structure.
- Tabu search has two main features. First, its adaptive memory, which allows the possibility to implement search procedures in different spaces to improve efficiency. Second, its responsive memory that explores new regions or promising solutions.
- Tabu search moves to the best neighbor solution that is not in the tabu list.
- Tabu list is the core component of the algorithm. It allows storing candidate solutions so it will not be considered again if the search returns to that point. In this way, the local optimal is less possible. The list is updated using short-term memory.

Considering the previous elements, the Tabu Search Algorithm (TSA) was defined as follows:

Algorithm 1. TSA

```
sBest ? s0
bestCandidate ? s0
tabuList ? []
tabuList.push(s0)
while (not stoppingCondition()){
  sNeighborhood ? getNeighbors(bestCandidate);
  for(sCandidate in sNeighborhood){
    if((not tabuList.contains(sCandidate)) and (fitness(sCandidate) >
```

Figure 6. Tabu Search

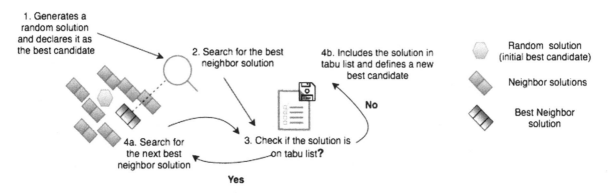

```
fitness(bestCandidate))){
    bestCandidate ? sCandidate;
  }
}
If (fitness(bestCandidate) > fitness(sBest)){
  sBest ? bestCandidate;
}
tabuList.push(bestCandidate);
if (tabuList.size > maxTabuSize){
  tabuList.removeFirst();
}
}
Return sBest
```

Constraints for Solution Generation

For model construction, two types of constraints were considered: problem constraints and quality constraints.

4. The problem constraints ensure that the solution complies with the defined network model:
   a. Ensure that each switch is managed by a single controller.
   b. Ensure that only one controller is deployed in a switch.
5. The quality constraints ensure a level of performance in the solution. The constraints defined are related to latency.
   a. The maximum flow setup latency is given by:

$$2 * l_{u,v}^p + \rho_c \leq \phi \tag{11}$$

That is, the bidirectional switch-to-controller latency plus the average time to serve the FER must be less than a predefined threshold.

b. The maximum latency for inter-controller communication is given by:

$$\sum_{u \in C} l_{u,v}^p * z_{u,v} \leq \eta \tag{12}$$

where C corresponds to the set of controllers and z is a binary variable that equals 1 if both $u,v \in C$, 0 otherwise.

That is, the inter-controller latency must be less than a predefined threshold.

c. Ensure that the switch to controller latency is zero when a controller is deployed in the same switch.

$$l_u = 0 \Leftrightarrow u \in C; \tag{13}$$

Neighbor Solution Generation

The Tabu search is made within the neighborhood with several iterations, going from one solution to another. Each solution has an associated neighborhood such that

Each solution $y' \in Y$ can be found from the initial solution applying an operator called "movement", that must generate a fully compliant solution. For this study, two types of "movements" were defined to generate new random solutions to help minimize the objective function (cost). Both types are defined in the subsections below.

Movement One: Controller Activation/Deactivation: In solution "" the active and inactive controllers are identified. One switch of each group is selected to change the state, i.e. the active controller goes to the inactive state while the inactive controller goes to an active state. Finally, the list of switches managed by the initially active controller is now assigned to the newly active controller. In any case, the constraints (3), (4), and (5) must be enforced.

Algorithm 2. Movement One

```
Movement One - Controller activation/deactivation
for (controller in obj_Controller){
  if(controller.get.State()==True){
  activecontrollers.append(controller);
  }
  else{
    unActiveControllers.append(controller);
  }
}
while(mynet.mynet.change_Controller == False){
  eject_movement;
}
self.unActiveControllers[x].controlledSW ? self.activecontroller[y].con-
```

```
trolledSW;
self.controllers[y].setState(False);
self.controllers[x].setState(True);
```

Movement Two: Link Variation: In the network, each pair of switches have multiple links between them, each offered by a different provider with different prices. However, in solution "y" there is only one active link. Movement two deactivates random links and activates an alternative.

Algorithm 3. Movement Two

```
lim_inf ← 0
while (mynet.mynet.change_links == False){
  eject_movement;
}
while (lim_inf < len(self.links)=3){
  hlp ← lim_inf;
  if (uniform(0,10 < 0.5){
    for (i in list(range(3))){
        if (i > 0){
            hlp ← hlp+1;
        }
        self.links[hlp].setState(False);
    }
    index = randint(lim_inf,lim_inf+2)
    sef.links[index].setState(True);
  }
  lim_inf ← lim_inf+3;
}
```

SIMULATION RESULTS AND ANALYSIS

To implement the algorithm and perform simulations, a Python application was developed using an object-oriented approach. The switches, controllers, and links were modeled according to the descriptions in section TABU-SEARCH MODEL FOR CONTROLLER PLACEMENT PROBLEM. A total of 19 real-life topologies from the Topology Zoo were used for simulations. A description of the topologies is presented in Table 2

To import the topologies in the algorithm, the GML format from the Topology Zoo was processed in the following way. Each pair of connected nodes were identified and the distance between them was obtained using Haversine formula (Robusto, C. C. 1957). The transmission latency of that link was then calculated considering optical fiber as the transmission medium. Later, the Dijkstra algorithm was used to calculate the shortest path between each pair of nodes of the topology. Finally, the links between each pair of connected nodes were generated, by the number of link providers.

The constraints φ and η were implemented as variables per iteration, to verify that the algorithm is operative and provide quality solutions.

The simulations were performed in parallel in two infrastructures: virtual machines t2.micro in AWS cloud and virtual machine in Oracle VM VirtualBox Manager on a PC with Ubuntu 16.04.3 LTS O.S, with 4GB RAM and 20 Gb of storage capacity.

Results

The results are considered from the 19 evaluated topologies. Below, there are the results for three topologies, presented as results.

Iteration_vs_Cost: Figure 7 shows the cost improvement for topologies ATMNet, YorkDataServices, and IntegraTelecom. It shows that the convergence of the cost presents around iteration 44.

Movement Quality: To evaluate the quality of movements, the objective function was evaluated with different percentages of controllers per total switches. Figure 8 shows the solutions as bubbles, in which the cost is represented as the size. The color represents the type of movement. The lowest cost was found in two different scenarios: for $\eta = 100$, the best movement was link variation. On the other hand, for $\eta = 150$, the best movement was Controller activation/deactivation.

Table 2. Description of the topologies for testing.

Topology	Nodes	Location	Type
AGIS	25	USA	Backbone
ATMnet	21	USA	Backbone
ATTNorthAmerica	25	USA	Backbone, Customer, Transit
Bbnplanet	27	USA	Backbone
Bellcanada	48	Canada, USA	Backbone, Customer, Transit
Bics	33	Europe	Backbone, Customer
CRL	33	USA	Backbone, Customer, Transit
Darkstrand	28	USA	Backbone, Customer
Digex	31	USA	Backbone, Customer
HurricaneElectric	24	Global	Backbone, Customer, Transit
IntegraTelecom	27	USA	Backbone, Customer, Transit
NetworkUsa	35	Louisiana, Texas, USA	Backbone
Packetexchange	21	Global	Backbone, Customer, Transit
PalmettoNet	45	North Carolina, South Carolina, USA	Backbone
PsiNet	24	USA	Backbone, Customer, Transit
Xeex	24	USA	Backbone, Customer, Transit
Xspedius	34	USA	Backbone, Customer
YorkDataServices	23	UK	Undefined
Claranet	15	Europe	Backbone, Customer, Transit

Figure 7. Cost vs iterations

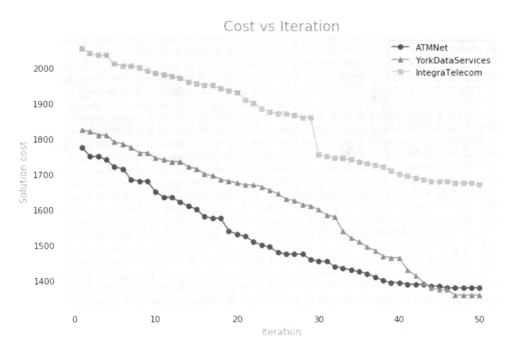

A total of 78.94% of the 19 topologies presented the best solution (lowest cost) using the link variation movement considering 10% of active controllers. However, the efficiency of this movement decreased to 63.15% if there was 30% of active controllers.

Solution Variation: To evaluate the quality of the generated solutions, the Coefficient of Variance (CV) was used. It is a statistical measurement of the data dispersion in a series, around the mean. It is defined as:

$$CV = \frac{\sigma}{\overline{U}} \tag{14}$$

where σ is the standard deviation, and $\overline{U}$ is the arithmetic mean. It is a useful method to compare the variation from one data series to another, even if the measures are highly different. For this study, the CV indicates if the algorithm produces solutions with a defined variation. The lower the CV, the better the algorithm, since the cost of the solutions are closer i.e. the algorithm can guarantee a level of quality of the generated solutions.

As presented in Figure 9, the CV is less than 4% for each obtained group of solutions. According to that, the algorithm provides homogeneous solutions. The highest variations were obtained from the Controller activation/deactivation movement.

Latency Constraints: The topologies that presented the most geographically disperse nodes (IntegraTelecom, HurricaineElectric, Packetexchange, and Darkstrand) required higher thresholds for and constraints.

Figure 8. Cost of each solution in Flow Setup Latency vs Percentage of controllers in ATMNet topology.
a) Intercontroller Latency = 100 ms. b) Intercontroller Latency= 150 ms

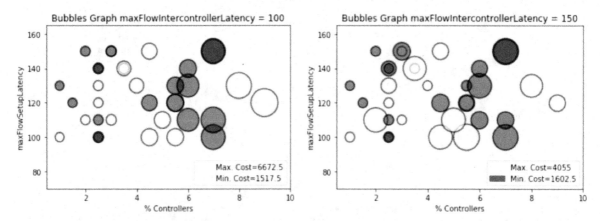

Computing Time: Using both of the platforms (virtual machine and AWS), all the topologies were processed in 78 hours for strict constraints (latencies of less than 150 ms). However, with less demanding constraints (200 to 250 ms), the computing time decreases by about 50%.

Cost Comparison: The last result is the cost comparison of the topologies. Considering Table 2, there is a direct relationship between the number of nodes and the cost of the deployment of the solution. The largest topology (Bellcanada with 48 nodes) has three times the cost of the smallest one (Claranet with 15 nodes), as presented in Figure 10.

CONCLUSION AND FUTURE WORKS

The CPP was studied using a network model in which IT resources (controllers and links) are provided as a service. For that purpose, a Tabu Search heuristic optimization method was proposed to minimize the operational cost of the network. The algorithm considered Flow Setup Latency, and Inter-Controller Latency constraints, as well as a predefined percentage of nodes acting as controllers. Simulations of the model and Tabu Search method were performed using Python with 19 real-life topologies from the Topology Zoo. The results of the CV showed that the algorithm provides homogeneous solutions and a reduction of the cost compared to the initial random solution. For the presented model the size of the network is the main contributor of the OPEX since it requires more controllers and links.

Regarding future work, there is room for improvement in the efficiency of the computing time of the algorithm, in such a way that it can be used in a dynamic network environment. Additionally, is important to study the effect of the switch assignment on the controllers. Finally, constraints such as reliability must be included in the model to use it in an operative environment.

Figure 9. CV in ATMNet topology. a) CV for movement one with 10% controllers per node. b) CV for movement one with 30% controllers per node. c) CV for movement two with 10% controllers per node. d) CV for movement two with 30% controllers per node

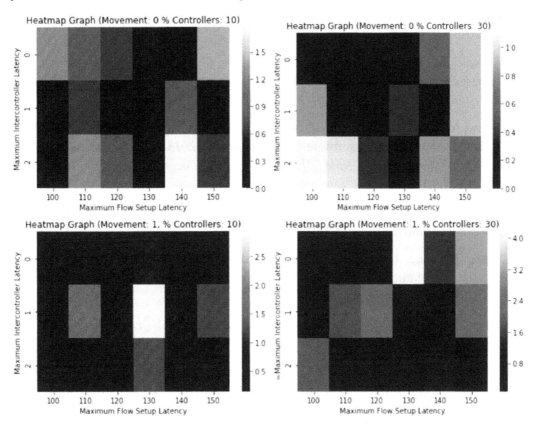

Figure 10. Cost results for 19 topologies

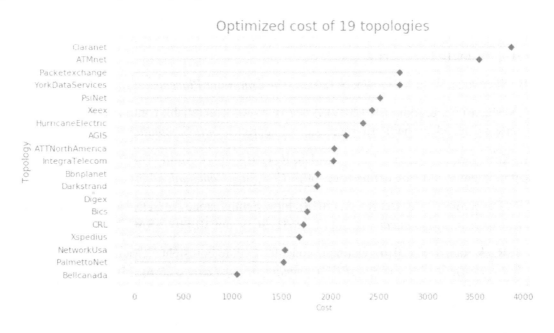

REFERENCES

Ahmadi, V., & Khorramizadeh, M. (2018). An adaptive heuristic for multi-objective controller placement in software-defined networks. *Computers & Electrical Engineering, 66*, 204–228. doi:10.1016/j.compeleceng.2017.12.043

Aloulou, N., Ayari, M., Zhani, M. F., Saidane, L., & Pujolle, G. (2017). Effective controller placement in controller-based Named Data Networks. *2017 International Conference on Computing, Networking and Communications, ICNC 2017*, 249–254. 10.1109/ICCNC.2017.7876134

Bari, M. F., Roy, A. R., Chowdhury, S. R., Zhang, Q., Zhani, M. F., Ahmed, R., & Boutaba, R. (2013). Dynamic controller provisioning in software defined networks. *2013 9th International Conference on Network and Service Management, CNSM 2013 and Its Three Collocated Workshops - ICQT 2013, SVM 2013 and SETM 2013*, 18–25.

Boussaïd, I., Lepagnot, J., & Siarry, P. (2013). A survey on optimization metaheuristics. *Information Sciences, 237*, 82–117. doi:10.1016/j.ins.2013.02.041

Chen, W., Chen, C., Jiang, X., & Liu, L. (2018). Multi-controller placement towards SDN based on louvain heuristic algorithm. *IEEE Access: Practical Innovations, Open Solutions, 6*, 49486–49497. doi:10.1109/ACCESS.2018.2867931

Diaz-Sanchez, A. F. (2014). *Cloud brokering: new value-added services and pricing models*. Retrieved from https://pastel.archives-ouvertes.fr/tel-01276552/

Diaz-Sanchez, F., Al Zahr, S., Gagnaire, M., Laisne, J. P., & Marshall, I. J. (2014). CompatibleOne: Bringing Cloud as a Commodity. *2014 IEEE International Conference on Cloud Engineering*, 397–402. 10.1109/IC2E.2014.62

Glover, F. (1989). Tabu Search—Part I. *ORSA Journal on Computing, 1*(3), 190–206. doi:10.1287/ijoc.1.3.190

Gong, S., Chen, J., Kang, Q., Meng, Q., Zhu, Q., & Zhao, S. (2016). An efficient and coordinated mapping algorithm in virtualized SDN networks. *Frontiers of Information Technology & Electronic Engineering, 17*(7), 701–716. doi:10.1631/FITEE.1500387

Heller, B., Sherwood, R., & Mckeown, N. (2012). *The Controller Placement Problem*. Academic Press.

Jalili, A., Keshtgari, M., & Akbari, R. (2019). A new framework for reliable control placement in software-defined networks based on multi-criteria clustering approach. *Soft Computing*, 1–20.

Jalili, A., Keshtgari, M., Akbari, R., & Javidan, R. (2018, July). Multi criteria analysis of controller placement problem in software defined networks. *Computer Communications*.

Khalili, R., Despotovic, Z., & Hecker, A. (2018). Flow Setup Latency in SDN Networks. *IEEE Journal on Selected Areas in Communications, 36*(12), 2631–2639. doi:10.1109/JSAC.2018.2871291

Killi, B. P. R., & Rao, S. V. (2016). Optimal Model for Failure Foresight Capacitated Controller Placement in Software-Defined Networks. *IEEE Communications Letters, 20*(6), 1108–1111. doi:10.1109/LCOMM.2016.2550026

Knight, S., Nguyen, H. X., Falkner, N. R. B., & Roughan, M. (2011). *The Internet Topology Zoo*. Retrieved January 18, 2019, from http://www.topology-zoo.org/

Liu, J., Liu, J., & Xie, R. (2016). Reliability-based controller placement algorithm in software defined networking. *Computer Science and Information Systems*, *13*(2), 547–560. doi:10.2298/CSIS160225014L

Liu, J., Shi, Y., Zhao, L., Cao, Y., Sun, W., & Kato, N. (2018). Joint Placement of Controllers and Gateways in SDN-Enabled 5G-Satellite Integrated Network. *IEEE Journal on Selected Areas in Communications*, *36*(2), 221–232. doi:10.1109/JSAC.2018.2804019

Lu, J., Zhang, Z., Hu, T., Yi, P., & Lan, J. (2019). A Survey of Controller Placement Problem in Software-Defined Networking. *IEEE Access: Practical Innovations, Open Solutions*, *7*, 24290–24307. doi:10.1109/ACCESS.2019.2893283

Mckeown, N., Anderson, T., Peterson, L., Rexford, J., Shenker, S., & Louis, S. (2008). Sigcomm08_Openflow. *Pdf.*, *38*(2), 69–74.

Open Networking Foundation. (2013). SDN Architecture Overview. *Onf*, (1), 1–5.

Open Networking Foundation. (2016). *SDN Architecture*. Author.

Ramya, G., & Manoharan, R. (2018). Enhanced Multi-Controller Placements in SDN. *International Conference on Wireless Communications*.

Rath, H. K., Revoori, V., Nadaf, S. M., & Simha, A. (2014). Optimal controller placement in Software Defined Networks (SDN) using a non-zero-sum game. *Proceeding of IEEE International Symposium on a World of Wireless, Mobile and Multimedia Networks 2014, WoWMoM 2014*.

Ren, W., Sun, Y., Luo, H., & Guizani, M. (2019). A novel control plane optimization strategy for important nodes in SDN-IoT networks. *IEEE Internet of Things Journal*, *6*(2), 3558–3571. doi:10.1109/JIOT.2018.2888504

Robusto, C. C. (1957). The cosine-haversine formula. *The American Mathematical Monthly*, *64*(1), 38–40. doi:10.2307/2309088

Ros, F. J., & Ruiz, P. M. (2014). Five nines of southbound reliability in software-defined networks. *Proceedings of the Third Workshop on Hot Topics in Software Defined Networking - HotSDN '14*, 31–36. 10.1145/2620728.2620752

Sallahi, A., & St-Hilaire, M. (2017). Expansion model for the controller placement problem in software defined networks. *IEEE Communications Letters*, *21*(2), 274–277. doi:10.1109/LCOMM.2016.2621746

Singh, A. K., & Srivastava, S. (2018). A survey and classification of controller placement problem in SDN. *International Journal of Network Management*, *28*(3), e2018. doi:10.1002/nem.2018

Tanha, M., Sajjadi, D., Ruby, R., & Pan, J. (2018). Capacity-Aware and Delay-Guaranteed Resilient Controller Placement for Software-Defined WANs. *IEEE eTransactions on Network and Service Management*, *15*(3), 991–1005. doi:10.1109/TNSM.2018.2829661

Wang, G., Zhao, Y., & Huang, J. (2017). *The controller placement problem in software defined networking: a survey*. Homepages.Dcc.Ufmg.Br.

Wu, S., Chen, X., Yang, L., Fan, C., & Zhao, Y. (2018). Dynamic and static controller placement in Software-Defined Satellite Networking. *Acta Astronautica*, *152*, 49–58. doi:10.1016/j.actaastro.2018.07.017

Zhang, B., Wang, X., & Huang, M. (2018). Multi-objective optimization controller placement problem in internet-oriented software defined network. *Computer Communications*, *123*, 24–35. doi:10.1016/j.comcom.2018.04.008

Chapter 6
A Study on Efficient Clustering Techniques Involved in Dealing With Diverse Attribute Data

Pragathi Penikalapati
Vellore Institute of Technology, India

A. Nagaraja Rao
Vellore Institute of Technology, India

ABSTRACT

The compatibility issues among the characteristics of data involving numerical as well as categorical attributes (mixed) laid many challenges in pattern recognition field. Clustering is often used to group identical elements and to find structures out of data. However, clustering categorical data poses some notable challenges. Particularly clustering diversified (mixed) data constitute bigger challenges because of its range of attributes. Computations on such data are merely too complex to match the scales of numerical and categorical values due to its ranges and conversions. This chapter is intended to cover literature clustering algorithms in the context of mixed attribute unlabelled data. Further, this chapter will cover the types and state of the art methodologies that help in separating data by satisfying inter and intracluster similarity. This chapter further identifies challenges and Future research directions of state-of-the-art clustering algorithms with notable research gaps.

INTRODUCTION

In this web 3.0 era, information is accelerating from the Big data and with the applicability of IoT devices. Each day, information is generated from the use of Social sites to wearable devices and IOT's. For instance, millions of google searchers, thousands of YouTube video uploads, data from service-based applications (like medical, transport, logistics, education, shopping sites, etc.), tweets, comments are being generated. With this information, researchers are trying to extract the patterns that help in analyzing and understand-

DOI: 10.4018/978-1-7998-1839-7.ch006

ing data. However, this raw data cannot be analyzed using any algorithm. Often in real-time situations, data is not available with any appropriate classifications and pre-defined labels. Consequently, there is a need to develop certain models of machine learning capable of precise classification of the new data, based on certain similarities in features. This process can be accomplished by 'Clustering', an algorithm of unsupervised learning. In machine learning () as well as data mining, this analysis of clustering is a very important technique. The objective of clustering analysis is to segregate an ensemble of undefined objects into diverse clusters in such a way that the data objects of a specific cluster are either different or similar to the data objects of another cluster. The applications of cluster analysis are numerous including the categorization of customers, setting market targets, analysis of social networks, bioinformatics, and analysis of scientific data (Han & Kamber, 2000). Segmentation of a specific dataset into a homogeneous collection is performed by an optimization model of partitioning depicted by a cost function, in such a way that there is a similarity among the observations inside a cluster, while dissimilarity among the observations of other clusters. Input: An unlabeled training set with attribute values $D=\{observations,$ $i=1,\dots,N\}$ with N objects described by d attributes where *observations*= {*Attribute_1, Attribute_2,...,* *Attribute_d*} $\hat{I}R^d$ K depicts the total number of initial clusters. Output: A set of K clusters C_1, C_2,..., C_k. The variations in size, shape, and density in the resultant clusters largely depend on the number of clusters K and the processes of clustering adopted. The prime characteristic feature of a good clustering is in its intense compactness, which means that the intra-cluster observations should be as proximate as can be possible, and isolation which means that the inter-cluster variations in observations should be as scattered as can be possible.

As illustrated in Figure 1, it can be noted that clusters C1 and C3 are different in shape but very compact, while C2 and C3 are comparatively not so compact. Certain observations are found to be secluded from the cluster's core. Such secluded observations could be the representations of the outlier and noise in the resultant clusters and may not be causing a negative influence on the comprehension towards the close of the process (Ben Salem, Naouali, & Chtourou, 2018).

d1 is the representation of the inter-distance existing between the observations associated with diverse clusters requiring to be maximized. This results in the procurement of isolated clusters. Similarly, d2 depicts the intra-distance existing between the observations pertaining to the same cluster requiring to be minimized. This results in the procurement of compact clusters.

Accuracy in finding the clusters and the high scalability are the major constraints to be considered in the design of new approaches of clustering, more specifically in the context of larger datasets. In general, there are certain major issues requiring to be addressed in the process of clustering such as discovering suitable measure (distance) of similarity for the optimization of objective function, implementation of competent iterative steps for the discovery of the most precise clusters and the derivation of an appropriate description for the contextualization of the elements in every individual clusters and according permission for the extraction of patterns (Ben Salem et al., 2018). An objective function's optimization can be represented as

$$Obj\_fun = \sum_{i=1}^{n} \left| Observation_j . \text{Centroid}_j \right|$$

where Centroid$_j$ is the center of the jth cluster, *Observation$_j$* is the ith object selected from Observations and $\|.\|$ is a distance metric.

Figure 1. Segmenting observations into groups based on inter and intra distance

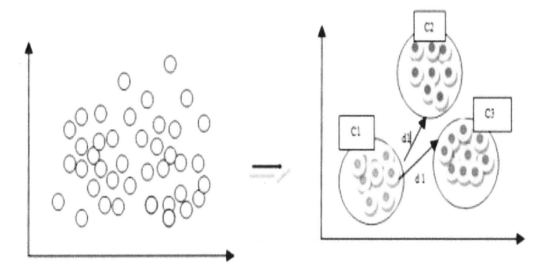

Distance Measures

Certain methods are required for classifying the observations into diverse groups for the computation of either dissimilarity or the distance existing between every individual pair of observations. The output of such calculation is largely understood as the matrix of dissimilarity or distance. A very crucial step in clustering is the selection of distance metric, as it defines the method of similarity computation between two elements x and y (Janjanam & Reddy, 2019). It also influences the cluster's shape. Hence, it is certainly a huge challenge to choose the appropriate metric of distance in clustering. Such a metric should be capable of maximizing the distance between dissimilar members and minimizing the distance between dissimilar members. As the datasets contain both the numeric and categorical features, the prime focus of the current researches on clustering is mostly on numeric datasets with a moderate focus on the categorical clustering. The metrics of distance are well illustrated in (Boriah, Chandola, & Kumar, 2008). The following are some of extensively used metrics of distance employed in clustering:

1. **L1 Norm (Manhattan):** Absolute value distance is the basis of the Manhattan distance. Consider the sum total of the absolute values pertaining to the differences of the coordinates. The following is the definition:

$$d(x,y) = \sum_{i=1}^{n} |x_i - y_i|.$$

2. **L2 Norm (Euclidean):** Euclidean distance is one of the largely employed conventional parameters in clustering. Consider a two-dimensional space and take two points representing two samples. The two points could be represented as x and y in the considered sample space corresponding to the real physical distance existing between the two points which may be computed employing Pythagora's formula with i=1…n corresponding to the considered samples, which is defined as

$$d(\mathrm{x},y) = \sqrt{\sum_{i=1}^{n}(x_i - y_i)^2} \ .$$

3. **Pearson Correlation Distance:** Distances based on correlation has been extensively used for the expression of a gene in data analyses. The degree and scale of the linear relationship existing between two profiles can be measured using Pearson correlation, defined as:

$$d(x,y) = 1 - \frac{\sum_{i=1}^{n}(x_i - \overline{x})(y_i - \overline{y})}{\sqrt{\sum_{i=1}^{n}(x_i - \overline{x})^2 \sum_{i=1}^{n}(y_i - \overline{y})^2}} \ .$$

4. **Jaccard Coefficient:** Several researchers have adopted the Jaccard Coefficient for the computation of similarity pertaining to the categorical data. This Jaccard index generally called the Jaccard similarity coefficient, originally framed by Paul Jaccard, is a statistic metric employed for the comparison of either similarity or diversity of considered sample sets. The Jaccard coefficient computes the similarity existing between a finite number of sample sets, and can be further illustrated as the intersection's size divided by the size of the combination of the sample sets, represented as:

$$J(\mathrm{x},y) = \frac{(\mathrm{x} \cap y)}{(\mathrm{x} \cup y)} \ |.$$

5. **Hamming Distance:** Hamming distance is another standard parameter used for the computation of similarity in categorical data clusters. Hamming distance computes the total number of disagreements existing between the vectors. For the computation of the Hamming distance that exists between two strings, their XOR operation can be performed, (a Å b), and subsequently, calculate the total number of 1s present in the output string.

6. **Gower Distance:** Gower Distance is a metric of distance, used for the computation of distance existing between two entities having mixed attributes, both numeric and categorical. Gower distance makes use of Manhattan for the computation of distance, existing between continuous data points and Dice for the computation of distance, existing between categorical data points. We find these Gower packages in R as well as Python. The Gower distance is calculated as the mean of partial dissimilarities existing among the individuals. Every individual partial dissimilarity (and thus Gower distance) has a range of [0 1]. Its formula can be represented as:

$$d(\mathrm{i},\mathrm{j}) = \frac{1}{p}\sum_{i=1}^{p} d_{ij}^{(\mathrm{f})} \ .$$

More details of clustering data employing Gower distance can be seen in the paper (Akay & Yüksel, 2018).

MIXED DATA CLUSTERING

The composition of mixed data could be numeric as well as categorical features. Mixed data sets frequently occur in several domains such as finance, marketing, and health. The mixed datasets are often applied to cluster in order to discover structures and to cluster objects of similarity for further investigations. But, the application of clustering approaches to mixed datasets can be very challenging as the direct application of mathematical operations including averaging and summation to the feature values pertaining to these datasets could be very complex and difficult. Most of the clustering algorithms are capable of handling only homogenous data containing exclusively either numeric values or categorical feature values. But, the computation of similarity existing between two data points becomes complex and difficult if the numerical, as well as the categorical features, exist in the datasets at the same time.

Table 1 depicts an illustration of a typical mixed dataset:

There are four features in this sample dataset with two numeric features of Height and Weight and two of categorical features of Gender and Blood group. A sample approach for identifying similarity existing between the two data points pertaining to this considered dataset is to break the numeric and categorical components and identify the Euclidean distance in between the specific two data points with regard to the numeric features and the Hamming distance with regard to the categorical features (Boriah et al., 2008). The similarity existing between the values of numeric and categorical features individually can be found through this process. The subsequent step is to integrate the two metrics in order to obtain a single value which depicts the distance existing between the two mixed data points. However, such integration of the two categories of distances could be non-trivial as it is not evident if both the distance metrics compute a 'similar' category of similarity or for that matter, if there is similarity in the scales considered in these distances.

At present, there are two standard types of approaches for processing the mixed data in the analysis of clustering. One type is based on the transformation of numerical features into categorical or categorical into numerical thereby, the methods of clustering either for numerical or clustering data could be employed. But, these approaches are ineffective as the metric of similarity of such transformed data fails to adequately reflect the similarity of the considered mixed data prior to the transformation. The other type is based on an extension of the algorithms of clustering to numerical as well as categorical data in such a way as to match with mixed data in order to enhance the results of clustering. In the literature, certain algorithms of clustering have been developed making use of these two categories of approaches (Zhao, Cao, & Liang, 2018).

Table 1. An example of a mixed dataset

Weight (kg)	Height (mts)	Blood Group	Gender
54	1.54	B+	M
65	1.68	O+	F
56	1.66	Nan	M
88	1.98	B+	M
78	1.73	A-	Nan
Nan	Nan	AB+	F
55	1.44	Nan	M

Types of Datasets

Clustering datasets can be classified into the following 3 sets:

1. **Numeric Dataset:** Numeric datasets include consideration of the real values of features such as distance, height and weight. Standard mathematical operations such as summation, mean, angles and distances can be applied to them for computation of similarity between the values of numeric features. Metrics of similarity based on distance are used in general for such numeric data points. In the context of numerical attributes, several studies (Kane, 2012) propose a clustering algorithm based on volume for determining the number of clusters that uses k-means for the purpose of clustering, starting from 1 and incrementing by 1, so that the total volume for every individual clustering can be computed. A comparison of the volume ratio of two adjacent clusters and the author defined threshold determine the optimal cluster number. In spite of its enhanced performance in instances of clear variance between the two clusters, the efficacy of the approach is in instances of non-spatial data as text.

2. **Categorical Dataset:** Any data that can be segregated into fixed categories on the basis of certain features such as profession, sex, color, race and blood group can be considered as categorical datasets. These attributes of categories are not organized by default, for instance as categorical values of blue or red. Consequently, a direct computation of the distance between two values of categorical features is not always possible. Hence, employing the similarity metrics based on the distance for the computation of categorical data becomes complex and challenging (Boriah et al., 2008). Due to several technical reasons, the standard algorithm of k-means cannot be directly applied o categorical data. With reference to categorical data, the sample space is discrete and without a natural origin. A function Euclidean distance applied to such a space does not really contain any meaning. K-modes is an existing variation of k-means advocated in (Zhexue Huang et al., 2006), which suits the requirements of categorical data. Researchers in (Kim, Lee, & Lee, 2004) have effectively made use of algorithms of fuzzy k-modes for the purpose of clustering categorical data, further extended by the representation of the clusters of categorical data along with fuzzy centroids substituting the normal practice of using the hard-type centroids in the original algorithm. This employment of fuzzy centroids enables it to comprehensively exploit the potential of the fuzzy sets in the representation of the uncertainty that always exists in classifying the categorical data.

3. **Mixed Dataset:** In most of the real-time applications, we come across mixed numerical and categorical features in the analyzed datasets, especially when they are integrated from diverse sources. In brief, data exists in a mixed-mode. Data in such mixed mode is likely to be encountered in instances of analysis of medical diagnosis, survey data, and images. For instance, the medical diagnosis data could include the attributes such as the details of age, sex, weight and blood pressure pertaining to patients, where the feature of sex alone is categorical while the others are numerical. The occurrence of mixed data is quite frequent in several of the applications related to medical, health and marketing (Ahmad & Dey, 2007). The clustering algorithms are primarily concerned with the datasets either of numerical or categorical features all through. The direct application of any conventional algorithms of clustering to mixed data proves difficult. Consequently, clustering of

mixed data proves to be complex, difficult and challenging at the same time drawing the attention of several researchers to the domains of machine learning and data mining (Zhao et al., 2018). A very common approach to the application of clustering to mixed data is to transform the datasets into a single data type and subsequently apply the standard algorithms of clustering to such transformed data. For instance, He et al. (He, Xu, & Deng, 2005) have experimented by considering a numerical feature as a category through the approach of discretization. They further extended their previous algorithm of clustering intended for categorical data to the mixed data also. Another option could be to design a common metric of similarity or distance for application on mixed data and then apply the same to the existing algorithms of clustering. K-prototype (Huang, 1998) is one such very popular algorithm, where the algorithms of the k-means and the k- modes are integrated by a definition of joint metric of dissimilarity, so that the mixed numerical and categorical features can be clustered.

Applications

Business and Marketing: The ever-evolving and swift strides in technology and the advent of cloud and mobile applications have been adversely affecting the finance industry posing severe threats in the areas of fraud detection on one side and enhanced digital threats on the other (West & Bhattacharya, 2016). The conventional approaches to these challenges are inadequate and inappropriate as they consume more time, drain more finances and are impracticable in many instances. Hence, an alternative approach specifically the emerging Artificial Intelligence, which can surmount these constraints, has become inevitable. The methods of clustering also come in handy for the purposes of marketing in grouping the all the available shopping items on the web into a well-defined group of unique products, thereby creating a platform for the users to make decisions. Clustering of mixed data can also be effectively used in diverse areas including prediction of income (adult data), approvals of credit, discovery of customer behavior pattern, segmentation of customers, marketing research, catalogue marketing, construction management and motor insurance the details of which are to be found in (Ahmad & Khan, 2018).

Medical Imaging: Analyzing medical images always has been a very challenging task in several medical applications. The suspicious patterns are outlined by extracting the most significant features from different clusters. In modern times, such techniques are used for disease identification or for the comprehension of diverse developments such as tumors or for the differentiation of the category of tissues (Ben Salem et al., 2018). (McParland & Gormley, 2016) have developed clustering algorithms for application on mixed data for investigating phenotypic numeric data of high dimensions and genotypic categorical data. The current investigations lead to an enhanced comprehension of the metabolic syndrome (MetS). Several researchers have been using different categories of approaches for clustering of mixed data related to heart ailments, severe inflammations, the life span of humans, age of abalone snails, dermatology, genetic regulation, medical diagnosis, analyzing biomedical datasets, and grouping of cancer samples. In literature, a detailed illustration of the diseases is furnished (Ahmad & Khan, 2019).

Information Retrieval and Pattern Recognition (IR&PR): Several applications based on facial recognition have their extensive utilization in diverse domains today. However, such applications are conditioned by a number of constraints such as illumination variations and expressional and postural differences requiring intelligent and robust systems of analysis.

METHODOLOGIES

The following are some of the chief methodologies of currently existing clustering algorithms:

1. **Partitional:** The algorithms of partitional clustering are extensively researched themes pertaining to the clustering of mixed data. These algorithms depict an interesting aspect regarding pattern recognition owing to their enhanced efficiency and scalability. One such is the k-means, which has been advocated in 1965. It demonstrates high efficiency when numeric clustering is considered but proves quite inadequate for categorical clustering. In order to address this, k-modes was proposed as supplementation of k-means in 1998 for clustering the categorical datasets. These algorithms primarily purpose to define (i) the center of a cluster which can reflect both numeric and categorical features, (ii) the metric of distance which can integrate both numeric and categorical features, and (iii) can iteratively minimize the cost function, which can negotiate the mixed data. Integrating the three concepts mentioned above, several partitional clustering algorithms iteratively optimize the cost function. The primary factors contributing to the initial adoption and large-scale adaptability of the partitional algorithms are their inherent linear nature pertaining to the number of data points, the capability to scale well in the context of large datasets and compatibility with parallelization frameworks such as MapReduce. (Akay & Yüksel, 2018) have suggested that clustering on mixed panel dataset can be implemented by using Gower's distance based on k-prototypes. The performance of these algorithms has been investigated on panel data comprising both numeric and categorical features. The efficacy of the algorithms is assessed by a comparative analysis using cluster accuracy. K-prototype is one such very popular approach, which inherits its fundamental concepts from k-means. Here, Euclidean distance is applied to numeric features and a distance function is described, so as to be included in the metric of the proximity between two specific objects. Pairs of objects having diverse categorical values have the tendency to increase the proximity between them. In comparison with the existing conventional algorithms of clustering, the clustering algorithm of improved k-prototypes applicable for mixed numeric as well as categorical data as proposed by (Ji, Bai, Zhou, Ma, & Wang, 2013) demonstrates more efficient results.

Cluster center initialization: The initialization of cluster center is a very common problem encountered by the algorithms of partitional clustering. In general, the initial cluster centers are randomly selected leading to diverse outcomes of clustering on different runs of the application of the algorithms seriously affecting the consistency in the outcome. Hence, the researchers cannot rely on such inconsistent clustering outcomes. A number of initiatives have been considered in order to address these issues of disadvantages in partitional clustering, the details of which have been employed by the researchers can be found in (Ahmad & Khan, 2018).

The number of clusters: Prior knowledge of the number of clusters is the basic assumption under which most of the algorithms of partitional clustering are applicable to numeric as well as categorical data, functions. This specific number may be either user-defined or derived from various other domains or computed through other algorithms. But still, most of these approaches fail to ensure a one-to-one correspondence between the selected number of clusters and the natural number of clusters present in the data. Similar problems occur in partitional algorithms also in the context of mixed data. The Authors in (Ahmad & Hashmi, 2016) have presented K-Harmonic means clustering algorithm for application on mixed data. Definitions for a center of cluster and distance metric also have been provided. The proposed algorithm makes use of these metrics of distance and cluster centers along with the cost function pertaining to the clustering algorithm of K-Harmonic means. Several experiments have been conducted using pure categorical datasets as well as mixed datasets. The output results demonstrate a total insensitivity of the clustering algorithm to the initialization problem of the cluster center. Authors in (Wei, Chow, & Chan, 2015) have tried clustering of diverse heterogeneous data employing the conventional algorithm of k-means through the approach of unsupervised feature transformation (UFT) based on mutual information (MI), which is capable of transforming the attributes from non-numerical to numerical without any loss of information. The UFT can substitute the original non-numerical attributes with numerical values, which simultaneously maintain the framework of the original non-numerical attributes along with the characteristics of continuous values. It is evident from the conducted experiments on datasets of real-world and the analysis of the results that the clustering algorithm obtained from the integration of K-means and UFT have better performance efficacy over the other existing clustering algorithms with reference to heterogeneous data of numerical as well as categorical attributes.

2. **Hierarchical:** Hierarchical clustering has been evolving as a possible substitute for the clustering algorithm based on a prototype. The prime advantage of Hierarchical clustering is that the number of clusters doesn't need to be specified, as it is capable of discovering the same by itself. Besides, the algorithm facilitates the dendrogram plotting as depicted in Figure 2. Dendrograms can be comprehended as the visual representation of a binary hierarchical clustering.

Bottom fusing observations are found to be similar mostly while the top fusing ones are completely different. The location and position of the vertical axis rather than the horizontal axis is the basis for drawing conclusions with dendrograms. We have the twin approaches to this category of clustering namely, Divisive and Agglomerative.

Figure 2. Hierarchical clustering

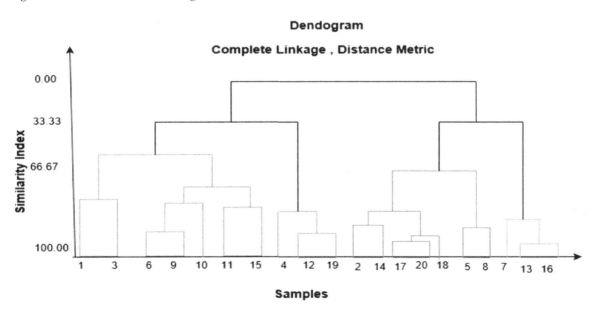

Divisive: All the considered data points are englobed in this approach initially; followed by iteratively splitting the clusters into still smaller ones until only one sample remains in each cluster.

Agglomerative: This approach begins with considering every sample as a different cluster and combines all the proximate clusters until only a single cluster remains. Agglomerative hierarchical clustering largely makes use of the single and complete linkage algorithms. Single Linkage: The agglomerative nature of the single linkage algorithm conditions it to assume that every single individual sample is a cluster by itself. Then, it starts computing the distances existing between the largely similar members pertaining to the individual cluster pairs in order to subsequently combine such two clusters having the smallest distance between their largely similar members.

Complete Linkage: This approach demonstrates exactly the opposite to that of the single linkage. It resorts to the comparison of the most dissimilar and different dataset points associated with a pair of clusters for conducting the operation of merging

Advantages of Hierarchical Clustering: The resultant hierarchical depiction and dendrograms can provide very interesting and informative visualization. Their significance is all the more powerful in instances of datasets containing actual hierarchical relationships.

Disadvantages of Hierarchical Clustering: Hierarchical Clustering is extremely sensitive to outliers with a significant decrease in the performance due to their presence. Besides, it proves to be very expensive in terms of computation.

Several of the algorithms of hierarchical clustering contain a large time complexity of $O(n^3)$ and requires $O(n^2)$ memory, where n is the number of data points. The authors in (Ahmad & Khan, 2019) have reviewed many evolving algorithms of hierarchical clustering capable of handling the mixed data.

3. **Density:** Clustering algorithms based on density are very efficient in precise identification of noise in data. DBSCAN is a very popular and largely employed clustering algorithm based on density. It depends on several points with a radius ε and a special label assigned to each individual data point. The following is the process of assigning the label: Minpts is a specified number of proximate points. If this Minpts falls in the ε radius, then a core point is allotted. A threshold point occurs within the ε radius of the core point, but contains fewer neighbors than the considered MinPts number. All the other points are noise points. The following is the logic of the algorithm: (i) Identification of a core point to form a group for everyone or for every inter-connected cluster of core points (if they comply with the metrics to be considered as core points). (ii) Identification and assignment of border points to the relevant core points. The commented notation and illustration of the process are clearly represented in Figure 3.

The following are the advantages of DBDSCAN Advantages:

(i) Cluster number specification is not required.
(ii) High flexibility in sizes and shapes for the clusters to choose from and adopt.
(iii) Highly useful in the identification and handling the outliers and noise data.

The following are the disadvantages of DBSCAN:

(i) Encounters several issues while handling the border points approachable by two clusters
(ii) Fails to properly identify different densities.

Figure 3. A typical clustering using DBSCAN

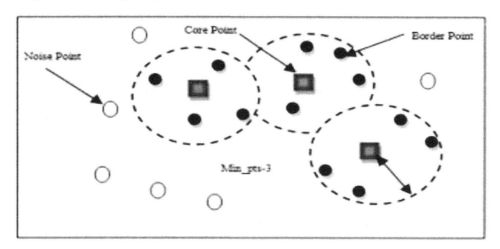

Several revisions to the algorithm of DBSCANs have been developed until now in order to enhance its performance. But the sensitivity of the parameter continues to pose a challenge for the DBSCAN algorithm for its extended application (Jinyin, Huihao, Jungan, Shanqing, & Zhaoxia, 2017). In order to address the issue of the determination of clustering center, the algorithms of fast density clustering have been advocated. But, its approach of mixed similarity computation relies on the inter-relationship of all the considered attributes making the computation highly complex. Also, its approach to determine the cluster center largely relies on such metrics, the prior setting of which is difficult. For instance, the functions of distance metrics intended for numerical values are incapable of capturing the similarity in data containing mixed attributes. In addition, the depiction of a cluster containing numerical values is generally described as the mean value pertaining to all the data objects existing in that cluster, which could be illogical with reference to the other features. Several researchers over a period of time have been employing a number of heuristics approaches some of which are based on density for mixed data clustering. A fast density clustering algorithm (FDCA) (Jinyin et al., 2017) was put forward, which is auto-determined by the center set algorithm (CSA). The authors have developed a novel metric of data similarity for clustering of data containing both numerical and categorical attributes. The very purpose of developing CSA is the auto-selection of cluster centers from among the data objects, thereby over-coming the previous constraint of setting the cluster center as experienced with several other clustering algorithms. The proposed approach has been verified and validated through a series of experimental simulations considering ten diverse mixed data sets and comparing them with diverse existing algo-rithms for the efficacy in performance, clustering purity and complexity of time. (Du, Fang, Huang, & Zeng, 2018) further extend these clustering algorithms based on density to the categorical realms and the domains containing mixed numeric and categorical features.

4. **Model-based and Fuzzy Centroid:** The conventional models of clustering as k-means and hier-archical clustering are heuristic by nature and do not rely on formal models. In addition, due to the common practice of random initialization of the k-means algorithm, the results yielded in different runs are not consistent and differ largely. Besides, the user needs to specify the optimal number of clusters in this k-means algorithm. Clustering based on the model could be an ideal alternative for

this, which regards the data as emerging from a distribution, which is a combination of two clusters or more (Fraley & Chris, 2002). Different from the k-means, the clustering based on model makes use of a soft assignment, in which each data point demonstrates the likelihood of belonging to each cluster. The data in model-based clustering can be considered as transmitted from a mixture of density. Every individual component or cluster, k is modeled after either the normal or Gaussian distribution approach, characterized by the metrics μ_k: -mean vector, Σ_k- covariance matrix and an associated probability in the mixture. All the relevant details pertaining to the mean vector, metric of covariance and the related probability can be perceived further from (Fraley, Chris, 2002). Each individual point has a strong probability of being associated with each cluster. The R package contains the entire library of the pre-defined model clustering. The use of Expectation Maximization (EM) duly initialized by the hierarchical clustering algorithm based on a model enables appropriate estimation of the parameters of the model. Every cluster k center at the means μ_k, with enhanced density for points proximate to the mean. The covariance matrix Σ_k determines each cluster's geometric attributes such as the orientation, shape, and volume. The mclust in R package contains the diverse possible parameterizations of Σ_k the model options as are available in the *mclust* package, depicted using appropriate identifiers such as EII, VII, EEI, VEI, EVI, VVI, EEE, EEV, VEV and VVV. The volume, shape, and orientation are represented by the first three identifiers, while E, V and I depict 'equal", "variable' and 'coordinate axes' respectively. For instance, the representation EVI depicts a model where all the cluster volumes are equal (E) with a possible variation in shapes (V) and with the orientation identity (I) as "coordinate axes". Similarly, EEE signifies the presence of the same volume, shape, and orientation in all the considered clusters in the space of p-dimensions and VEI represents the clusters with variable volume but with similar shape and the orientation identical to coordinate axes.

The above plot suggests a minimum of 3 clusters in the considered mixture. All the three clusters have almost identical and elliptical shape suggesting three possible bivariate normal distributions. There is a strong scope to predict that the mixture's all the three components could contain similar homogeneous matrices of covariance in the context of all of them having almost identical terms of volume, shape, and orientation. Several researchers have applied these model-based clustering algorithms to mixed datasets. In particular, one study by Damien et.al (McParland & Gormley, 2016), considers clustering of mixed data using the technique of clustMD. This clustMD uses the structure of a parsimonious covariance for the latent variables, resulting in a suite of six models of clustering which are different in complexity and furnishes an accurate and unified technique for mixed data clustering.

An algorithm of Expectation Maximization (EM) is employed for the estimation of clustMD. There is a requirement for the algorithm of Monte Carlo EM, when nominal data is present. This model of clustMD is well illustrated through mixed-type data simulated by clustering pertaining to patients of prostate cancer on whom the recording of mixed data has been done. Bettina (Grün, 2018) provides the theoretical information associated with the mixed data clustering based on a model. Bettina well justifies the model-based clustering pertaining to the general methods of heuristic clustering and furnishes a very general overview of diverse means of specifying the cluster model. It has been commonly experienced that several of the clustering algorithms pose problems of optimization, where the functions of internal cluster validity have been used as the objectives in order to discover the optimal partitions. However, several of these approaches consider only one criterion for mere application for finding the specific structure/ data distribution. Uncertainties can be ideally handled employing fuzzy logic as well as fuzzy set theory. The models based on fuzzy clustering have furnished specifically ideal solutions in a vast range of areas (Kim et al., 2004). (Zhu & Xu, 2018) have proposed a novel clustering algorithm based on multiple objective fuzzy centroids for application on categorical data employing non-dominated sorting genetic algorithm based on the reference point, which optimizes a number of indices of cluster validity simultaneously. They have employed an effective algorithm of fuzzy centroids in their work for designing the proposed approach, which varies from the other contestant approaches of k -modes-type. In this instance, the approach of fuzzy memberships has been employed for the representation of chromosome which fuses with a novel genetic operation in order to produce novel solutions. Besides, a scheme of changeable length encoding has been developed for discovering the clusters without having any prior knowledge. Diverse experiments held on a variety of datasets illustrate the performance superiority of the proposed algorithm Experiments on several data sets demonstrate the superior performance efficacy of the proposed algorithm in comparison with the other existing state-of-the-art approaches in terms of stability and accuracy of clustering. The authors in (Lam, Wei, & Wunsch, 2015) have applied the unsupervised feature learning (UFL) on the data of mixed-type for obtaining sparse representation, while they have applied fuzzy adaptive resonance theory (ART) for clustering. UFL in conjunction with fuzzy ART (UFLA) accomplishes enhanced results of clustering, through the elimination of the differences in the treatment of categorical as well as numeric features. The self-evident advantages of this process are illustrated against a number of real-time datasets of ground truth such as heart ailments, evaluation of teaching assistant and approval of credit. This method has also been demonstrated in instances of noisy and mixed-type data pertaining to the petroleum industry and has been established to have accomplished good results.

5. **Ensemble-based:** Figure 4 depicts a very general framework of ensemble clustering, having three major components of a generation of ensemble members, consensus function and finally the evaluation. As is evident, in the context of the framework of clustering ensemble, the input is a specific data requiring to be clustered and the output is the final clustering result obtained from the considered dataset. Further details can be comprehended from the illustrations in (Alqurashi & Wang, 2018). For a dataset of n objects: $X=\{x_1,x_2,\ldots,x_n\}$ let $P_q = \{c_1^q, c_2^q, \ldots c_{kq}^q\}$ be a clustering result of k_q clusters produced by a clustering algorithm as q^{th} a partition, so that $c_i^q \cap c_j^q = \varnothing \cup_{j=1}^{kq} c_j^q = X$ and A clustering ensemble ϕ can then be built with m partitions $T=\{P_1, P_2, \ldots, P_m\}$ and a consensus function F, and denoted by $\phi(F,T) = F(P_1, P_2, \ldots, P_m) = F(T)$. It should be noted that the members may not necessarily have the same number of clusters in their partitions, that is, k_q may not be equal to a pre-set value k. The task of a clustering ensemble is to find a partition P∗ of dataset X by combining the ensemble members $\{P_1, P_2, \ldots, P_m\}$ with F without accessing the original features so that P∗ is probably better in terms of consistency and quality than the individual members in the ensemble.

Just as the functioning of ensemble methods in the environment of supervised learning, the approaches of ensemble clustering function in two distinct steps of generation and combination of clustering. The two principal factors influencing the performance efficacy of ensemble clustering approaches are the diversity and the quality of the base clustering. Consequently, a number of heuristics have been projected for the purpose of generation of diverse clustering approaches for a specific dataset, which can be categorized into the following three methods:

Figure 4. A generic clustering ensemble framework

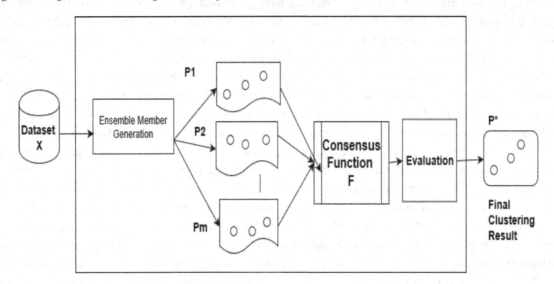

Homogeneous method: It is a method of generating base clustering by repeated execution of a single algorithm of clustering with diverse initializations, for instance as the cluster number or number of cluster centers (Kuncheva & Vetrov, 2006).

Data subspaces/subsamples method: It is a method of obtaining a bunch of results of base clustering through the projection of data onto diverse sub-spaces, selecting diverse subsets of attributes or data sampling (Yu, Li, Liu, Zhang & Han, 2015).

Heterogeneous method: It is a method of producing base partitions with a range of diverse algorithms of clustering on a specifically considered dataset (Yu, 2013). In the subsequent step, considering a base clustering set, a consensus function is employed for integrating them towards the final results of clustering.

The recent past decade witnessed the emergence of a number of approaches of ensemble clustering as can be seen in (Boongoen & Iam-On, 2018). Broadly considering, the above methods can be categorized into the following four groups namely, the methods based on features, similarity, graph and relabeling. Numerous novel technologies for ensemble clustering have been emerging to address these specific issues. For instance, the approach of ensemble selection clustering can certainly enhance the quality of clustering by a process of evaluation and selection of a subset of base partition corresponding to its contribution in the process of integration (Boongoen & Iam-On, 2018).

EVALUATION METRICS

In order to obtain a comprehensive picture of results, three largely popular external metrics have been considered for the performance evaluation and validating the efficacy of clustering algorithms namely, Clustering Accuracy (CA), Normalized Mutual Information (NMI) and Adjusted Rand Index (ARI), which comparatively analyze and compute the agreement between the ground truth and the clustering results generated through the application of an algorithm. Presume that $C=\{c_1,c_2,...,c_k\}$ and $P=\{p_1,p_2,...,p_k\}$ depict the results of clustering and the pre-defined dataset classes having N number of objects, respectively. Let k be the number of clusters C, and classes P, $N_{i,j}$ the number of common objects existing in the cluster C_i and pre-defined class P_j. N_i^c and N_j^p be the number of data points in cluster C_i and class P_j, then, the following are the three very popular and common external criteria:

1. **Clustering Accuracy:** Clustering Accuracy (CA) calculates the percentage of the exactly classified data points existing in the clustering solution in comparison against the pre-defined labels of class. The following can be the definition of the CA:

$$CA = \frac{\sum_{i=1}^{k} \max_{j=1}^{k} N_{i,j}}{N}$$

2. **Normalized Mutual Information (NMI):** NMI is a very common metric of external clustering validation which assesses the clustering quality pertaining to specific class labels of the data. To be more specific, NMI is capable of efficient quantification of the shared statistical information among the random variables reflecting the assignments of the clusters as well as the pre-defined label assignments pertaining to the respective objects. The following formula can well define and compute the NMI:

$$NMI = \frac{\sum_{i=1}^{k} \sum_{j=1}^{k} N_{i,j} \log \frac{N \cdot N_{i,j}}{N_i^c \cdot N_j^p}}{\sqrt{\sum_{i=1}^{k} N_i^c \cdot \log \frac{N_i^c}{N} \cdot \sum_{j=1}^{k} N_j^p \cdot \log \frac{N_j^p}{N}}}$$

3. **Adjusted Rand Index (ARI):** The ARI is concerned with the number of objects present either in the same cluster or different clusters and is generally defined as:

$$ARI = \frac{\binom{N}{2} \sum_{i=1}^{k} \sum_{j=1}^{k} \binom{N_{i,j}}{2} - \left[\sum_{i=1}^{k} \binom{N_i^c}{2} \sum_{j=1}^{k} \binom{N_j^p}{2} \right]}{\frac{1}{2} \binom{N}{2} \left[\sum_{i=1}^{k} \binom{N_i^c}{2} + \sum_{j=1}^{k} \binom{N_j^p}{2} \right] - \sum_{i=1}^{k} \binom{N_i^c}{2} \sum_{j=1}^{k} \binom{N_j^p}{2}}$$

The maximum value of the three external criteria is 1. In instances where the result of the clustering is very near to the true class distribution, their values become high. The enhanced performance of the clusters is directly in proportion to the higher level of the values pertaining to the three metrics for a clustering result.

LIMITATIONS OF MIXED DATA CLUSTERING

- The prior knowledge as to the number of clusters is presumed in several algorithms of partitional clustering for their functioning on numeric as well as categorical data. This assumed number could be user-defined or derivations from the domain of computational output from the other algorithms. The problem is recurrent in partitional algorithms employed for mixed data.

- Most of the algorithms of hierarchical clustering depend on the computation of the similarity matrix, based on which cluster construction becomes possible. But, this similarity matrix relies on a proper description of similarity/distance. As has been stated already, the distance existing between two objects of mixed data is not always self-explanatory requiring further probe.

- Diverse models of clustering and data information and several research efforts have been exploited in the area of ensemble clustering. However, these approaches do not always comprehensively deal with the correlation existing among the diverse base clustering during their generation process, which is a prominent aspect in obtaining qualitative and various clustering decisions.

FUTURE RESEARCH DIRECTIONS

- A typical mixed dataset as shown in Table 1 from section Mixed Data Clustering, contains some 'Nan' values which are missing values or not valid numbers. The existing algorithms are inefficient to cope up with such data; thus, they can be handled by using preprocessing techniques before processing through algorithm. Concentrating on clustering algorithm that handles missing values in mixed dataset can be initiated.

- All the attributes pertaining to the mixed datasets may not be significant. Elimination of insignificant attributes certainly could enhance the results of clustering. Several aspects related to a methodical exploration of the selection of unsupervised features in the context of mixed datasets have not been seriously considered so far, opening vast scope for research to study the datasets with multiple features.

- Due to the inadequacies in the methods of data acquisition or certain inherent constraints in data acquisition, uncertain data is present in several application areas of mixed datasets, such as socio-economic and medical applications. It is obvious that appropriate approaches capable of handling this category of datasets are scarce still. This clustering of uncertain mixed datasets in several domains could be another prominent research direction.

- As of today, very few methods have been developed to convert mixed datasets into pure and un-mixed numeric datasets in such a way that algorithms of clustering intended for pure datasets can also be extended to mixed datasets. This is certainly a novel aspect worthy of serious research consideration pertaining to the complex problem of clustering of mixed data. Further, it is evident that there could be serious loss information when the mixed data is transformed into numeric data. Hence, the scope is wide open for developing such algorithms capable of considerable reduction in the undesirable effects occurring during data transformation.

- As discussed in the Methodology section the algorithms of Ensemble clustering are prone to work well when compared to standard algorithms of clustering. Since the prior objective of ensemble clustering is to enhance the quality of clustering based on certain criteria, research on ensemble clustering would outperform standard clustering techniques (Zhao et al., 2018).

CONCLUSION

With the enormous rise of information from diverse sources, the available datasets in present days are mixed type and are very challenging to recognize patterns out of that data. One such admired advent reason is the compatibility issues among the characteristics of data involving numerical as well as categorical attributes usually called mixed data. Clustering is often used to group identical elements and to find structures out of data. However, clustering categorical data poses some notable challenges. Particularly Clustering diversified (mixed) data constitute bigger challenges, because of its range of attributes. Computations on such data are merely too complex to match the scales of numerical and categorical values due to its ranges and conversions. This chapter presents different details of algorithms involved in dealing with mixed data. The literature models are highlighted to make the researches familiar with on-going ideas that would be helpful in determining new methodologies. Research gaps and future research direction are discussed based on the reviewed mixed data literature models.

REFERENCES

Ahmad, A., & Dey, L. (2007). A k-mean clustering algorithm for mixed numeric and categorical data. *Data & Knowledge Engineering*, *63*(2), 503–527. doi:10.1016/j.datak.2007.03.016

Ahmad, A., & Hashmi, S. (2016). K-Harmonic means type clustering algorithm for mixed datasets. *Applied Soft Computing*, *48*, 39–49. doi:10.1016/j.asoc.2016.06.019

Ahmad, A., & Khan, S. S. (2018). Survey of state-of-the-art mixed data clustering algorithms. *IEEE Access : Practical Innovations, Open Solutions*, *7*, 1–16. doi:10.1109/ACCESS.2019.2903568

Ahmad, A., & Khan, S. S. (2019). Survey of State-of-the-Art Mixed Data Clustering Algorithms. *IEEE Access : Practical Innovations, Open Solutions*, *7*(i), 31883–31902. doi:10.1109/ACCESS.2019.2903568

Akay, Ö., & Yüksel, G. (2018). Clustering the mixed panel dataset using Gower's distance and k-prototypes algorithms. *Communications in Statistics. Simulation and Computation*, *47*(10), 3031–3041. doi:10.1080/03610918.2017.1367806

Alqurashi, T., & Wang, W. (2018). Clustering ensemble method. *International Journal of Machine Learning and Cybernetics*, *10*(6), 1227–1246. doi:10.1007/s13042-017-0756-7

Ben Salem, S., Naouali, S., & Chtourou, Z. (2018). A fast and effective partitional clustering algorithm for large categorical datasets using a k-means based approach. *Computers & Electrical Engineering*, *68*, 463–483. doi:10.1016/j.compeleceng.2018.04.023

Boongoen, T., & Iam-On, N. (2018). Cluster ensembles: A survey of approaches with recent extensions and applications. *Computer Science Review*, *28*, 1–25. doi:10.1016/j.cosrev.2018.01.003

Boriah, S., Chandola, V., & Kumar, V. (2008). Similarity Measures for Categorical Data: A Comparative Evaluation. Proceedings of the 2008 SIAM International Conference on Data Mining, 243–254. doi:10.1137/1.9781611972788.22

Du, H., Fang, W., Huang, H., & Zeng, S. (2018). MMDBC: Density-Based Clustering Algorithm for Mixed Attributes and Multi-dimension Data. Proceedings - 2018 IEEE International Conference on Big Data and Smart Computing, BigComp 2018, 549–552. 10.1109/BigComp.2018.00093

Fraley, C., & Raftery, A. E. (2002). Model-Based Clustering, Discriminant Analysis, and Density Estimation. *Journal of the American Statistical Association*, *97*(458), 611–631. doi:10.1198/016214502760047131

Grün, B. (2018). Model-Based Clustering. doi:10.100700357-016-9211-9

Han, J., & Kamber, M. (2000). *Data Mining Concepts and Techniques*. Elsevier.

He, Z., Deng, S., & Xu, X. (2006, August). Approximation algorithms for k-modes clustering. In *International Conference on Intelligent Computing* (pp. 296-302). Springer.

He, Z., Xu, X., & Deng, S. (2005). Scalable algorithms for clustering large datasets with mixed type attributes. *International Journal of Intelligent Systems*, *20*(10), 1077–1089. doi:10.1002/int.20108

Huang, Z. (1998). Extensions to the k-means algorithm for clustering large data sets with categorical values. *Data Mining and Knowledge Discovery*, *2*(2(3)), 283–304. doi:10.1023/A:1009769707641

Janjanam, P., & Reddy, C. P. (2019). Text Summarization: An Essential Study. *2019 International Conference on Computational Intelligence in Data Science (ICCIDS)*, 1-6.

Ji, J., Bai, T., Zhou, C., Ma, C., & Wang, Z. (2013). An improved k-prototypes clustering algorithm for mixed numeric and categorical data. *Neurocomputing*, *120*, 590–596. doi:10.1016/j.neucom.2013.04.011

Jinyin, C., Huihao, H., Jungan, C., Shanqing, Y., & Zhaoxia, S. (2017). Fast Density Clustering Algorithm for Numerical Data and Categorical Data. *Mathematical Problems in Engineering*, *2017*, 1–15. doi:10.1155/2017/6393652

Kane, A. (2012). Determining the number of clusters for a k-means clustering algorithm. *Indian Journal of Computer Science and Engineering*, *3*(5), 670–672.

Kim, D., Lee, K. H., & Lee, D. (2004). Fuzzy clustering of categorical data using fuzzy centroids. *Pattern Recognition*, *25*(11), 1263–1271. doi:10.1016/j.patrec.2004.04.004

Kuncheva, L. I., & Vetrov, D. P. (2006). Evaluation of the stability of k-means cluster ensembles with respect to the random initialization. *IEEE Transactions on Pattern Analysis and Machine Intelligence*, *28*(11), 1798–1808. doi:10.1109/TPAMI.2006.226 PubMed

Lam, D., Wei, M., & Wunsch, D. (2015). Clustering Data of Mixed Categorical and Numerical Type With Unsupervised Feature Learning. *IEEE Access : Practical Innovations, Open Solutions*, *3*, 1605–1616. doi:10.1109/ACCESS.2015.2477216

McParland, D., & Gormley, I. C. (2016). Model-based clustering for mixed data: clustMD. *Advances in Data Analysis and Classification*, *10*(2), 155–169. doi:10.1007/s11634-016-0238-x

Wei, M., Chow, T. W. S., & Chan, R. H. M. (2015). Clustering heterogeneous data with k-means by mutual information-based unsupervised feature transformation. *Entropy (Basel, Switzerland)*, *17*(3), 1535–1548. doi:10.3390/e17031535

West, J., & Bhattacharya, M. (2016). Intelligent financial fraud detection: A comprehensive review. *Computers & Security*, *57*, 47–66. doi:10.1016/j.cose.2015.09.005

Yu, Z. (2013). Hybrid Fuzzy Cluster Ensemble Framework for Tumor Clustering from Biomolecular Data. Computational Biology and Bioinformatics, IEEE/ACM Transactions On, 10(3), 657–670.

Yu, Z., Li, L., Liu, J., Zhang, J., & Han, G. (2015). Adaptive noise immune cluster ensemble using affinity propagation. *IEEE Transactions on Knowledge and Data Engineering*, *27*(12), 3176–3189. doi:10.1109/TKDE.2015.2453162

Zhao, X., Cao, F., & Liang, J. (2018). A sequential ensemble clusterings generation algorithm for mixed data. *Applied Mathematics and Computation*, *335*, 264–277. doi:10.1016/j.amc.2018.04.035

Zhu, S., & Xu, L. (2018). Many-objective fuzzy centroids clustering algorithm for categorical data. *Expert Systems with Applications*, *96*, 230–248. doi:10.1016/j.eswa.2017.12.013

Chapter 7
Digital Detection of Suspicious Behavior With Gesture Recognition and Patterns Using Assisted Learning Algorithms

Nancy E. Ochoa Guevara
Fundación Universitaria Panamericana, Colombia

Wilmar Calderón Torres
Fundación Universitaria Panamericana, Colombia

Andres Esteban Puerto Lara
https://orcid.org/0000-0002-3818-5667
Fundación Universitaria Panamericana, Colombia

Laura M. Grisales García
Fundación Universitaria Panamericana, Colombia

Ángela M. Sánchez Ramos
Fundación Universitaria Panamericana, Colombia

Nelson F. Rosas Jimenez
Fundación Universitaria Panamericana, Colombia

Omar R. Moreno Cubides
Fundación Universitaria Panamericana, Colombia

ABSTRACT

This chapter presents a study to identify with classification techniques and digital recognition through the construction of a prototype phase that predicts criminal behavior detected in video cameras obtained from a free platform called MOTChallenge. The qualitative and descriptive approach, which starts from individual attitudes, expresses a person in his expression, anxiety, fear, anger, sadness, and neutrality through data collection and feeding of some algorithms for assisted learning. This prototype begins with a degree higher than 40% on a scale of 1-100 of a person suspected, subjected to a two- and three-iterations training parameterized into four categories—hood, helmet, hat, anxiety, and neutrality—where through orange and green boxes it is signaled at the time of the detection and classification of a possible suspect, with a stability of the 87.33% and reliability of the 96.25% in storing information for traceability and future use.

DOI: 10.4018/978-1-7998-1839-7.ch007

INTRODUCTION

Bogota in Colombia, according to Chamber of Commerce of Bogota 2018 respondents (4,780) perceived insecurity in the city has increased. Over 30% of people who suffer from some form of aggression do not report because it is very difficult and there is little confidence that count for something. The(Mayor of Bogota, 2017) He reported that by 2016, the city had 532 video surveillance devices and only 302 were in operation, however, theft individuals increased by 64% by 2017 (Office of Information Analysis and Strategic Studies - OAIEE, 2017).

Computer vision has become a technology that properly applied, can help detect suspicious behavior in environments covered surveillance video. In most cities in Colombia the main causes of victimization are simple theft, bike theft, murder and quarrels among others, violating the integrity of its inhabitants, as well as spreading terror and fear.

Therefore, the question arises how the digital recognition of suspicious behavior by identifying gestures and patterns, will predict criminal behavior detected in video cameras using assisted learning algorithms? Based on the premise put forward by (Goffmann, 1971)"A person can stop talking but can not stop communicating with your body." In order to identify classification techniques and digital recognition suspicious behavior by detecting gestures and patterns, using algorithms assisted learning, so that through a prototype phase, to predict criminal behavior detected in video obtained from a free platform. With the support of the qualitative methodological approach and a descriptive study, which starts from the individual expressing attitudes a person in his expression, happiness, sadness and fear among others, through data collection and feeding of some algorithms for assisted learning, through a process simulation and testing with a control group (CG) and a experimental group (GE). This is able to build a reliable prototype,

BACKGROUND

Problematic

According to Norza, Peñaloza and Rodriguez (2016) perpetrators of crimes against life and personal integrity they increased by 49.61% during 2016, compared to the immediately preceding year. A total of 147,891 cases in 2015, he went on to have 221,256 in 2016. Likewise, we can see that being Bogota the capital of the heads republic with a 16.31% registration of occurrences of crimes nationwide. Theft 146,643 people had records, noting an increase of 45% over 2015. The 50.30% of the total was presented at: Bogota (25.84%), Antioquia (14.44%) and Valle (10, 01%).

This shows that criminal behavior and perceived insecurity in the cities especially in Bogota, is not only in perception; but they are a reality and are a problem that can be tackled from several disciplines. The certain suspicious attitudes that precede the crime early, could serve to generate the necessary alarms and take measures to prevent its occurrence. With the information provided by the OAIEE in 2019, evidence that measures such as increasing police in critical areas such as Transmilenio (transport stations), parks and school zones have not been very effective in cities, especially in Bogota.

Problem Formulation

How the digital recognition of suspicious behavior by identifying gestures and patterns, will predict criminal behavior detected in video cameras using assisted learning algorithms?

Overall Objective

Identify classification techniques and digital recognition of suspicious behavior by detecting gestures and patterns, using algorithms assisted learning, so that through a prototype phase, to predict criminal behavior detected in video obtained from a free platform.

Specific Objectives

1. Identify the profile of an individual by applying and combining different classification methodologies and gesture recognition and patterns.
2. Analyze various types of assisted learning algorithms, so that to allow profiling combine and apply the methodologies used for the development of the study.
3. Develop a functional prototype that integrates assisted learning algorithms and digital gesture recognition and selected patterns.
4. Assess the functionality of the software prototype developed with a control group and one experimental, through a simulation process.

Applied Studies

Deep-anomaly: Fully convolution neural network for fast anomaly detection in crowded scenes, M. Sabokrou, M. Fayyaz, M. Fathy, Z. R. Klette and Moayed. 2018. Institute for Research in Fundamental Sciences, Tehran, Iran. This study presents an efficient method for the detection and localization of abnormal behaviors in videos. By using neural networks and temporary data completely convolution (FCN). Where FCN previously trained and supervised transferred to unsupervised FCN ensuring anomaly detection (global) scene. This architecture based FCN addressed two major tasks, the representation of features and outlier detection cascade. The experimental results in the two reference points suggest that the proposed method overcomes the existing methods in terms of accuracy with respect to the detection and location. Given that one of the objectives of the proposed system is the use of RNA, the results of this study, and the algorithm HoG and FCN will be used during the development stage of this research, mainly by the degree accuracy in detecting patterns.

Abnormal behavior detection using hybrid agents in crowded scenes, Sang-Hyun Cho and Kang-Bong Hang. 2014. Catholic University of Korea. This study proposes a hybrid agent method to detect abnormal behavior in a busy scene. In their model, they classify the behavior of people in individual behavior and interactive group behavior. Experimental results show that the proposed method efficiently detects abnormal behavior in such scenes. Given that the proposed system aims to analyze and classify the behavior of people in crowded environments, the authors consider to use the techniques proposed in this study in order to compare their performance against that of a classification system based on FCN.

Detection and tracking of objects moving cameras, Hector Lopez. 2011. Autonomous University of Madrid, Spain. The study was to conduct a compilation of existing methods for solving problems of detecting and tracking patterns and presented an own surveillance system applied to moving targets in video sequences. The method employed for detecting movement consists in applying a measure of image quality, structural comparison, and with respect to tracking algorithm used Kalman filter. This filter will be considered by the authors of the study on the development of the tracking module of individuals of the proposed system by allowing filtering and predict linear systems more efficient way.

Gesture Recognition and Patterns

Gestures and positions of the human body is one of the most basic forms of communication has; as expressed by (Goffmann, 1971)"A person can stop talking but can not stop communicating with his body, he can tell the truth or lies, but can not stop saying something" and, according to this premise, we note that from a young age humans They express feelings through various events. bodily expressions such as raised eyebrows and open your eyes to express surprise, raise the cheeks, moving the edge of the lips and squinting to generate a smile or a cry for fear show. Where are the first forms of communication and as the individual grows involuntary reactions such as sweating, a slight movement of the lips or cross your arms in a conversation, showing more things that characterizes it. In Figure 1, it is observed that M

SPC to which an individual reacts daily in response to more or less intense stimuli, classified into three groups in order to understand better how they are exaggerated signals, unsatisfying signals, meta, with some attitudes mentioned by the author Anta (2012) as is fear, sadness, contempt, anger and signs of deception.

Image Processing Phases

In Figure 2, the different phases are applied to image processing in the studio is displayed; as is the acquisition, pre-processing, segmentation, description and comparison characteristics and definitions of each object is analyzed during the process.

Histogram of Oriented Gradients (HOG)

As expressed by (Intel Software, 2018) the HOG, is a descriptor features, which was used to detect objects in computer vision and image processing. The technique of the descriptor HOG counts the occurrences of the orientation of the gradient in localized portions of a window detection image or ROI in Figure 3, the implementation of the algorithm is presented HOG descriptor.

Classifier Cascade (Haar)

This classifier is used to recognize faces in images. It is based on decision trees that handle supervised training. Figure 4 shows that when an image analysis is performed to find a face image is traversed entirely by making small windows of the original image. Most of these windows or portions of the image corresponding to portions in which there are faces, so a simple method is to analyze each image portion and if not a face, is discarded and not further processed, thus it is optimizing the time it takes to check the possible part of a face that is in the whole picture.(Knight Barriga, 2017).

Figure 1. Groups of SC in an individual
Source: (Rebel, 2009). SC group that can occur on a individual. Design built by the authors

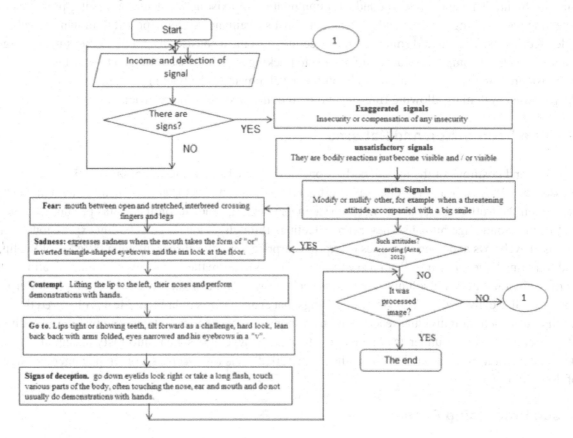

Figure 2. Image processing
Source: (Medina, 2018). Essential elements in the processing of images
Design constructed by the authors

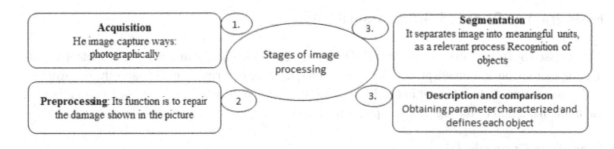

The features detailed in this example are two; one shows how the eye region is darker than that of the cheeks and nose. The other shows that the eye region is darker than the part of the bridge of the nose, thanks to the operation of Haar, this classifier will be part of the face detection for the correct operation of the prototype build.

Figure 3. Algorithm descriptor HOG
Source: (Intel Software, 2018). HOG algorithm instructions. Design constructed by the authors

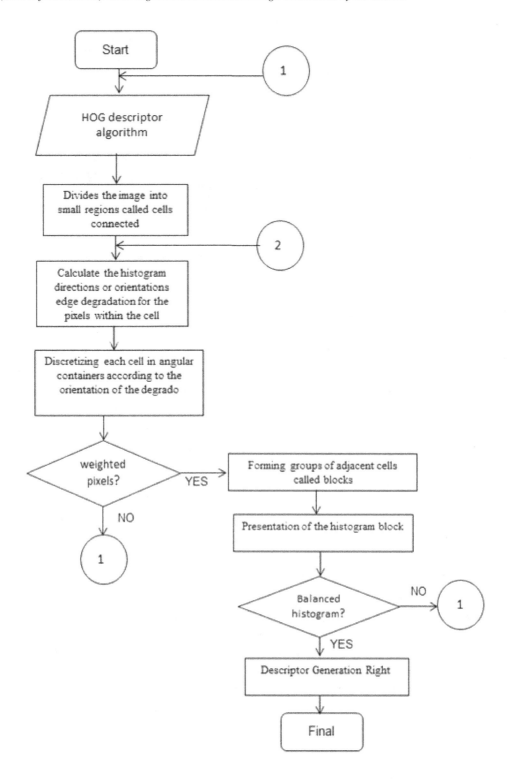

Figure 4. Characteristics of Haar type applied to a face detection
Source: (Knight Barriga, 2017)

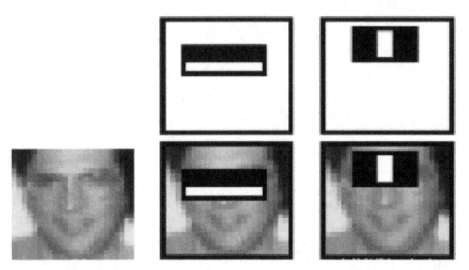

Aided Learning Algorithm (AAA)

In an AAA, instructions are received for a process that get results based on the information provided is sought.In Figure 5, shown as architecture within a local environment receives the information from the camera, in this process of image collection, the information is processed in the different modules have, which obtain information already stored in the database and leave records about new learned behaviors and create new records, which then returns the same response behavior to the core module which returns to the main user of the query. This process is done through scheme detection and monitoring performed in three aspects: pattern classification, advents generators, management and monitoring with the records generated and stored in the database (Wilmer Asanza Rivas, 2018).

Legal Delimitations

The Congress of Colombia, develops the aforementioned constitutional provision to the issuance of the Statutory Law 1581 of 2012, partially regulated by Decree National 1377 of 2013, whose scope shall apply to personal data recorded in any database that are processed by public entities or private nature, it should be noted that the definition of personal data is stipulated in the Act referred to as "Article 3, any information related or may be associated with one or more natural persons determined or determinable." (Law 1581, 2012). In the same way the Constitutional Court of Colombia argues that video surveillance systems are considered intrusive to the privacy tools to involve as monitoring and observing the activities that people do throughout the day. In this regard, it is stated that before making the decision to implement such systems should take into account the need to use (Constitutional Court, 2018)

Figure 5. Interactivity elements for detecting attitudes
Source: authors

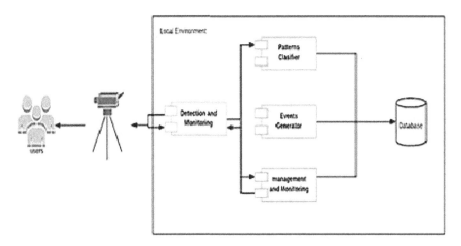

Methodology

The research is based on a qualitative approach, where practice allows sorting and ordering information expressions by voluntary or involuntary individuals, such as anxiety, fear, sadness, anger and neutrality, which will be analyzed in order to determine possible behaviors that precede a criminal act. The study was explorative was just kind, which aims to investigate the impact of modalities or levels of one or more variables in the study population (Fernandez & Baptista, 2014).

Population, Sample and Observation

Given the characteristics of the study, this research focuses on people of different ages, genders, contextures and statures; therefore, the target population corresponds to people who have been captured on video, available for free on the platform (MOTChallenge)and some students of the University Unipanamericana Foundation. It works with a random sample for GC four (4) students Unipanamericana Foundation and the GE will be taken from free videos provided by(MOTChallenge)with an amount of 160 traveling through public areas, three processes of observation time intervals and pointing to variables gender, body type and approximate age. There is achieved a special population identified for the study as shown in Figure 6 the characterization of this population observed:

In Table 1, evaluation levels were allocated for the reliability of the proposed system (prototype), which helped determine the allocation of appropriate information to ensure rigorous evaluation of the system are appreciated. However, for reasons of improvement, each thread with less than 100 score will be reviewed on a weekly basis margin established to verify if the score rises as the capture and processing more data.

Figure 6. Characterization of the population of the study
Source: authors

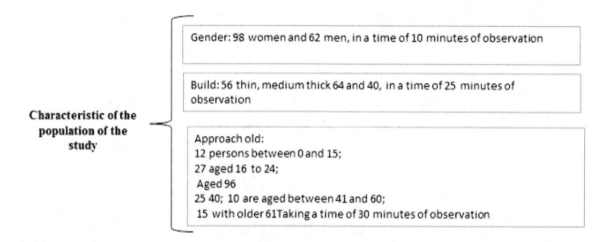

Table 1. Levels of reliability of the prototype

Score	Actuate
0-50	Not valid
51-84	Unstable
85-100	Stable

Source: authors

Materials

Hardware Components

- **Computer Laptop Model**: Toshiba Satellite S55; Processor: Intel Core i7 5500U 2.4GHz (3.0 GHz) with Turbo; RAM: 16GB; Hard Drive: 1TB SSD; AMD Radeon R7 M260 2GB DDR3;: Video Card Connectivity: Gigabit Ethernet card (RJ-45), Wi-Fi 802.11b / g / n / Intel® Dual Band Wireless-AC 3160 AC 1x1 (433Mbps) and Bluetooth 4.0.
- **Video Camera**: Model: Anran AR-PTD22-WIFI; Type: Domo; Connectivity: IP / Network - Wireless; Color mode: Color (Day), B & W (night); Features: Outdoor / Weatherproof, Vandal Proof, Pan / Tilt (Electronic), Infrared / Night Vision.
- **Fast Ethernet Switch Model**: Netgear GS108-400NAS; Ports: 8 Gigabit Ethernet.

Software Components

Use of tools for developing application logic (back-end), the presentation layer (front-end) and for the persistence layer is made:

- **Python 3.6.4**: Anaconda distribution installed from 4.6.14, which comes in open source version.

- Operating System: Linux Mint 19.1 Tessa.
- 4.1.0 OpenCV (Open Source Computer Vision) is a library of computer vision and machine learning open source.
- **Angular 7.2.15**: Is a JavaScript framework, free and open source, created by Google and designed to facilitate the creation of web applications such SPA (Single Page Application).
- **MongoDB 4.0.8**: A system NoSQL database-oriented documents, is free to use under the public license the server side (SSPL) v1.
- **PyCharm 2019.1.2 Community Edition**: an integrated development environment (IDE) used in programming, specifically for the Python language. Community version is released under the Apache License.

Proposed System (Prototype)

It consists of four independent functional modules. In Figure 7, the modules of the proposed system, where the first is charged with MACA is observed; the second is the MDS subjects of interest; the third receives the results of the above and based on a historical record of a few minutes or seconds, analyzes and compares the MPOPD; and the fourth allows event generation, which evaluates the results of the previous module and generates corresponding alerts (MGEA) if present.

FOCUS MAIN CHAPTER[1]

Enlistment Process

The process begins with the collection of photos of persons with suspicious behavior from various angles and a distance of no more than 20 meters, as well as videos with minimal resolutions of 1280 x

Figure 7. Diagram of the system components generally proposed
Source: authors

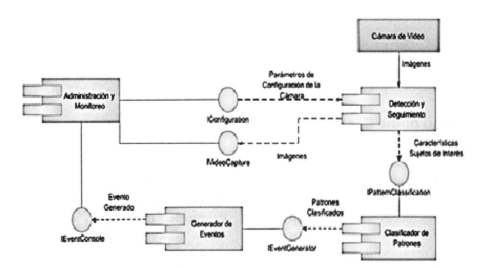

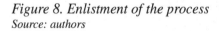

Figure 8. Enlistment of the process
Source: authors

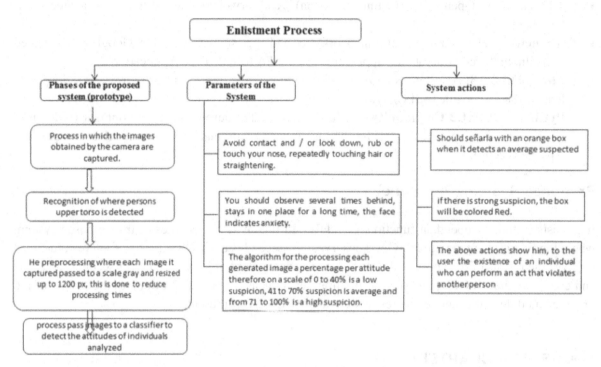

720 pixels. In the case of these videos is required that the image is fixed, which means that the camera should not move and additionally having the capture of several people simultaneously. This training must include individuals with: the face uncovered, with cap, hatless, wearing glasses, without glasses, or elements covering the face so that the system is able to identify attitudes regardless of the implements that hinder the full view of facial appearance. The process of gathering information is obtained from classified as: anxiety, fear, anger,

In Figure 8, the four phases shown implemented in the system, the relevant parameters in the construction of the algorithm and the actions that must be followed to achieve pattern recognition detecting gestures

To this end initiates the development of the software with the description of the requirements according to the analysis of its purpose, which is to have the ability to detect suspicious behavior by recognizing gestures and patterns in a given population.

Implementation of the System

As a first step the requirements and technical definition are specified, as shown in Figure 9, according to the general functionalities to apply four different packages, monitoring (Operator Monitoring and Management and Monitoring) detection and tracking are generated, pattern classifier and event generator; Each package includes use cases to help visualize the complete interaction of the software.

Figure 9. Case utility system
Source: authors

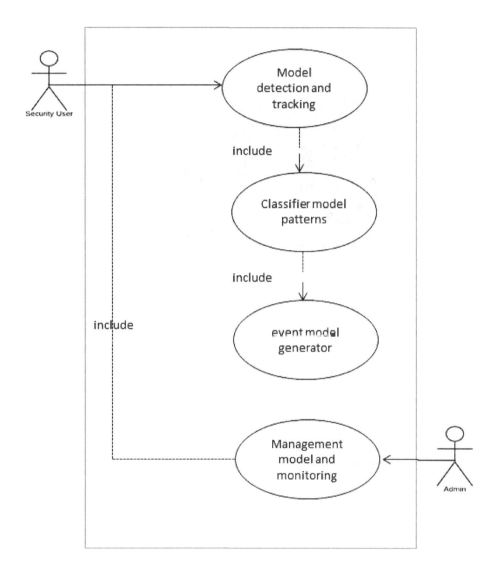

Simulation and Analysis

It was carried out through experimentation with GC and GE, with the following characteristics: The GC was formed by researchers of the study, who pretended suspicious behavior such as anxiety, which was determined at the time of the experiment with in order to measure variables defined with learning algorithms. GE does not have a default selection of people, since this validation was performed on the results obtained with the GC. In Figure 10, the process that took place during the simulation process with GC, in which they asked participants to undertake suspicious gestures, so that the algorithm can demarcate orange or red, as the case may be observed and well, gestures that do not generate any kind of warning from the prototype.

Figure 10. Process simulation with research participants
Source: authors.

For understanding the study took into account the connection with a video camera, which can be integrated into a device or native is necessary for the execution of the software. This video capture gestures and emotions of people and send information at runtime to the computer; this, run software designed performing the analysis of algorithms, information collection and sent the video comparison with already saved and generated. Thus, it returns a result of the analysis evaluated person giving a color according to the level of suspicion reached by the software. In contrast, software and information are already stored feed back.

The collected data are stored so that these can be reused for future analysis system interactions. In Figure 11, the process of ordering a learning algorithm is generally observed, in which various images are entered that must be interpreted and ordered to define what emotion and gesture each one has

Figure 11. Process ordering of the learning algorithm
Source: authors

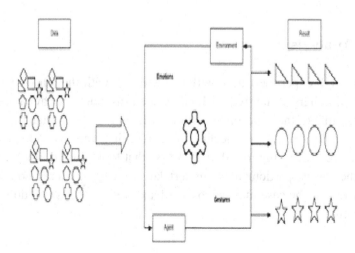

Communication Between Modules

As illustrated in the DGC in Figure 11, the communication between each of the modules back-end, is done by the consumption of the interfaces arranged for each. And for communication between the presentation layer and drivers will make use of API's Rest when required to process information from the persistence layer. A Rest API approaches the HTTP protocol to provide the CRUD operations (Create, Read, Update and Delete) of an information entity.

Implementation Details

This process is carried out under the scheme (3) proposed iterations. Then the implementation of each of them is detailed.

ITERATION 1

Focal length: capture and pre-processing segmentation image from a video camera parameters work, 8 mm; height: between 2 and 3 meters; angle: 33 °; range: 20 to 30 m; Resolution: 720 P (1280x720); format: H.264 or MPEG-4 frames per second: 30 fps.

- **Detection and Monitoring Module:** It must be borne in mind that there is an inverse relationship between the focal length and the angle of view of the same, ie, the smaller the focal length of the camera, an image will have a wider viewing angle. Also, the larger the focal length, a lower viewing angle is obtained, but this means that a greater depth gain. Similarly, there must be adequate lighting for capturing images. This is very important when extracting and analyzing the desired characteristics include good lighting conditions, ie light should come from the light source, reflected in the focused objects and be collected by the lens device. In Figure 12, the step is seen with the help of the openCV and function "Cv2.VideoCaptra.

- **Pre-Processing:** Then, in order to improve the process of image analysis is applied on each image obtained filter gray scale, for this the "cv2.cvtColor" OpenCV function will be used in this process, you will get a copy of the original image in gray scale, which will be used in the following steps; similarly, if the image has a width greater than 400 pixels, it is reduced to this size, otherwise it is left in the original size. This is done to reduce the detection time and improve accuracy. Next, they are shown in Figure 13 step by step.

- **Segmentation:** This process is divided into two main aspects: First, the descriptor characteristics HoG applied, along with a linear SVM pre-trained next with openCV for detection of persons within the image obtained in step pre processing. During this stage the positions of each of the people detected in the image are obtained. In Figure 14, the shape is illustrated as the operation was performed

Finally, having obtained the vector positions of persons in the above process, it proceeds to extract the images corresponding to each person of the input image to subsequent one by one with LBP, along with a CAAC based in Haar features pre-trained openCV providing, as shown in Figure 15.

Figure 12. Capture process openCV
Source: authors

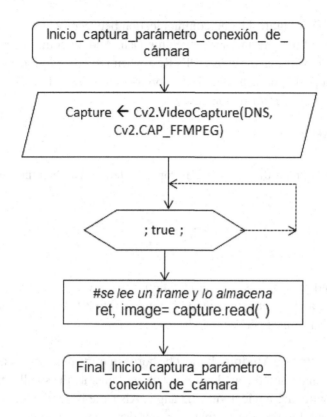

Figure 13. Filter gray scale OpenelCV
Source: authors

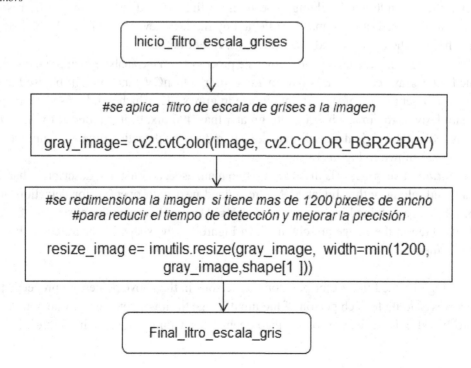

Figure14. Model HoG + linear pre-trained SVM - openCV
Source: authors

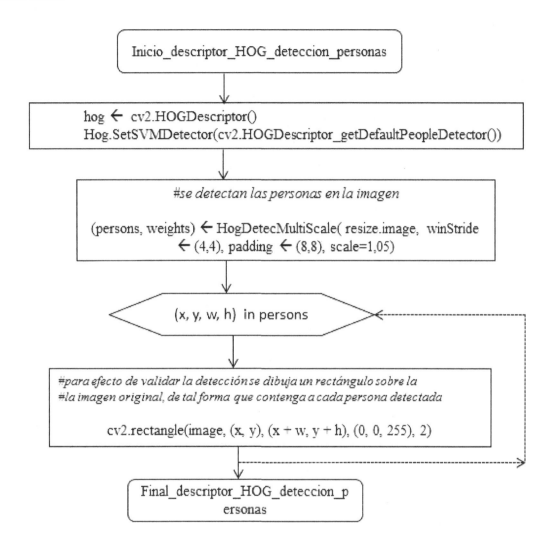

- **Tracing:** This process records the position coordinates for each of the people detected in the previous stage. For that, each detected person is identified and recorded in the history component positions. In order not to saturate the database with duplicate information or with very little variation, the registration process takes place every 2 seconds.

ITERATION 2

- **MCP:** This module is responsible for advancing the feature extraction processes of each of the upper body images obtained during segmentation step for subsequent classification as to the categories defined suspected. To meet the above objective, the class will be used: "cv2.createFisherFaceRecognizer ()" of OpenCV, specialized in recognizing facial expressions and will be used

Figure 15. Sorter upper trunk
Source: authors

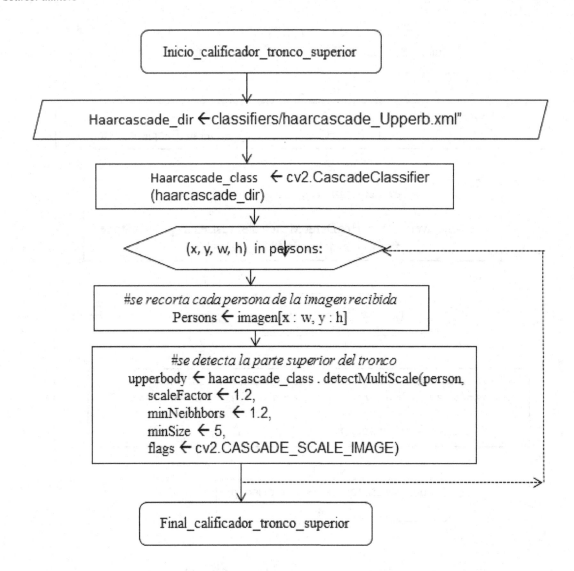

for training assisted learning algorithm and its subsequent use for the detection and classification suspicious attitudes. These processes will be implemented taking into account the project developed by (Gent, 2016), Making the necessary adjustments and adjusted to the needs of the study.

- **PED:** First, we proceed to prepare the data set that meets the requirements of the proposed system; hooded people, people with helmet, hat people, people who show anxiety and people who demonstrate a neutral expression for which, images that can be classified into the following categories were obtained. For each category we were able to collect 100 images of the area of the head and part of men of different people and at different angles capture. Subsequently, each photograph was applied to a manual process of preparation and standardization, which consisted of performing a scaling to a dimension 350x350 pixels and gray scale transformation by applying the corresponding filter. This is done because the classifier function in a better way if the images used for the training stage and classification are of the same size. After applying the standardization process, all photographs were arranged in a directory corresponding to the category it represents; in this case, they were appointed as follows: hood, helmet, hat, anxiety and neutral.

- **EC:** Having prepared the data required to train the classifier to learn to identify the categories defined suspicion. For this, the images in each category are divided into two groups; the first containing the images to be used in the training process and with which the learning algorithm and the second will contain the images to be used for the classification process, with which the performance of the classifier is estimated. The proportion of images in each group is given as follows: 80% for training and 20% for the classification, which are taken randomly. The training process is executed in 20 opportunities, for which success rates were obtained between 56.26% and 87.5%, with an overall average success of 66.56%. This shows that the classifier is working acceptably. While these results are encouraging, it is still considered to be possible to improve the degree of classification accuracy. This increase was achieved in 100 (double) the number of images in each category, applying the same standards process done previously.

After doing this, we proceeded to run the training process in 20 opportunities, for which success rates were obtained between 65.87% and 93.86%, with an overall average of 82.96% accuracy. This resulted in an improvement in the results of the 14.51%. This result is quite acceptable, even taking into account the complexity of recognizing suspicious attitudes through still images. It is important to note that the correct recognition of an expression or suspicious attitude depends largely on the context in which it is observed.

- **CP:** This component is responsible for receiving the images of the upper body of people and classifies providing a degree of suspicion by using pre-trained classifier generated in the previous stage.

ITERATION 3

- **GeV:** This module receives data sent by the pattern classifier and compares them with the scales set for alert generation. If the weight of the classification received, to be coupled with the weight of the classification of the movement of the person is within the following values, the corresponding event is generated: 0 to 40: not generate event; between 41 and 70: Generates orange alert event; Greater than 71: Generates event red alert
- **A.M:** Through this module, they are carried out all tasks of the system configuration (parameters, users, cameras, motion patterns and descriptors). In addition, images from the camera and alerts that occur are displayed.
- **CM:** This component is in charge of receiving the images from the camera and in the case of receiving an event from the generator print events a warning message on the image being displayed at that time and an audible alert is generated to call you care operator monitoring.

SOLUTIONS AND RECOMMENDATIONS

- **MDS:** At this stage several tests aimed at measuring and evaluating the execution times of the steps of pre-processing and segmentation are performed; for which a set of 606 images was used belonging database DrivFace created by(Diaz, Hernandez, & Lopez, 2016)Where each photo is in JPEG format and a size of 640 × 480.
- **Pre-processing:** To test this component, a set of 100 images was used DrivFace database and application times were measured filter grayscale and resizing the image (500 pixels); whereby, time pre-processing on each image and the average and total time over the entire data set was obtained. Table 2 shows the performance of this step.

Analyzing the data shows that the process of applying the filter scale only impacts gray 12% to the pre-processing and the resizing process image is the most expensive weighing 88%. In general the pre-processing is quite efficient because average execution time is 0.0020 seconds and the data set of 100 images took 0.2012 seconds.

- **Segmentation:** During the segmentation step tests aimed at evaluating the effectiveness of the pre-trained classifier suspicious behavior detection they were performed.
- **Training the Classifier:** During the training stage two cycles were performed, with groups 100 and 200 images per category. In the first cycle worked with 100 images per category, enn Figure 16, one success rates observed between 56.26% and 87.5%, with an overall average success of 66.56%, indicating an effectiveness rate in the detection of a moderately acceptable form. While in the second cycle increased the images to 200 by category.

In Figure 17, it was possible that the data obtained in this new cycle (training) were percentages of much better effectiveness, where they have success rates between 65.87% and 93.86%, with an overall average of 82.87% accuracy. This result is quite acceptable, even taking into account the complexity of recognizing suspicious attitudes through still images.

- **People Recognition:** To test the operation of the detection of persons; First, tests were performed with four different still images; which they recorded 6, 10, 21 and 26 persons respectively, obtained through a process of direct observation on the same image. In Table 3, this is observed most people present in a photograph, the percentage of decrease detection and processing time increases.

Table 2. Summary tests preprocessing stage

Process	Minimum	Maximum	Average	Total	Weight
grayscale	0.0001	0.0015	0.0002	0.0235	12%
Resize	0.0044	0.0053	0.0018	.1777	88%
Cumulative Process			0.0020	.2012	100%

Source: authors

Figure 16. Effectiveness training 1
Source: authors

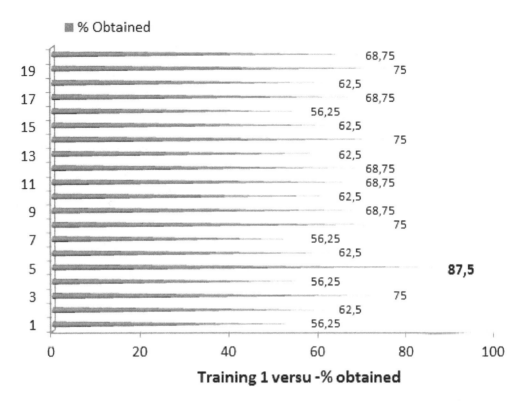

Table 3. Results of people still image recognition

Cacrtacterísticas	Photography No.1	Photograph No.2	Photography No.3	Photo No.4
People	6	10	twenty-one	26
detected	6	9	19	2. 3
failures	0	1	two	3
Detection%	100%	90%	90.48%	88.46
Detection time	0.14 sec	0.19 sec	0.27 sec	0.31 sec

Source: authors

In Figure 18, a sample of the output of the process of identifying persons occurs. Where analyzing the data of Table 3, it is evident that the detection rate is affected mainly by the following factors: occlusion (each other close together people), the distance of the person with respect to the first plane of the camera and resolution image at the time of analysis. It should be borne in mind that if the image is too large is applied to a process of downsizing, doing that during this process details are lost in the picture

At some point, a test was attempted without applying the process of re-sizing; however, the results were not satisfactory, mainly due to high processing times of working with high-resolution images. It must be rescued, that in this scenario improved effectiveness in detecting, but response times increased considerably.

Figure 17. Effectiveness training 2
Source: authors

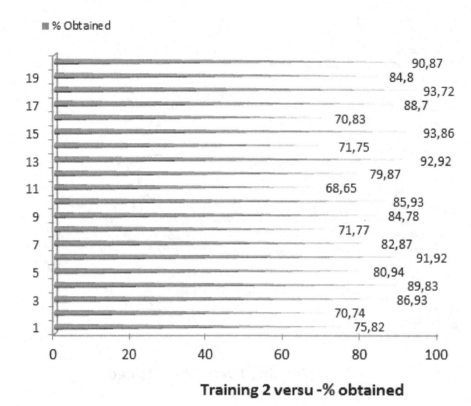

Figure 18. Detection of persons
Source: Design constructed by the authors

Comprehensive Tests: For these tests, a set of 3 videos (MOT17-02, MOT17-04 and MOT17-08) was used obtained from (MOTChallenge); which they were loaded one by one and analyzed by the system, using each of its modules and components. The aims these tests are: Ensure integration of all components and modules System; Measure the performance of the detection process of people; Measure effectiveness in detecting potential people with some degree of suspicion; and measure overall system performance. In Table 4, the results are presented in obntenidos these tests.

Where it can be concluded that: The greater the shooting distance, the lower the effectiveness of the process detection and classification, as seen in the video MOT17-04, where an average of people was calculated by frame 45.3 and applying to the detection process of the system we gave an average detection 20.07, it is a percentage of effectiveness of 44%. In the video MOT17-02, it has a proportion of 31 people per frame calculated in the video, versus an average of 16.61 detected by the system, which delivers a 54% effectiveness and finally the video MOT17-08, it has a ratio of 33.8 people on average per frame and the system detected an average of 23.54, to an effectiveness rate of 70%.

In Figure 19, a higher concentration of people in the scene, processing times are increased and detection; Thus a concentration of 31 people per frame in the video MOT17-04, an average processing time of 0.96 sec, for video MOT17-08 where having a concentration of 33.8, was obtained processing time was obtained average and 1.05 sec for video MOT17-06, where it has a concentration of 45.3 was obtained an average processing time of 1.47 sec. These results cause that level monitoring console delay is taken into the image display from 1 to 1.5 seconds, giving the effect of slow motion.

In Figure 20, a sample is presented with the results obtained with the source of the video MOT17-04 where regarding the detection of suspicious behavior; it was found that the classification system in some frames generated false positives, as highlighted people who, in the opinion of the authors, should not be classified in any category of suspicion. Also, in others recorded degrees of suspicion for some people in

Table 4. Comprehensive testing results

Characteristics	MOT17-02	MOT17-04	MOT17-08
FPS	30	30	30
Resolution	1920x1080	1920x1080	1920x1080
Duration	00:20	00:35	00:21
Number of frames	600	1050	625
Shooting distance	HALF	HIGH	NEAR
Average number of persons x frame	31.0	45.3	33.8
Average people detected	16.61	20.07	23.54
Average quiet people	2.49	16.04	4.55
Average people walking	14.12	14.03	18.99
Average people running	0	0	0
potential suspects	0	two	3
Average detected suspicious	0	0.92	2.52
Average processing time	0.96 sec	1.47 sec	1.05 sec

Source: authors

Figure 19. Detection of persons in video, using as source video MOT17-02
Source: https://motchallenge.net/. Design constructed by the authors

Figure 20. Detection of suspicious behavior with the source of the video MOT17-04
Source: https://motchallenge.net/. Design constructed by the authors

the scene that the authors had overlooked in the visual identification of possible suspects, which, to the focus on the identified subject, it is observed that records certain characteristics common to the images of the suspect category in which he qualified.

In Figure 21, the captured image has been manipulated by the authors in order to identify a letter between A and D to each of persons classified as potentially suspicious by the system. In the carefully observe both the image and the video to the detected individuals, it was observed that the subjects A and B had been listed as suspects potentially by the authors due to stay long in one place and it looked like they were waiting the opportunity to act in some way.

It is indicating that the system classified in appropriate ways, even taking into account the degree of assigned suspicion is low (41.47% and 41.01%) compared with the percentage of effectiveness delivered by the classifier in the training stage, which corresponded a 82.87%.

With respect to the subject C, it was observed that higher weights delivered by the system corresponded to categories cap, where 35.83% weighing and motion category "walking" with a weight of 10% was given. Authors, the video review, observe and understand the reason for the classification by the system, especially in relation to the category "cap" which was the most important.

If you look closely, the subject is wearing a hat and considering that in this category there are 27 pictures of people wearing hat, it understands why he gave more weight to the classification in this category. The remaining 10% is allocated by category of movement "walking", which delivers 10% in terms of the degree of suspicion.

Figure 21. Mark and percentage of detection of suspicious attitudes in MOT17-04 video
Source: https://motchallenge.net/. Design constructed by the authors

With respect to the subject D, a similar situation is observed, the higher weights delivered by the system correspond to the categories of movement "static", weighing 30% and assigned category "cap", which gave 13.29%. Authors, the video review, observe and understand the reason for the classification by the system, especially in relation to the category of movement, which was the most important. If you look closely, the subject is all the time in the same place and wear a hat, hence the reason for the weights assigned. While this situation at some point could be seen as a possible flaw in the standings, analyzing depth and reviewing the weights delivered by the system it follows that it made its classification process according to the characteristics recorded in the classifier.

Analysis of Facial Gestures Based on Images

This analysis is based on six criteria:

- **For Training 1**: The results were: receiving information 100; loss of image quality data 100; Lost data communication device 100; execution of the algorithm 64.46; detecting the process exceptions unevaluated 35.54 and analysis results 66.56, for a total of 466.56 with an overall rating of 77, 76%, indicating that it is unstable operation of the overall software for image analysis in processing the algorithm.
- **For Training 2**: The results were: receiving information 100; loss of image quality data 100; Lost data communication device 100; execution of the algorithm 100; detecting the process exceptions unevaluated 100 and analysis results 81.45, for a total of 581.45 with an overall rating of the 96.90%, indicating that stable operation of the overall software for image analysis processing the algorithm.

It is therefore possible to observe the sum of the criteria so that all the information is saved for traceability and future development have a result of 385 points (from receiving information 100, data storage 100, execution of automated processes 85; review of corrupted data 100) therefore, no adjustment is needed to have the storage process, with a percentage of reliability of the 96.25% in the stability of their processes.

List of Acronyms

- **AM**: Administration and Monitoring.
- **API's Rest**: Representational State Transfer, or Representational State Transfer. Is an architectural style API which conforms to certain rules or restrictions.
- **CAAC**: machine learning classifier cascade.
- **CM**: Monitoring Console.
- **CP**: Pattern Classification.
- **CRUD**: Create, read, update and delete.
- **DGC**: General Component Diagram.
- **EC**: Training and classification.
- **GeV**: Event Generator.
- **HTTP**: Hypertext Transfer Protocol is a transfer protocol where a system of different services and clients used in web pages.

- **LBP**: Application Descriptor Features.
- **MCP**: Modulo pattern classifier proposed (prototype).
- **PED**: Preparation and Data Standardization.
  SVM: supervised learning classifier.

Future Directions in Research

It is convenient in future work to try to extend the range of the scope of information capture and resolution of the video camera used because it is necessary to have a component of clear capture in order to more provide accuracy when assessing the gestures of passersby.

A level of interaction, for work with experimental and control groups would be important to work directly with a larger number of images of passersby for more combinations and thus achieve make a more detailed in recognition of gestures analysis. Unfortunately in this study it was limited to few combinations with the GC and it was not possible to know in detail its curve behavior patterns.

Finally, it is important to achieve solutions to get good access data through REST AdvanReader because in this way you will not need .csv files directly but could work with the data that interest almost instantaneously. This would contribute to better time in electronic data processing and testing of trial and error

As illustrated in the DGC in Figure 6, the communication between each of the modules back-end, is done by the consumption of the interfaces arranged for each. And for communication between the presentation layer and drivers will make use of API's Rest when required to process information from the persistence layer. A Rest API approaches the HTTP protocol to provide the CRUD operations (Create, Read, Update and Delete) of an information entity.

CONCLUSION

The proposed system (prototype), even with its limitations in terms of processing capabilities and does not have a database large enough to make the training process; which is estimated should be about 10,000 images per category to achieve an optimal level of classification, generates alerts impartially simply using the characteristics with which he was trained, where it is the responsibility of the operator monitoring, do the analysis appropriate and take the decisions it deems appropriate. Still, the authors consider it viable and functioning as a first phase of implementation in compliance with the objectives of the study

Within the overall level of performance of the prototype, even with the possible misclassification that were detected in the tests and that can arise in use, it is important to review the operation of the input video, communication with the computer that is running level software and lost data (video interruptions).

It is important to note that the video input videographic or materials to the algorithm does not affect how it behaves, and from this evaluation can generate frames for feeding and imaging system from them. It should be remembered that these are stored dynamically and allow a better response from the software through time, thus giving an impression of results according to what the user expects.

It is relevant to note that in the training process shows that the number of images allows more precise to the expected results, but the same can help create a larger gap if it has no control over the images entered to the algorithm, then, it is necessary to set specific parameters to know what criteria the images entering and processing time for each defined as well as the quality and hardware capable of running the process of these workouts. It should be noted that growth can not be defined as exponential and could not give you a formula determined growth as more thorough studies up to one thousand images per category would be needed.

For high response times, it is required that the processing of video images get a lot of resources for interpretation, therefore, the use of an algorithm that can take a long time in the process of analyzing the information would not be the ideal for the development of this type of study.

REFERENCES

Alcaldía Mayor de Bogotá. (2017). *Logros del plan de gobierno de Peñalosa en 2017*. Obtenido de http://www.bogota.gov.co/logros-del-plan-de-gobierno-de-penalosa-en-2017/mejora-la-seguridad-en-bogota.html

Anta, J. A. (2012). Análisis verbo-corporal (AVC): Su utilidad en los secuestros. *Revista de Criminologia e Ciencias Penitenciárias*, 9-14.

Barbona Ivana, B. C. (2016). Método de clasificación supervisada support vector machine: Una aplicación a la clasificación automática de textos. *Revista de Epistemología y Ciencias Humanas*, 37-42.

Caballero Barriga, E. R. (2017). *Aplicación Práctica de la Visión Artificial para el Reconocimiento de Rostros en una Imagen*. Utilizando Redes Neuronales y Algoritmos de Reconocimiento de Objetos de la Biblioteca OPENCV. Obtenido de http://hdl.handle.net/11349/6104

Cámara de Comercio de Bogotá. (2018). *Encuesta de Percepción y Victimización de seguridad en Bogotá*. Bogotá: Author.

Cho, S.-H., & Kang, H.-B. (2014). Abnormal behavior detection using hybrid agents in crowded scenes. Pattern Recognition Letters, 44, 64-70. doi:10.1016/j.patrec.2013.11.017

Constitución Política de Colombia. (1991). Legis.

Corte Constitucional, Sentencia, T - 114 (3 de Abril de 2018).

Dalal, N., & Triggs, B. (2005). *Histograms of oriented gradients for human detection*. Obtenido de https://lear.inrialpes.fr/people/triggs/pubs/Dalal-cvpr05.pdf

Diaz, K., Hernández, A., & López, A. (2016). A reduced feature set for driver head pose estimation. Applied Soft Computing, 45, 98 - 107. doi:10.1016/j.asoc.2016.04.027

Dollar, P., Wojek, C., Schiele, B., & Perona, P. (2012). Pedestrian Detection: An Evaluation of the State of the Art. *IEEE Transactions on Pattern Analysis and Machine Intelligence*, *34*(4), 743–761. doi:10.1109/TPAMI.2011.155 PMID:21808091

Felzenszwalb, P., Girshick, R., McAllester, D., & Ramanan, D. (2010). Object Detection with Discriminatively Trained Part-Based Models. *IEEE Transactions on Pattern Analysis and Machine Intelligence*, *32*(9), 1627–1645. doi:10.1109/TPAMI.2009.167 PMID:20634557

Fernández, C., & Baptista, P. (2014). *Metodología de la investigación*. McGraw Hill Education.

Garlan, D., & Shaw, M. (1994). An Introduction to Software Architecture. Pittsburgh, PA: School of Computer Science, Carnegie Mellon University.

Gent, P. (2016). *Emotion Recognition With Python, OpenCV and a Face Dataset. A tech blog about fun things with Python and embedded electronics.* Obtenido de http://www.paulvangent.com/2016/04/01/emotion-recognition-with-python-opencv-and-a-face-dataset/

Gill, V. (2010). *BBC News.* Obtenido de https://www.bbc.com/mundo/ciencia_tecnologia/2010/05/100528_vigilancia_defensa_pl

Goffmann, E. (1971). *Conducta en situaciones sociales.* Academic Press.

Gómez, C. M. (2016). *Análisis comparativo de los algoritmos fisherfaces y lbph para el reconocimiento facial en diferentes condiciones de iluminación y pose, tacna.* Guia de Protección de Datos Personales de la Superintendencia de Industria y Comercio. Obtenido de http://www.sic.gov.co/sites/default/files/files/Nuestra_Entidad/Guia_Vigilancia_sept16_2016.pdf

Intel Software. (2018). *Developer Reference for Intel® Integrated Performance Primitives 2019.* Obtenido de https://software.intel.com/en-us/ipp-dev-reference-histogram-of-oriented-gradients-hog-descriptor

Kalal, Z., Matas, J., & Mikolajczyk, K. (2009). Online learning of robust object detectors during unstable tracking. In *IEEE 12th International Conference on Computer Vision Workshops* (pp. 1417–1424). IEEE.

Leach, M., Sparks, E., & Robertson, N. (2014). Contextual anomaly detection in crowded surveillance scenes. *Pattern Recognition Letters, 44,* 71–79. doi:10.1016/j.patrec.2013.11.018

Leonardo, D., & Dominguez, A. J. (2016). Herramientas para la detección seguimiento de personas a partir de cámaras de seguridad. *Congreso Argentino de Ciencias de la Computación, 22,* 251–260.

Ley 1581 (Congreso de la República 17 de Octubre de 2012).

Lopez, H. (2011). *Detección y seguimiento de objetos con cámaras en movimiento.* Madrid: Universidad Autónoma de Madrid.

Lopez Garcia, J. C. (2009). *Algoritmos y Programación para Docentes.* Fundación Gabriel Piedrahita Uribe.

Mannini, A., & Sabatini, A. M. (2010). *Machine Learning Methods for Classifying Human Physical Activity from On-Body Accelerometers.* Obtenido de mdpi.com: https://www.mdpi.com/1424-8220/10/2/1154/htm

Br. Medina Olivera, V. J. (2018). *Desarrollo de un Sistema Automatizado Basado en Procesamiento Digital de Imágenes para mejorar el control de Videovigilancia en empresas de Trujillo.* Academic Press.

MOTChallenge. (n.d.). Obtenido de Multiple Object Tracking Benchmark: https://motchallenge.net/

Norza, E., Peñalosa, M. J., & Rodríguez, J. D. (2016). *Exégesis de los registros de criminalidad y actividad operativa de la Policía Nacional en Colombia.* Academic Press.

Oficina de Análisis de Información y Estudios Estratégicos (OAIEE). (2017). *Boletín mensual de indicadores de seguridad y convivencia.* Bogotá: Author.

Oficina de análisis de información y estudios estratégicos (OAIEE). (2019). *Boletín mensual de indicadores de seguridad y convivencia.* Bogotá: Author.

Oficina de Naciones Unidas contra la Droga y el Delito. (2010). *12° Congreso de las Naciones Unidas sobre Prevención del Delito y Justicia Penal.* Obtenido de https://www.un.org/es/events/crimecongress2010/pdf/factsheet_ebook_es.pdf

Personería de Bogotá, D. C. (2018). *Alarmante situación de inseguridad en Bogotá.* Bogotá: Author.

Piccardi, M. (2004). Background subtraction techniques: a review. *2004 IEEE International Conference on Systems, Man and Cybernetics.* 10.1109/ICSMC.2004.1400815

Potthast, J. (2011). *Sense and Security. A Comparative View on Access Control at Airports.* Science, Technology & Innovation Studies.

Rebel, G. (2009). *El lenguaje corporal. Lo que expresanlas actitudes, las posturas, los gestos y su interpretación.* Madrid: EDAF, S.L.

Rodriguez, M., Laptev, I., & Audibert, J. (2011). Density-aware person detection and tracking in crowds. In *IEEE International Conference on Computer Vision* (pp. 2423-2430). IEEE. 10.1109/ICCV.2011.6126526

Roger, S., & Pressman, P. (2010). *Ingeniería del software un enfoque práctico.* McGraw Hill.

Sabokrou, M., Fayyaz, M., Fathy, M., Moayed, Z., & Klette, R. (2018). Deep-anomaly: Fully convolutional neural network for fast anomaly detection in crowded scenes. *Computer Vision and Image Understanding, 172,* 88–97. doi:10.1016/j.cviu.2018.02.006

Somoza Castro, O. (2004). *La muerte violenta. Inspección ocular y cuerpo del delito.* Madrid: Grefol, S.L.

Soriano, R. R. (2002). *Investigación social teoría y praxis.* Editorial Plaza y Valdés, S.A. de C.V.

SuperDataScience. (2017). *SuperDataScience.* Obtenido de https://www.superdatascience.com/blogs/opencv-face-recognition

Superintendencia de Industria y Comercio. (n.d.). *Oficina Asesora Jurídica.* Obtenido de http://www.sic.gov.co/sites/default/files/files/Nuestra_Entidad/Guia_Vigilancia_sept16_2016.pdf

Universidad de California & Departamento de Policía de los Ángeles. (2010). *Predpol.* Obtenido de Predpol: https://www.predpol.com/technology/

Valera, M., & Velastin, S. (2005). Intelligent distributed surveillance systems: a review. *IEE Proceedings Vision, Image and Signal Processing.*

Viola, P., & Jones, M. (2001). *Rapid Object Detection using a Boosted Cascade of Simple.* Obtenido de https://www.cs.cmu.edu/~efros/courses/LBMV07/Papers/viola-cvpr-01.pdf

Wilmer Rivas Asanza, B. M. (2018). *Redes neuronales artificiales aplicadas al reconocimiento de patrones.* Editorial UTMACH.

Zhao, Chellappa, Rosenfeld, & Phillips. (2003). *Face Recognition: A Literature Survey.* Academic Press.

ADDITIONAL READING

Ansari, F. J. (2017). Hand Gesture Recognition using fusion of SIFT and HoG with SVM as a Classifier. *International Journal of Engineering Technology Science and Research*, *4*(9), 913–922.

Paredes, D. (2009). *Seguimiento y Caracterización de Componentes del Rostro para la Detección de Expresiones Faciales*. Tesis de Maestría, Centro Nacional de Investigación y Desarrollo Tecnológico.

Platero, C. (2009). Apuntes de visión artificial, Departamento de Electrónica, Automática e Informática Industrial, Editor Universidad Politécnica de Madrid.

Priego-Pérez, F. (2012). *Reconocimiento de Imágenes del Lenguaje de Señas Mexicano*. Tesis de Maestría en Ciencias de la Computación, Instituto Politécnico Nacional, Centro de Investigación en Computación.

Seijas, L. (2011). *Reconocimiento de patrones utilizando técnicas estadísticas y conexionistas aplicadas a la clasificación de dígitos manuscritos*. Tesis de Doctorado, Universidad de Buenos Aires.

ENDNOTE

[1]   First module of Administration and Configuration of the Application (MACA) of the proposed system (prototype) Second Detection and Monitoring Module (MDS) of the proposed system (prototype) Third module of Patterns Obtained with the Previously Defined (MPOPD) of the proposed system (prototype) Fourth module Event Generation and Alerts (MGEA) of the proposed system (prototype).

Chapter 8
Recent (Dis)similarity Measures Between Histograms for Recognizing Many Classes of Plant Leaves:
An Experimental Comparison

Mauricio Orozco-Alzate

Universidad Nacional de Colombia, Manizales, Colombia

ABSTRACT

The accurate identification of plant species is crucial in botanical taxonomy as well as in related fields such as ecology and biodiversity monitoring. In spite of the recent developments in DNA-based analyses for phylogeny and systematics, visual leaf recognition is still commonly applied for species identification in botany. Histograms, along with the well-known nearest neighbor rule, are often a simple but effective option for the representation and classification of leaf images. Such an option relies on the choice of a proper dissimilarity measure to compare histograms. Two state-of-the-art measures—called weighted distribution matching (WDM) and Poisson-binomial radius (PBR)—are compared here in terms of classification performance, computational cost, and non-metric/non-Euclidean behavior. They are also compared against other classical dissimilarity measures between histograms. Even though PBR gives the best performance at the highest cost, it is not significantly better than other classical measures. Non-Euclidean/non-metric nature seems to play an important role.

INTRODUCTION

Biologists and ecologists have turned their attention to engineering, particularly to image processing and pattern recognition techniques, with the aim of developing visual recognition systems that give them support for the time-consuming and often ambiguous activities demanded by the species identification processes of individual organisms. Such identification processes are crucial in several tasks including

DOI: 10.4018/978-1-7998-1839-7.ch008

—among others— refinement of taxonomic classification (Seeland et al., 2019), biodiversity estimation (Peng et al., 2018), detection of invaders (Pyšek et al., 2013), monitoring of endangered species (Omer et al., 2016) and assessment of ecosystem health (Li et al., 2014). Exemplar cases of the above-mentioned visual recognition systems are those developed by interdisciplinary endeavors of botanists and engineers for the automated classification of plant leaves.

According to Agarwal et al. (2006), the following reasons motivate the development of visual recognition systems for the leaf-based identification of plant species: (i) providing field botanists with a tool for rapid *in situ* comparison of their collected specimens against prototype ones from reference collections stored in herbaria; (ii) accelerating the process for identifying potentially novel species because the destruction of habitats is also occurring at dramatic speeds and (iii) easily accessing to descriptive information and meta-data of the species. Many different techniques have been used for leaf recognition systems; see for instance the ones reviewed by Cope et al. (2012) and Wäldchen et al. (2018) as well as other studies not found there such as Larese et al., 2014, Fan et al., 2015, Grinblat et al., 2016, Wäldchen & Mäder, 2018, Nguyen Thanh et al., 2018 and Lorieul et al., 2019.

Leaf recognition can be roughly decomposed into the four stages of a conventional pattern recognition system, namely *acquisition, preprocessing, representation* and *classification*. The first stage typically corresponds to a scanner or a camera that provides high quality photographs; the second one is aimed at adequating the images —via operations as filtering, segmentation and binarization— such that the subsequent stage becomes easier; the third one consists in extracting a set of features (characteristic measurements stored as a vector) from the preprocessed images that are known to be discriminative; examples of these features include geometric descriptors such as shape, contour, textures, color and histograms. The last stage is fed with a collection of examples such that a classification algorithm, either based on probabilities or (dis)similarities, assigns a class label (the name of the species) to the image of the examined leaf.

In the last stage —classification— almost all the available classifiers have been used for leaf recognition, from simple ones such as the nearest neighbor rule, decision trees and Bayesian classifiers (Rahmani et al., 2016) to complex ones such as support vector machines and, more recently, convolutional neural networks (Barré et al., 2017; Nguyen Thanh et al., 2018; Lorieul et al., 2019). However, in spite of the current availability of many sophisticated methods, the nearest neighbor classifier is still a competitive tool in terms of accuracy and it is very intuitive, allowing thereby an easy understanding of the motivation of the class label assignment (Duin et al., 2014). Moreover, the performance of the nearest neighbor classifier can be significantly boosted by choosing a proper dissimilarity measure according to the nature of the representation that is derived from the raw input images.

Histograms, including spectra, have proven to be useful for visually representing objects in general as well as for characterizing leaf images in particular (Mallah et al., 2013). Their popularity is explained not just by the fact of their effectiveness to capture discriminative information from the images but also because a number of specific dissimilarity measures have been proposed to compare histograms. The collection of those measures ranges from classical ones such as the Jeffrey's divergence (*JD*), the chi-squared (χ^2) histogram and the histogram intersection (*HI)* dissimilarity to modern ones such as the Weighted Distribution Matching (*WDM*) similarity (Correa-Morris et al., 2015) and the Poisson-Binomial Radius (*PBR*) distance measure (Swaminathan et al., 2017). The latter ones, in particular, are claimed to be more accurate or convenient than the classical ones. However, to the best of the author knowledge, *WDM* and *PBR* have not been compared to each other yet, neither in general terms nor in particular for the task of leaf recognition.

This chapter is therefore aimed at presenting an experimental comparison of the *WDM* and the *PBR* measures between histograms for recognizing many classes of plant leaves. Other classical dissimilarity measures between histograms are also included in the comparative study, particularly those that were also considered in both (Correa-Morris et al., 2015) and (Swaminathan et al., 2017), namely: the L_1 distance, *JD*, the χ^2 distance and *HI*. The Euclidean distance was also included in the study because it is the baseline for any comparison among dissimilarity measures. For the sake of research reproducibility (Kerautret et al., 2019), a public data set, originally delivered by Mallah et al. (2013), is used for the experiments. This data set contains examples of one-hundred plant species with 16 examples per class. Moreover, the comparison is made not just in terms of classification performance but also considering computational costs and non-conventional characterization indices (Duin & Pękalska, 2009) including the asymmetry coefficient, the non-metricity fraction and the negative eigenfraction. General recommendations are given to the practitioner interested in simple, effective and explainable (Holzinger, 2018) plant leaf recognition systems.

BACKGROUND

Histograms

Histograms can be regarded as a fundamental representation for both signal/image analysis and recognition. They model probability distributions by dividing the variable domain into bins and, afterwards, counting occurrences of data points (observations) that fall in each bin. A final step consists in dividing all counts by the number of observations such that a normalized model for the probability density is obtained (Bishop, 2006, p. 120). In images, particularly, histograms describe the distribution of the intensities (Burger & Burge, 2009), that is, the number of pixels per either each intensity value or each bin defined for the intensity range. Even though spatial information is lost when computing histograms directly from raw images, they are widely employed as a convenient representation for subsequent processing. Alternatively, histograms can also be computed not from the images themselves but from a secondary representation, typically a transform of the raw data or a feature vector extracted from them. Mallah et al. (2013), for instance, built a data set to recognize many classes of plant leaves by computing histograms from both texture and leaf margin descriptors. That data set is used in this chapter, but only with the histograms computed from textures. Further details about the data set are given below.

The Nearest Neighbor Classifier

Many different pattern classification rules have been proposed along the years. Some of them are simple, well-established an intuitive while others are complex, exotic and difficult to grasp. As pointed out by Hand (2006), what is understood by either simple or complex depends on the individual criterion. One of those criteria —the one adopted here— is that a simple classification rule is one with none or a few parameters to be tuned and a clear interpretation of its working principle.

It is somehow a consensus in the pattern recognition community that the nearest neighbor classifier (1-NN) is the *de facto* simplest classification rule under the above-mentioned criterion. It assigns an unseen (unlabeled) object with the class label of its closest one (i.e. its nearest neighbor) from a set of labeled examples. The closeness is estimated with a dissimilarity measure *d*, typically but not necessarily a distance: e.g. the Euclidean one. The following is a mathematical formulation of the 1-NN rule to classify an unlabeled object x, given a set of N labeled examples $\mathcal{T} = \left\{ \left(x_1, \theta_1 \right), \ldots, \left(x_N, \theta_N \right) \right\}$ where θ_i are the class labels:

$$\hat{\theta} = \theta_j, \text{ where } j = arg \min_{i=1,\ldots,N} d\left(x, x_i \right) \tag{1}$$

Most dissimilarity measures are symmetrical and, therefore, it is not relevant which vector is the reference to compute the distance. In case of asymmetrical measures, however, it is important to take into account the order of the arguments when computing the dissimilarity measure.

When comparing the effectiveness of other stages of a pattern recognition system —in terms of their effects on the classification performance— the 1-NN rule is often chosen for the evaluation because it does not involve tunable parameters (except for the dissimilarity measure that is chosen) as well as because it provides easily interpretable results. Moreover, it exhibits a good classification performance provided that a sufficiently large and, therefore, representative training set is available (Theodoridis & Koutroumbas, 2009). Such a good behavior is guaranteed by the fact that the asymptotic error of the 1-NN rule is never larger than twice the Bayes rate (the error when the true underlying probability distributions are known); see (Hastie et al., 2009, p. 465) for further details about this theoretical property of 1-NN.

Yet another reason to prefer the 1-NN rule over other more sophisticated methods - such as support vector machines (SVMs) and artificial neural networks (ANNs)—is that the former does not suffer neither from excessive computational complexity nor from parameter optimization issues in small-sample cases as the two latter ones. In fact, even though SVMs are still considered as state-of-the-art classification methods, they are unfortunately not suited for many-class problems —as the one we are considering in our chapter, which includes one-hundred classes of plant leaves— due to their very time-consuming character as well as because sufficient examples per class are needed for the proper optimization of the parameters. The reason is that multi-class problems when using SVMs are cast into several two-class problems, which are typically solved in a one-against-rest approach (Wang & Xue, 2014, p. 26). Consequently, using SVMs in a one-hundred-class problem implies the training of one-hundred two-class SVMs. Moreover, a leave-one-out cross-validation procedure scales this by the cardinality of the dataset. In the particular case of the one-hundred plant leaves, since there are 1600 objects in the dataset, the complete experiment using SVMs in a leave-one-out fashion would imply the computationally prohibitive training and evaluation of 160000 two-class individual SVMs, each one requiring the optimization of the kernel parameters. Besides, a proper optimization of the kernel requires a sufficient amount of examples per class and, in the considered dataset, only 16 examples per class are available.

Dissimilarity Measures Between Histograms

Let $x=(x_1,x_2,\ldots,x_R)$ and $y=(y_1,y_2,\ldots,y_R)$ be R-dimensional vectors containing the counts of two different histograms having R bins each. The following are the descriptions corresponding to the two dissimilarity measures to be compared (*WDM* and *PBR*) along with other classical and most commonly used measures between histograms:

The Weighted Distribution Matching (*WDM*) Dissimilarity Measure

The Weighted Distribution Matching (*WDM*) measure was originally proposed by Correa-Morris et al. (2015) as a similarity measure. However, as suggested by Duin (2015), any similarity measure (let's denote it by *s*) can be converted into a dissimilarity one (denoted by *d*) by applying a monotonic and decreasing function. Examples of the simplest transformations of that nature are the following ones:

- **Negative Transformation**: *d=-s*
- **Complement Transformation**: *d=1-s*
- **Inverse Transformation**: *d=1/s*

The first transformation produces negative dissimilarities that, in spite of being counterintuitive if interpreted as distances, are not problematic for ranking-based classifiers as the 1-NN rule. The two latter must be carefully employed by taking into account the dynamic range of the similarities and the eventual indetermination produced if *s*=0, respectively. In those cases, a proper scaling parameter must be introduced.

Due to those reasons, the first transformation is the one used in this chapter to convert the original formulation of *WDM* into a dissimilarity measure, as follows:

$$d_{WDM} = -\sum_{i=1}^{R-1} \left(\omega(x_i, y_i)\delta(x_i, y_i) - v(x_i, y_i)(1 - \delta(x_i, y_i)) \right) \qquad (2)$$

where $\delta=1$ if $\text{sgn}(x_{i+1} - x_i) = \text{sgn}(y_{i+1} - y_i)$ and 0, otherwise. Therefore, ω is a reward for the coincidence of either an increasing or a decreasing behavior in neighboring bins of both *x* and *y*. Consequently, *v* is a penalty for the case of noncoincident behaviors. Several versions of d_{WDM} can be considered according to the functions chosen for the ω and the *v*. The following ones are proposed by Correa-Morris et al. (2015), particularly using $v=1 - \omega$.

The WDM$_{mm}$ distance. This variant of the *WDM* dissimilarity measure corresponds to the case of considering $\omega = \dfrac{\min(x_i, y_i)}{\max(x_i, y_i)}$. According to that, Eq. (2) can be rewritten as follows:

$$d_{WDM_{mm}} = -\sum_{i=1}^{R-1} \left(\frac{\min(x_i, y_i)}{\max(x_i, y_i)}\delta_i(x, y) - \left(1 - \frac{\min(x_i, y_i)}{\max(x_i, y_i)}\right)(1 - \delta_i(x, y)) \right) \qquad (3)$$

The WDM$_\sigma$ distance. In this variant, $\omega = e^{\frac{-|x_i - y_i|}{\text{Å}}}$. In this case, Eq. (2) is rewritten as follows:

$$d_{WDM_{\text{Å}}} = -\sum_{i=1}^{R-1} \left(e^{\frac{-|x_i - y_i|}{\text{Å}}}\delta_i(x, y) - \left(1 - e^{\frac{-|x_i - y_i|}{\text{Å}}}\right)(1 - \delta_i(x, y)) \right) \qquad (4)$$

Notice that, in contrast with Eq. (3), the *WDM* variant in Eq. (4) includes a free parameter, σ, that must be tuned by the user and according to the application. Correa-Morris et al. (2015) applied the *WDM* measure to three exemplar recognition problems, namely: face verification, face identification and image retrieval.

The Poisson-Binomial Radius (*PBR*) Dissimilarity Measure

This measure was recently proposed by Swaminathan et al. (2017), using the Poisson-binomial distribution as starting point of their derivation. In contrast with d_{WDM}, this measure does not involve variants nor parameters to be tuned. The *PBR* dissimilarity measure between two histograms x and y is given by:

$$d_{PBR}(x, y) = \frac{\sum_{i=1}^{R} z_i (1 - z_i)}{R - \sum_{i=1}^{R} z_i} \tag{5}$$

where:

$$z_i = x_i \log\left(\frac{2x_i}{x_i + y_i}\right) + y_i \log\left(\frac{2y_i}{x_i + y_i}\right) \tag{6}$$

The proponents of *PBR* also tested their dissimilarity measure in a leaf recognition task, particularly using the Swedish leaf data set (Söderkvist, 2001) which contains examples for 15 tree species. The problem considered in this chapter, however, is much more challenging since the data set of Mallah et al. (2013) includes many more species with less examples per class. In addition to the leaf recognition problem, Swaminathan et al. (2017) also tested their dissimilarity measure for texture classification, material recognition, scene recognition and ear biometrics.

L_1 Distance

This distance is known under several names, among them *City-block distance*, *Manhattan distance* and *boxcar distance*. It is computed as follows:

$$d_{L_1}(x, y) = \|x - y\|_1 = \sum_{i=1}^{R} |x_i - y_i| \tag{7}$$

L_2 (Euclidean) Distance

It is assumed, unless specified otherwise, that the distance between two points in the space is computed by using the Euclidean distance. It can be computed in terms of the dot product of the difference vector, as follows:

$$d_{L_2}(x, y) = \|x - y\|_2 = \sqrt{(x-y)^\top (x-y)} \tag{8}$$

The Euclidean distance measures proximity in conventional vector spaces; i.e. in spaces where the well-known Pythagorean theorem holds. Besides, the Euclidean distance is often considered the baseline to contrast with any novel dissimilarity measure.

Jeffrey Divergence (JD)

The name *divergence* is traditionally applied to functions that estimate the dissimilarity between two probability distributions. As discussed above, a histogram is a count-based model of a probability distribution. Thereby, discretized versions of a divergence function can be applied to estimate the dissimilarity between histograms. Among the most common ones is the *Jeffrey divergence*, which is given by:

$$d_{JD}(x, y) = \sum_{i=1}^{R} \left(x_i \ln \left(\frac{2x_i}{x_i + y_i} \right) + y_i \ln \left(\frac{2y_i}{x_i + y_i} \right) \right) \tag{9}$$

In the experiments, we used the implementations of d_{JD} provided by Schauerte (2013).

χ² Distance

Several versions of this distance between histograms appear in the literature. Most of the variants only differ in the scalar outside the sum. In this chapter, we use the following version:

$$d_{\chi^2}(x, y) = \frac{1}{2} \sum_{i=1}^{R} \frac{(x_i - y_i)^2}{x_i + y_i} \tag{10}$$

Histogram Intersection (HI)

The common area below two partially overlapping functions is limited by the minimum values between them. Such a limit is, therefore, the intersection between the functions. A histogram is just a particular discrete function in which values are counts in bins. As a result, the intersection between two histograms is given by:

$$d_{HI}(x, y) = \sum_{i=1}^{R} \min(x_i, y_i) \tag{11}$$

Characterization Indices of Dissimilarity Matrices

Consider a set $\{x_1, \dots, x_N\}$ of N histograms and a dissimilarity measure d. In addition, let $D = [d(x_i, x_j)]$ be the $N \times N$ matrix of all pairwise dissimilarities between them. Duin & Pękalska (2009) proposed a number of indices to characterize D particularly emphasizing in properties related to the asymmetric, non-metric and non-Euclidean nature that some dissimilarity measures might exhibit. A brief definition of three of those indices is given here; namely *asymmetry coefficient*, *non-metric fraction* and *negative eigenfraction*. The reader is also referred to (Duin et al., 2013, 2014) for additional indices and further details.

Asymmetry Coefficient (Asym)

This coefficient serves to estimate the asymmetric behavior of D. Such an estimation is computed as:

$$Asym = \sum_{i \neq j} \frac{|d(x_i, x_j) - d(x_j, x_i)|}{|d(x_i, x_j) + d(x_j, x_i)|} \tag{12}$$

Non-Metric Fraction (NMF)

In a $N \times N$ dissimilarity matrix D the number of triangle inequalities to be checked is $N(N-1)(N-2)$ i.e. the number of triplets. Let τ be the number of triplets that violate the triangular inequality, the *NMF* index is given by:

$$NMF = \frac{\tau}{N(N-1)(N-2)} \tag{13}$$

The larger the *NMF*, the less metric the dissimilarity measure is.

Negative Eigenfraction (NEF)

If D is either symmetric or has been symmetrized, it can be eigendecomposed to find its eigenvalues and eigenvectors. If D exhibits a Euclidean behavior, all its eigenvalues will be positive. Conversely, if D is heavily non-Euclidean, most of its eigenvalues will be negative. Let λ_i be the eigenvalues of D the *NEF* is given by:

$$NEF = \frac{\sum_{i:\lambda_i < 0} |\lambda_i|}{\sum_i |\lambda_i|} \tag{14}$$

EXPERIMENTAL SETUP AND RESULTS

As stated in the introduction, the comparison of the dissimilarity measures between histograms is made in this chapter not just in terms of classification performance but also considering computational costs and non-conventional characterization indices. For the sake of notation simplicity, dissimilarity measures are referred hereafter by their names or acronyms but not as subscripts of d. All experiments were performed by using a publicly available data set of plant leaves that is described in the subsequent section.

Data Set

The so-called *one-hundred species leaves data set* was released by Mallah et al. (2013) and is freely available to be downloaded from the UCI Machine learning repository (https://bit.ly/2xxuiS7). It consists in a one-hundred class problem having 16 leaf images per class (plant species). The data set contains segmented and binarized versions of the images as well as their corresponding feature vectors for i) shape descriptors, ii) texture histograms and iii) margin histograms; each of them having a length of 64 entries. Sample images of one leaf per class are shown in Fig. 1.

The shape descriptor in this dataset consists in the so-called centroid contour distance curve, which is invariant to translations and rotations. The second descriptor —the texture histograms— are computed from the responses of five filters applied to small portions (windows) of the original image. Finally, the third descriptor —margin histogram— is estimated from several intermediate features computed for overlapped windows of the smoothed image, namely: magnitude, gradient, curvature.

Even though Mallah et al. (2013) proposed to use the three above-mentioned descriptors to build a recognition system, only texture histograms are used in this chapter since the aim is to make a comparison —for this particular application of leaf recognition— of several dissimilarity measures between histograms and not to combine different representations to boost the classification performance. No further details regarding the computation of the texture histograms are given here due to space constraints. The reader is referred to the cited paper for further information.

Figure 1. One sample image per class from the one-hundred species leaves data set

Experiments and Results

Comparison of Classification Performances

There are several methods for evaluating classification performance on a dataset with N objects. Bramer (2016) mentions the following ones: i) repeated train and test; ii) k-fold cross-validation and iii) leave-one-out (LOO) cross-validation. The two former methods involve a random partition of the design set into training and test parts and, therefore, the estimated performances may differ among experiments. The latter, in contrast, is not subject to randomness since all possible partitions —into a training set of N–1 instances and a singleton (the remaining object) for testing the performance— are considered. Even though LOO is costly since N classifiers must be trained and evaluated, it is often preferred to guarantee that other researchers repeating the experiments get the very same results. That is why LOO was chosen as the method to evaluate the performance. Remember that the standard error for the LOO estimation is given by $\sqrt{p(1-p)/N}$ where p stands for the performance estimation, either accuracy or classification error.

A first aspect to take into account before comparing LOO classification performances is that one of the dissimilarity measures contains a tunable parameter, namely σ in WDM_σ. For the sake of a fair comparison for this measure, a range of values for σ was explored trying to minimize the LOO error of 1-NN. The behavior shown in Fig. 2 was observed for values of σ ranging from 0.001 to 0.1. It is important to say that no separate validation set was considered for these experiments. Notice that lower values of σ quickly deteriorate the classification performance (i.e. the error increases). Larger ones also deteriorate the performance but a slower rate. The minimum LOO 1-NN error occurs at $\sigma=0.0275$. In consequence, hereafter we use $\sigma=0.0275$ and denote $WDM_{0.0275}$ simply as WDM_σ.

LOO 1-NN results, in descending order of classification error and for all the dissimilarity measures under comparison, are shown in Table 1. For the sake of an easier comparison, LOO errors are also shown in Fig. 3 including their corresponding error bars (using the standard errors). Notice that *PBR*— the newest dissimilarity measure (Swaminathan et al., 2017)— is indeed the best one closely followed by two classical ones: *JD* and χ^2. In fact, there is no significant difference between the performances of these three dissimilarity measures because the standard error whiskers horizontally overlap.

Following the same previous analysis, we can clearly distinguish four groups of dissimilarity measures according to their classification performances. From left to right in Fig. 3, a first group includes L_2 and WDM_{mm} which account for the largest classification errors around 24%. This is somehow expected for L_2 because the Euclidean distance is not designed to compare histograms and, therefore, it does not take into account shape information; however, it is surprising that WDM_{mm} accompanies L_2 in this group because it was originally proposed as being convenient to compare histograms. Moving to the right, WDM_σ appears alone (second group) with a classification error of 21.451%. A third group corresponds to *HI* and L_1, whose classification errors are around 19.3%. Finally, the fourth group is composed by the best performing dissimilarity measures that were already discussed in the precedent paragraph.

Classification error, as well as other global performance measures, give an idea of the general classification performance. However, in multi-class problems, it is often important to inspect the performances per class. Two performance measures are particularly relevant, namely *sensitivity* and *specificity* when considering one class as the target (also referred as the *positive* class) and all the remaining ones as the *negative* class. Remember that *sensitivity* is the true positive rate; that is, the proportion of positive objects

Figure 2. Exploration of the classification error for tuning the parameter σ in WDM$_σ$

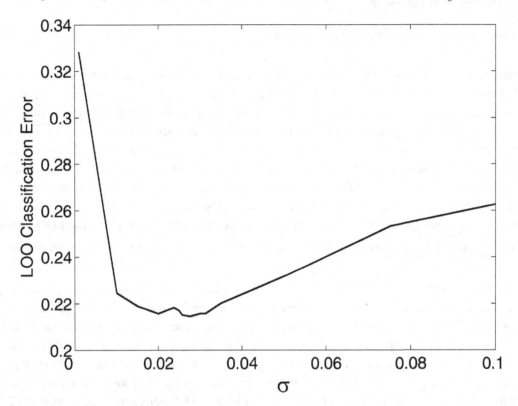

Table 1. Leave-one-out classification errors, along with their corresponding standard errors, for the dissimilarity measures under comparison

Measure	LOO Error	± std Error
L_2	24.203	1.071
WDM_{mm}	23.827	1.065
$WDM_σ$	21.451	1.027
HI	19.325	0.987
L_1	19.199	0.985
x^2	19.323	0.924
JD	16.010	0.917
PBR	15.947	0.916

that are correctly recognized as positive. Conversely, *specificity* is the true negative rate: the proportion of negative objects that are correctly recognized as negative. Following the suggestion by Wang et al. (2013) to intuitively illustrate the performance, sensitivities and specificities per class are compared by presenting them in a scatter plot; see Fig. 4, Fig. 5 and Fig. 6. These comparisons are only made for the two state-of-the-art dissimilarity measures, i.e. to compare *PBR* vs. *WDM$_σ$* (the best performing version of *WDM*) as well as to contrast both of them against L_2.

Figure 3. Comparison of LOO 1-NN errors, with their corresponding error bars, for the dissimilarity measures under comparison

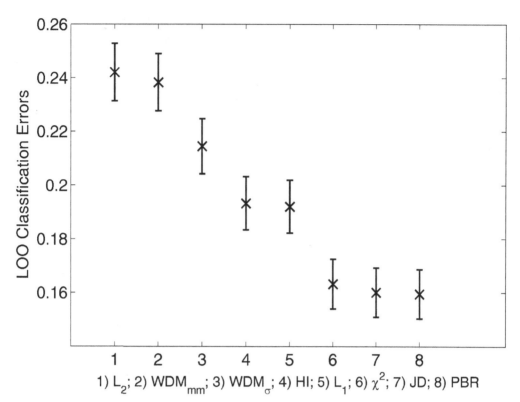

The inspection of Fig. 4 reveals that, as expected, *PBR* is better than WDM_σ in most classes, including several ones whose results appear on top of each other because the performance pairs are exactly the same. However, it is interesting to note that WDM_σ has a better sensitivity than *PBR* for a few classes and, similarly, it is a bit more specific than *PBR* only for a small subset of the one-hundred classes. From this figure, as well as from Fig. 5, it is noticeable that *PBR* accounts for perfect sensitivities and specificities in several classes while, in contrast, both WDM_σ and L_2 only exceptionally reach perfection in distinguishing positive and negative examples.

Regarding the comparison of both *PBR* and WDM_σ against the Euclidean distance, notice that sensitivities for L_2 cover the whole possible range while they are always greater than 0.25 for *PBR* (See Fig. 5a) or greater than 0.05 for WDM_σ (See Fig. 6a). In practice, this reveals that there were no cases in which *PBR* and WDM_σ completely failed in recognizing classes but some in which L_2 does. The comparison of Fig. 5b and Fig. 6b confirms that *PBR* is much more specific than L_2 and that the latter, in turn, is not significantly different than WDM_σ in terms of specificity.

Comparison of Computational Costs

The elapsed times for computing the dissimilarities between one-thousand pairs of histograms were measured under *MATLAB* R2013a, running on a 2.83 GHz Intel Core 2 Quad Q9550 processor with 2 GB in RAM. Afterwards, the resulting times were divided by 1000 in order to report more reliable

Figure 4. Scatter plots of PBR vs. WDM$_\sigma$ in terms of sensitivity and specificity

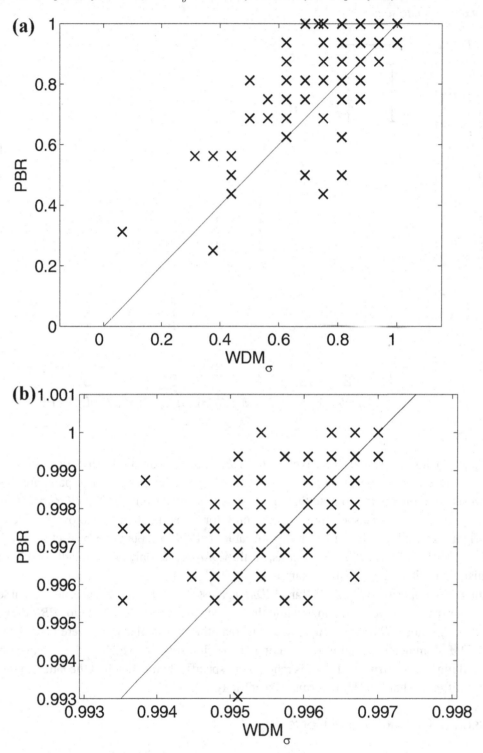

Figure 5. Scatter plots of PBR vs. L_2 in terms of sensitivity and specificity

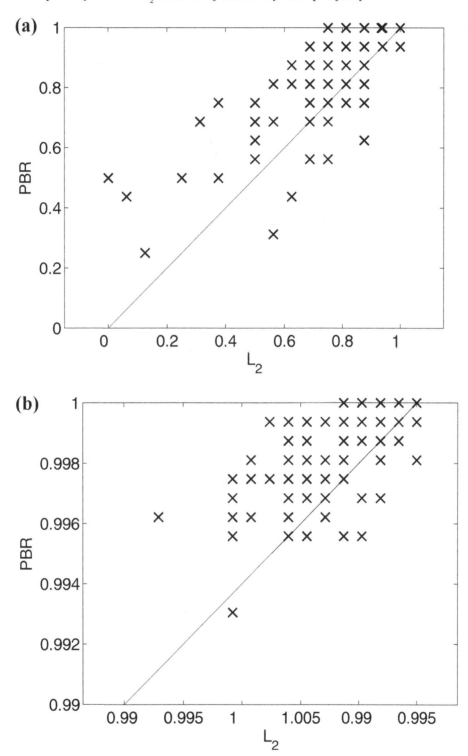

Figure 6. Scatter plots of WDM$_\sigma$ vs. L$_2$ in terms of sensitivity and specificity

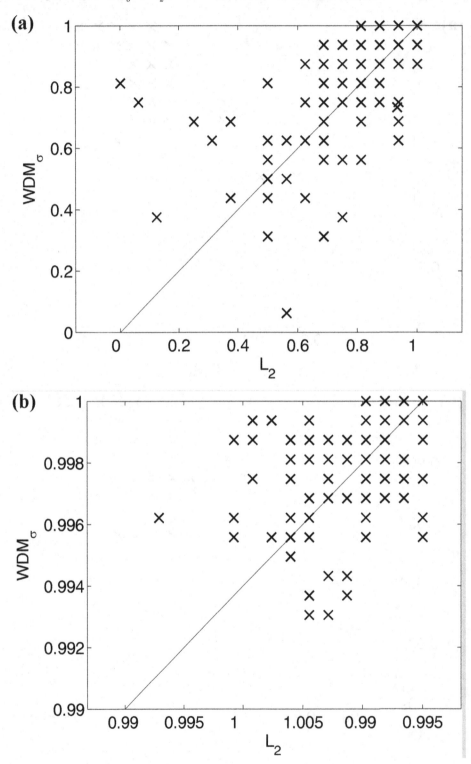

values for the elapsed times than the ones that would result from timing a single execution. This was done because, even though *MATLAB* was the only program launched in the computer, there could be other hidden processes running on the machine at the same time that might introduce spurious variations to the timing.

Runtimes, in microseconds, are reported in Table 2 and depicted in Fig. 7 in order to facilitate the interpretation. Notice that *PBR* is the most computationally demanding dissimilarity measure while, in contrast, L_2 and L_1 are the cheapest ones. In the middle, in descending order of computational cost, two groups of dissimilarity measures are found. The first one contains χ^2, *JD* and *HI*; the second one corresponds to the two variants of *WDM*: WDM_σ and WDM_{mm} respectively in descending order of computational cost.

Computational costs were also examined as the length of the histograms increases. Remember that the length of the histograms from the *one-hundred species leaves data* set is 64. Therefore, elapsed computation times were measured for histograms having lengths from 2 to 64. The resulting curves are shown in Fig. 8. Several trends can be distinguished, among them the following: the cost of *PBR* grows much faster than those of the other dissimilarity measures while, in contrast, the cost of χ^2 seems to be the most constant one. The cost of the other dissimilarity measures increases a bit faster than the one of χ^2 but their growths are, anyway, significantly slower compared with the one of *PBR*.

The groups of comparable measures that were previously commented when discussing Table 2 and Fig. 7 are also clear at the very end of the curves in Fig. 8. However, in this case, it is more evident that i) L_1, L_2, WDM_σ and WDM_{mm} belong to the same group, ii) *HI*, χ^2 and *JD* exhibit a similar computational behavior and, iii) *PBR* clearly behaves differently to the other dissimilarity measures. Finally, it is particularly interesting that *PBR* is cheaper than *HI*, χ^2 and *JD* in the first portion of the incremental curves but becomes more expensive than all of them at about half of the analyzed range of the histogram lengths.

Comparison of Characterization Indices

The three above-described characterization indices —namely *Asym*, *NMF* and *NEF*- were computed for the dissimilarity measures under comparison. For a proper analysis of the results, it is important to remember that metricity (fulfillment of the triangle inequality) is a stricter condition than symmetry and, in turn, more relaxed than Euclideaness. In other words, Euclideaness is the strictest condition for a dissimilarity measure while symmetry is the most relaxed one. Some authors also include the reflexivity

Table 2. Elapsed times, in microseconds, for computing the dissimilarity between two 64-length histograms

Measure	Elapsed Time
PBR	662.232
χ^2	641.987
JD	637.969
HI	634.831
WDM_σ	623.218
WDM_{mm}	619.848
L_1	615.735
L_2	615.678

Figure 7. Elapsed times, in microseconds, for computing the dissimilarity between two 64-length histograms

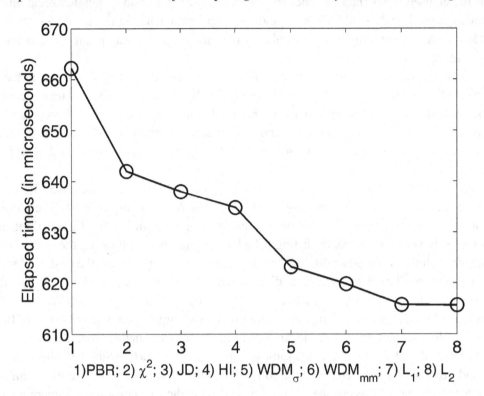

constraint (i.e. $d(\mathbf{x},\mathbf{x})=0$) as the very first condition to be fulfilled by a dissimilarity measure; nonetheless, the most common consideration is that reflexivity —along with non-negativity— is inherent to the concept of dissimilarity measure; see for instance (Pękalska & Duin, 2005, p. 168).

Values of the three characterization indices are shown in Table 3. All results are scaled by 100, i.e. reported as percentages, in order to present more digits in the less significant places. From the table, the following observations can be highlighted:

- None of the dissimilarity measures are asymmetric, as confirmed by the all zeros reported for the *Asym* coefficient.
- All the values for *NMF* are smaller than those for *NEF*, as expected because —as just said— Euclideaness is harder to satisfy.
- The two variants of *WDM* are heavily non-metric and non-Euclidean, specially WDM_σ whose *NEF* is almost 100. Such a strange behavior, however, is explained by taking into account that those dissimilarity measures were generated from their original similarity counterparts by using the negative transformation ($d=-s$). Any of the other transformations would also introduce a distortion to both metricity and Euclideaness but probably with less intensity than the negative transformation. Remember, however, that the other transformations must also be carefully treated in order to guarantee that the original similarities are restricted to the [0,1] range and that divisions by zero are avoided.

Figure 8. Elapsed computation times as the lengths of the histograms increase

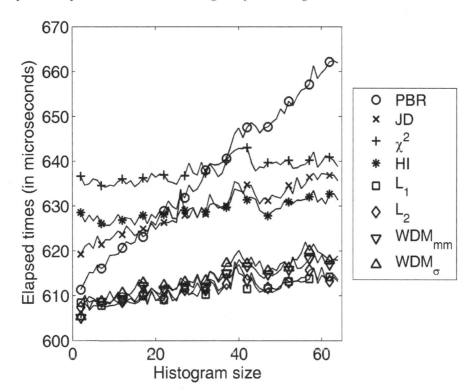

- *HI* and L_1 are non-Euclidean but metric. In contrast, χ^2, *JD* and *PBR* are not just non-Euclidean but also non-metric.
- The two best performing dissimilarity measures —*JD* and *PBR* - are the two most non-Euclidean and non-metric ones (without taking into account the *WDM* variants due to the reasons explained above). This fact confirms that, as pointed out by Pękalska et al. (2006), non-Euclideaness and non-metricity can be informative in some contexts. Indeed, other studies have shown that non-metricity is particularly useful in domains such as information retrieval (Skopal & Bustos, 2011) and visual recognition (Scheirer et al., 2014).

SOLUTIONS AND RECOMMENDATIONS

Results in this study suggest that practitioners dealing with the task of plant leaf recognition could still consider classical dissimilarity measures, such as *JD* and χ^2, to accurately compare pairs of histograms. The state-of-the-art *PBR* measure is slightly better and could also be considered alternatively if additional computational cost is not an important issue, particularly for short histograms.

This comparative study also serves to show that —in spite of the current boom of deep learning— the usage of simple but powerful strategies to compare pairs of histograms, such as the 1-NN rule together with an accurate dissimilarity measure, may be recommended to obtain competitive recognition performances when explainable results are also wanted (Holzinger, 2018).

Table 3. Values of the characterization indices (presented as percentages) for the square dissimilarity matrices of the compared measures

Measure	Asym	NMF	NEF
L_2	0.000	0.000	0.000
WDM_{mm}	0.000	36.138	83.037
WDM_σ	0.000	64.013	99.970
HI	0.000	0.000	20.585
L_1	0.000	0.000	20.585
χ^2	0.000	0.296	22.952
JD	0.000	0.854	23.564
PBR	0.000	0.658	23.881

FUTURE RESEARCH DIRECTIONS

As future work, it is planned to compare both *WDM* and *PBR* dissimilarity measures against the so-called bin ratio-based distance (*BDR*) proposed by Hu et al. (2014) as well as to study possibilities to combine the state-of-the art dissimilarity measures between histograms by applying, for instance, the strategies suggested by Ibba et al. (2010) and Duin & Pękalska (2012). In addition, research efforts on the parallelization of the dissimilarity measures —specially the most recent ones— are worthwhile to avoid the computational burden when comparing long histograms. A first attempt in that direction was already done for WDM_{mm}; see (Orozco-Alzate et al., 2017).

CONCLUSION

Even though *PBR* was the most accurate dissimilarity measure to compare histograms in the evaluated task of recognizing many classes of plant leaves, both *JD* and χ^2 achieved almost the same performance and at a more economical cost. It was also shown that the computational cost of *PBR* increases faster than the one of all the other dissimilarity measures considered in the study. Moreover, the fact that the best performing dissimilarity measures are those having the most non-metric and non-Euclidean behaviors confirms that both non-metricity and non-Euclideaness play an important role for capturing discriminative information from the histograms. *WDM*, in two of its variants, did not exhibit competitive classification results neither with respect to *PBR* —the other recently proposed dissimilarity measure— nor against the classical dissimilarity measures between histograms that were included in the study as a baseline for the comparisons. *PBR*, therefore, can be recommended when computational cost is not a decisive issue; otherwise, the classical *JD* and χ^2 are still the best options to recognize plant leaf images through the comparison of their associated histograms.

REFERENCES

Agarwal, G., Belhumeur, P., Feiner, S., Jacobs, D., Kress, J. W., Ramamoorthi, R., ... White, S. (2006). First steps toward an electronic field guide for plants. *Taxon*, *55*(3), 597–610. doi:10.2307/25065637

Barré, P., Stöver, B. C., Müller, K. F., & Steinhage, V. (2017). LeafNet: A computer vision system for automatic plant species identification. *Ecological Informatics*, *40*, 50–56. doi:10.1016/j.ecoinf.2017.05.005

Bishop, C. M. (2006). *Pattern recognition and machine learning*. New York, NY: Springer. Retrieved from https://www.microsoft.com/en-us/research/publication/pattern-recognition-machine-learning/

Bramer, M. (2016). Estimating the Predictive Accuracy of a Classifier. In *Principles of data mining* (3rd ed.; pp. 79–92). London, UK: Springer; doi:10.1007/978-1-4471-7307-6_7

Burger, W., & Burge, M. J. (2009). Histograms. In *Principles of digital image processing: Fundamental techniques* (pp. 37–54). London, UK: Springer; doi:10.1007/978-1-84800-191-6_3

Cope, J. S., Corney, D., Clark, J. Y., Remagnino, P., & Wilkin, P. (2012). Plant species identification using digital morphometrics: A review. *Expert Systems with Applications*, *39*(8), 7562–7573. doi:10.1016/j.eswa.2012.01.073

Correa-Morris, J., Martínez-Díaz, Y., Hernández, N., & Méndez-Vázquez, H. (2015). Novel histograms kernels with structural properties. *Pattern Recognition Letters*, *68*, 146–152. doi:10.1016/j.patrec.2015.09.005

Duin, R. P. W. (2015). *Distools examples: Classifiers in dissimilarity space*. Available at: http://37steps.com/distools/examples/disspace-classification/

Duin, R. P. W., Bicego, M., Orozco-Alzate, M., Kim, S.-W., & Loog, M. (2014, Aug). Metric learning in dissimilarity space for improved nearest neighbor performance. In P. Fränti, G. Brown, M. Loog, F. Escolano, & M. Pelillo (Eds.), *Structural, Syntactic and Statistical Pattern Recognition: Proceedings of the joint IAPR international workshop, S+SSPR 2014* (Vol. 8621, p. 183-192). Berlin: Springer. 10.1007/978-3-662-44415-3_19

Duin, R. P. W., & Pękalska, E. (2009, Nov). *Datasets and tools for dissimilarity analysis in pattern recognition* (Tech. Rep. No. 2009 9). SIMBAD (EU, FP7, FET). Retrieved from http://simbad-fp7.eu/techreports.php

Duin, R. P. W., & Pękalska, E. (2012). The dissimilarity space: Bridging structural and statistical pattern recognition. *Pattern Recognition Letters*, *33*(7), 826–832. doi:10.1016/j.patrec.2011.04.019

Duin, R. P. W., Pękalska, E., & Loog, M. (2013). Non-Euclidean Dissimilarities: Causes, Embedding and Informativeness. In M. Pelillo (Ed.), *Similarity-based pattern analysis and recognition* (pp. 13–44). Springer London. doi:10.1007/978-1-4471-5628-4_2

Fan, J., Zhou, N., Peng, J., & Gao, L. (2015, November). Hierarchical learning of tree classifiers for large-scale plant species identification. *IEEE Transactions on Image Processing*, *24*(11), 4172–4184. doi:10.1109/TIP.2015.2457337 PMID:26353356

Grinblat, G. L., Uzal, L. C., Larese, M. G., & Granitto, P. M. (2016). Deep learning for plant identification using vein morphological patterns. *Computers and Electronics in Agriculture*, *127*, 418–424. doi:10.1016/j.compag.2016.07.003

Hand, D. J. (2006). Classifier technology and the illusion of progress. *Statistical Science*, *21*(1), 1–14. doi:10.1214/088342306000000060 PMID:17906740

Hastie, T., Tibshirani, R., & Friedman, J. (2009). *The Elements of Statistical Learning: Data Mining, Inference, and Prediction* (Vol. 64). New York: Springer; doi:10.1007/978-0-387-84858-7

Holzinger, A. (2018, Aug). From machine learning to explainable AI. In *2018 World Symposium on Digital Intelligence for Systems and Machines (DISA)* (p. 55-66). 10.1109/DISA.2018.8490530

Hu, W., Xie, N., Hu, R., Ling, H., Chen, Q., Yan, S., & Maybank, S. (2014, December). Bin ratio-based histogram distances and their application to image classification. *IEEE Transactions on Pattern Analysis and Machine Intelligence*, *36*(12), 2338–2352. doi:10.1109/TPAMI.2014.2327975 PMID:26353143

Ibba, A., Duin, R. P. W., & Lee, W.-J. (2010). A study on combining sets of differently measured dissimilarities. In *20th International Conference on Pattern Recognition (ICPR 2010)* (p. 3360-3363). 10.1109/ICPR.2010.820

Kerautret, B., Colom, M., Lopresti, D., Monasse, P., & Talbot, H. (Eds.). (2019). *Reproducible Research in Pattern Recognition - Second International Workshop, RRPR 2018, Beijing, China, August 20, 2018, revised selected papers* (Vol. 11455). Springer. doi: 10.1007/978-3-030-23987-9

Larese, M. G., Namías, R., Craviotto, R. M., Arango, M. R., Gallo, C., & Granitto, P. M. (2014). Automatic classification of legumes using leaf vein image features. *Pattern Recognition*, *47*(1), 158–168. doi:10.1016/j.patcog.2013.06.012

Li, Z., Xu, D., & Guo, X. (2014). Remote sensing of ecosystem health: Opportunities, challenges, and future perspectives. *Sensors (Basel)*, *14*(11), 21117–21139. doi:10.3390141121117 PMID:25386759

Lorieul, T., Pearson, K. D., Ellwood, E. R., Goëau, H., Molino, J.-F., Sweeney, P. W., ... Joly, A. (2019). Toward a large-scale and deep phenological stage annotation of herbarium specimens: Case studies from temperate, tropical, and equatorial floras. *Applications in Plant Sciences*, *7*(3), e01233. doi:10.1002/aps3.1233 PMID:30937225

Mallah, C., Cope, J., & Orwell, J. (2013, Feb). Plant leaf classification using probabilistic integration of shape, texture and margin features. In L. Linsen, & M. Kampel (Eds.), *Proceedings of the 14th IASTED international conference on computer graphics and imaging (CGIM 2013) / track 798: Signal processing, pattern recognition and applications* (p. 1-8). Calgary, Canada: Acta Press. 10.2316/P.2013.798-098

Nguyen Thanh, T. K., Truong, Q. B., Truong, Q. D., & Huynh, H. X. (2018). Depth learning with convolutional neural network for leaves classifier based on shape of leaf vein. In N. T. Nguyen, D. H. Hoang, T.-P. Hong, H. Pham, & B. Trawiński (Eds.), *Intelligent information and database systems - 10th Asian Conference, ACIIDS 2018, Dong Hoi city, Vietnam, March 19-21, 2018, Proceedings, part I* (p. 565-575). Cham: Springer International Publishing. 10.1007/978-3-319-75417-8_53

Omer, G., Mutanga, O., Abdel-Rahman, E. M., & Adam, E. (2016). Empirical prediction of Leaf Area Index (LAI) of endangered tree species in intact and fragmented indigenous forests ecosystems using WorldView-2 data and two robust machine learning algorithms. *Remote Sensing, 8*(4), 1–26. doi:10.3390/rs8040324

Orozco-Alzate, M., Villegas-Jaramillo, E.-J., & Uribe-Hurtado, A.-L. (2017, Jun). A block-separable parallel implementation for the weighted distribution matching similarity measure. In S. Omatu, S. Rodr'ıguez, G. Villarrubia, P. Faria, P. Sitek, & J. Prieto (Eds.), *Distributed computing and artificial intelligence, 14th international conference, DCAI 2017, Porto, Portugal, 21-23 June, 2017* (Vol. 620, p. 239-246). Cham, Switzerland: Springer. doi: 10.1007/978-3-319-62410-5_29

Pękalska, E., & Duin, R. P. W. (2005). *The dissimilarity representation for pattern recognition: Foundations and applications* (Vol. 64). Singapore: World Scientific. doi:10.1142/5965

Pękalska, E., Harol, A., Duin, R. P. W., Spillmann, B., & Bunke, H. (2006). Non-Euclidean or nonmetric measures can be informative. In D.-Y. Yeung, J. Kwok, A. Fred, F. Roli, & D. de Ridder (Eds.), Structural, Syntactic, and Statistical Pattern Recognition Joint IAPR International Workshops, SSPR 2006 and SPR 2006 (Vol. 4109, p. 871-880). Springer. doi:10.1007/11815921_96

Peng, Y., Fan, M., Song, J., Cui, T., & Li, R. (2018). Assessment of plant species diversity based on hyperspectral indices at a fine scale. *Scientific Reports, 8*(4776), 1–11. doi:10.103841598-018-23136-5 PMID:29555982

Pyšek, P., Hulme, P. E., Meyerson, L. A., Smith, G. F., Boatwright, J. S., Crouch, N. R., … Wilson, J. R. U. (2013). Hitting the right target: taxonomic challenges for, and of, plant invasions. *AoB Plants, 5*(plt042), 1-25. doi:10.1093/aobpla/plt042

Rahmani, M. E., Amine, A., & Hamou, R. M. (2016). Supervised machine learning for plants identification based on images of their leaves. *International Journal of Agricultural and Environmental Information Systems, 7*(4), 17–31. doi:10.4018/IJAEIS.2016100102

Schauerte, B. (2013). *Histogram distances*. Available at: https:// www.mathworks.com/matlabcentral/fileexchange/39275-histogram-distances

Scheirer, W. J., Wilber, M. J., Eckmann, M., & Boult, T. E. (2014). Good recognition is non-metric. *Pattern Recognition, 47*(8), 2721–2731. doi:10.1016/j.patcog.2014.02.018

Seeland, M., Rzanny, M., Boho, D., Wäldchen, J., & Mäder, P. (2019, January). Image-based classification of plant genus and family for trained and untrained plant species. *BMC Bioinformatics, 20*(4), 1–13. doi:10.118612859-018-2474-x PMID:30606100

Skopal, T., & Bustos, B. (2011, Oct). On nonmetric similarity search problems in complex domains. *ACM Computing Surveys, 43*(4), 34:1-34:50. doi:10.1145/1978802.1978813

Söderkvist, O. J. O. (2001). *Computer vision classification of leaves from Swedish trees* (Master's thesis). Linköping University, Linköping, Sweden. Retrieved from http://www.cvl.isy.liu.se/en/research/datasets/swedish-leaf/

Swaminathan, M., Yadav, P. K., Piloto, O., Sjöblom, T., & Cheong, I. (2017). A new distance measure for non-identical data with application to image classification. *Pattern Recognition*, *63*, 384–396. doi:10.1016/j.patcog.2016.10.018

Theodoridis, S., & Koutroumbas, K. (2009). *Pattern recognition* (4th ed.). Burlington, MA: Academic Press. doi:10.1016/B978-1-59749-272-0.X0001-2

Wäldchen, J., & Mäder, P. (2018). Machine learning for image based species identification. *Methods in Ecology and Evolution*, *9*(11), 2216–2225. doi:10.1111/2041-210X.13075

Wäldchen, J., Rzanny, M., Seeland, M., & Mäder, P. (2018, April). Automated plant species identification— Trends and future directions. *PLoS Computational Biology*, *14*(4), 1–19. doi:10.1371/journal.pcbi.1005993 PMID:29621236

Wang, X., Mueen, A., Ding, H., Trajcevski, G., Scheuermann, P., & Keogh, E. (2013). Experimental comparison of representation methods and distance measures for time series data. *Data Mining and Knowledge Discovery*, *26*(2), 275–309. doi:10.100710618-012-0250-5

Wang, Z., & Xue, X. (2014). Multi-Class Support Vector Machine. In *Support Vector Machines Applications* (pp. 23–48). Cham, Switzerland: Springer. doi:10.1007/978-3-319-02300-7_2

ADDITIONAL READING

Duin, R. P. W., & Pekalska, E. (2015). *Pattern Recognition: Introduction and Terminology*. Delft, The Netherlands: 37 Steps. Retrieved from http://37steps.com/documents/printro/

Lei, B., Xu, G., Feng, M., Zou, Y., van der Heijden, F., de Ridder, D., & Tax, D. M. J. (2017). *Classification, Parameter Estimation and State Estimation: An Engineering Approach Using MATLAB* (2nd ed.). Chichester, UK: Wiley. doi:10.1002/9781119152484

KEY TERMS AND DEFINITIONS

Asymmetry: A dissimilarity measure is said to be asymmetric if $d(x,y) \mathbin{!} d(y,x)$.

Computational Cost: The amount of time required to complete certain operation. Even though computational cost has to do also with several computer resources (memory, power supply, etc.), it typically refers to computation time.

Dissimilarity Measure: A measure to judge nearness or closeness between either the objects themselves or their representations. It should, at least, fulfill the reflexivity condition.

Histogram: Count-based model of a probability distribution over a discrete number of bins.

Leave-One-Out Classification Error: Estimation of the performance by training the classifier N times, in such a way that each object is used one time for testing and the remaining $N - 1$ objects for training.

Non-Euclideaness: Deviation from the Euclidean behavior. In practice, such a deviation occurs when triplets of dissimilarities do not obey the relation established by the Pythagorean theorem. For a dissimilarity matrix, the degree of non-Euclideaness can be estimated by the so-called negative eigenfraction.

Non-Metricity: Violation of the triangle inequality by a triplet of dissimilarities. For a dissimilarity matrix, the degree of non-metricity can be estimated by the so-called non-metric fraction.

Plant Leaf Recognition: The task of identifying plant species based on discriminant properties of their leaves.

Chapter 9

Weed Estimation on Lettuce Crops Using Histograms of Oriented Gradients and Multispectral Images

Andres Esteban Puerto Lara

https://orcid.org/0000-0002-3818-5667

Fundacion Universitaria Panamericana, Colombia

Cesar Pedraza

Universidad Nacional de Colombia, Colombia

David A. Jamaica-Tenjo

Universidad Nacional de Colombia, Colombia

ABSTRACT

Each crop has their own weed problems. Therefore, to understand each problem, agronomists and weed scientists must be able to determine the weed abundance with the most precise method. There are several techniques to scouting, including visual counting for density or estimations for coverage of weeds. However, this technique depends by the evaluator subjectivity, performance, and training, causing errors and bias when estimating weeds abundance. This chapter introduces a methodology to process multispectral images, based on histograms of oriented gradients and support vector machines to detect weeds in lettuce crops. The method was validated by experts on weed science, and the statistical differences were calculated. There were no significant differences between expert analysis and the proposed method. Therefore, this method offers a way to analyze large areas of crops in less time and with greater precision.

DOI: 10.4018/978-1-7998-1839-7.ch009

INTRODUCTION

The interest in precision agriculture has increased since it facilitates the agronomical, economical and environmental management of a crop. Currently, the agriculture in Colombia has not appropriately developed (Serrato & Castillo, 2018), considering 70% of the country is not even technified yet (Junguito, Perfetti, & Becerra, 2014; Rodríguez, Plaza, & Gil, 2008). The competition between crops and weeds over nutrients – in addition to the spatial interference caused by weed – represents a huge problem within agriculture, as global estimates of crop losses due to weeds are approximately 34% (Oerke, 2006), and in certain crops, the losses could reach 100% (Zimdahl, 2007). Therefore, it is crucial to quantify, characterize and identify accurately the weeds abundance, floristic composition, biodiversity, functional components and the spatial distribution, in order to suggest improved management plans with the least possible impact, or having more accurate research data (Westwood et al., 2018). There are conventional assessment techniques (Rew & Cousens, 2001) to estimate the coverage (area covered by plants per unit area, expressed as percentage) and density (number of plants per unit area) of these plants, but it entails a subjectivity problem that depends on the evaluator's abilities and training.

Nowadays, there are multiple advances in many areas that allows to improve many agricultural processes (Kim, Kim, & Chung, 2019). Even including modern artificial intelligence techniques such as convolutional neural networks (dos Santos, Freitas, Silva, Pistori, & Folhes, 2017) despite of their extra hardware requirement (Umamaheswari, Arjun, & Meganathan, 2018). Some of these tools can be used in order to estimate weed in crops.

Despite some works apply machine learning techniques in weed detection, there are a lack of tools to estimate weeds in crops (Murawwat, Qureshi, Ahmad, & Shahid, 2018), for example vegetables like lettuce. In consequence, it would be appropriate to design and develop software tools based in artificial intelligence, to measure the amount of weed present in a crop with more precision (Sujaritha, Annadurai, Satheeshkumar, Sharan, & Mahesh, 2017).

The main objective of this chapter is related to introduce a method to estimate weed in vegetables crops. The method is based on the implementation of multi-spectral images, built upon techniques to extract features using histogram of oriented gradients (HOG) and conduct further processing using support vector machines in order to implement a learning model that enables the proper detection of weed in a specific crop.

This chapter is organized as follows: a background section with a brief compilation about conventional sampling, in addition to a brief summary emphasizing on some studies on precision agriculture that use digital image processing techniques. Also the problem addressed is explained. Then, the method proposed to detect weed using multi-spectral images in a lettuce crop is explained in main focus section taking into account that this method is more accurate that human visual assessment, an experiments section is included in order to test the methodology proposed. Finally, issues, results and conclusions of the method proposed are exposed.

BACKGROUND

Conventional Sampling

The conventional sampling of weeds in crops is a technique that allows to identify, quantify and characterize weed populations in a field. In the precision agriculture paradigm it is required to have a large amount of data, the conventional process is to sample determined number of quadrats randomly distributed in the field or defining rectangular or squared grids of different resolutions according to the objective proposed (Ambrosio, Iglesias, Marin, & Del Monte, 2004), in each of these quadrats it is usually estimated the density or the coverage of the weeds, among others.

These processes of estimation and removing processes is conducted manually and at the discretion of the capability of the person performing each process (Sakthi & Yuvarani, 2018). Besides, this technique requires a lot of time, being exposed to different climatic conditions and comprises error probabilities depending of the evaluator's experience. An experienced evaluator may not have significant errors when the test is performed, meanwhile an amateur evaluator may have significant errors in the coverage estimation causing incorrect results in the weed sampling study. Another kind of problem lies on underestimation or over estimation when a coverage test is performed. Depending of each evaluator and the amount of weed on the crop, the percentage of area could be exaggerated or attenuated causing high variance related with the real value of coverage. In Colombia, this type of sampling is used for research in different crops. For instance, performing experiments in potato, spinach and sugar cane crops using conventional sampling, Jamaica, (2013) demonstrates that weeds distributes in patches had different abundance assessments using different size of quadrats (0.5×0.5, 2×2, 4×4 and 8×8 meters).

Image Processing in Agriculture

There are a set of solutions for the remote detection of weed using image processing (Lehoczky, Riczu, Mazsu, Viktória Dellaszéga, & Tamás, 2018). However, López, (2011), in her revision of the methodologies and techniques to detect weed using image processing, highlights the lack of interdisciplinary and research groups in this regard in addition to the high costs generated by these studies. (Jadhav & Patil, 2014; Thorp & Tian, 2004) suggest using hyper spectral cameras, taking advantage of the fact that spectral signatures are easy to identify due to the large number of bands in the cameras. However, this equipment is very expensive and requires specific expertise for its operation.

Additionally, multi spectral sensors have also specific characteristics that differ from RGB sensors and are less expensive than hyper spectral sensors. Normally, these sensors can collect images in the near infrared band (790nm). Moreover, there is a chance to use the Normal differenced vegetation index (NDVI) (Whiting, Ustin, Zarco-Tejada, Palacios-Orueta, & Vanderbilt, 2006) (equation 1) as an identifier of photosynthetic activity and also as a background estimator in a context where the objects of interest are vegetables (Alchanatis, Ridel, Hetzroni, & Yaroslavsky, 2005; Feyaerts & Van Gool, 2001).

$$NDVI = \frac{NIR - RED}{NIR + RED} \tag{1}$$

Since multi-spectral cameras have a subtle band spectrum, their spectral signature per pixel does not provide enough information unlike the continuous spectrum of a hyper spectral camera. Some authors suggest changing this type of spectrum using image processing. For example, Thorp and Tian, (2004) points out the hyper spectral sensors are capable to create information related to biomass and the light reflective capacity due to their ability to obtain different values according to the large number of bands present at these sensors, while a multi-spectral sensor only has a few bands extra to the ones in the visible spectrum – specially the near infrared and red edge – thus, its spectral resolution is very limited.

Most of the techniques highlighted by some of the authors are based on common algorithms to process images in areas unrelated to precision agriculture. These techniques are mainly supported by the extraction of physical features in the objects involved once the background has been estimated using a specific index (NDVI)(Louargant et al., 2018).

Perez *et al.* (2000) suggests using color filters to segment the background first, based on the background estimation using the NDVI index. Then the features are extracted, such as capacity, main axis of the plant and centroid; whilst the identification uses algorithms such as the k-nearest neighbors algorithm (K-NN) or Bayesian methods. According to Piron (Piron, Leemans, Lebeau, & Destain, 2009), one way to complement these methodologies is using colored probability densities, besides extracting other types of features from each object such as area, perimeter and height. Additionally, Bossu *et al.* (2009) implements useful techniques in his work using Wavelet transforms to extract features and Gabor filters for textures. Finally, Camps-Valls and Bruzzone, (2005) research was used as reference. They assessed different methods to detect weed using machine-learning techniques (neural networks, adaboost, support vector machines) and assessed the performance of each system. Lottes *et al.* (2017), assessed a series of crops, first attempting to segment the seed rows through spatial relations and orientations to subsequently eliminate them from the scene, thus extracting the plants features to estimate the amount of weed in sugar beet crops.

It is imperative to compare the weed scouting between human and computer estimations. In order to establish if the computer estimations is at least the same as the conventional. To compare the results between a human weed coverage test and the possibility of perform a computer weed coverage test, a machine learning method was developed here in order to get coverage measures and establish a statistical comparison between human and computers results. the hypothesis tested were that humans overestimate weed coverage when they asses big quantities of weed, by other way they underestimate weed coverage when the quantities tend to be low. In addition, computers calculations brings more specific estimations due to capabilities of give a lot of significative figures instead of humans that usually give integer number in weed coverage estimation.

MAIN FOCUS OF THE CHAPTER

This work focuses on collecting features from the objects pertaining to the crop, which will be classified later as weed or lettuce. Fig 1 shows how each image was processed to subsequently detect weed within a specific lettuce crop. The phases of the proposed method are described below (Wang, Zhang, & Wei, 2019).

Figure 1. Block diagram for the main procedure

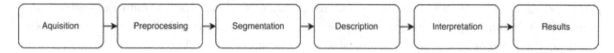

Preprocessing

The method proposed was designed for images acquired with a multispectral camera with four bands: red, green, near infrared and red edge with 1Mpx of resolution by spectral sensor. The images are shown in green, near infrared and red to obtain a false green. Preprocessing phase is based on projective and geometrical transformations that were made to correct the camera lens defects, such as fish-eye and misalignment between bands as Fig 2 shows.

Actually, every channel is provided by the sensor independently. The issue is that every channel is misaligned in relation to the other one (Fig 2a)., so an affine transformation was required to align channels between them. In that way the input image will not be blurred anymore. Finally a lens distortion correction was applied by software using a set of distortion coefficients and camera matrix, (Fig 2b).

Segmentation

Once the image has been corrected, a procedure is performed in order to separate the background from the objects of interest; in this case, the soil is the object that must be eliminated from the scene (Irias Tejeda & Castro Castro, 2019). This was achieved by using the NDVI index and adjusting it manually or adaptively through a threshold to create a mask. A value between 0 and the maximum value that

Figure 2. a) Original image b) Corrected image.

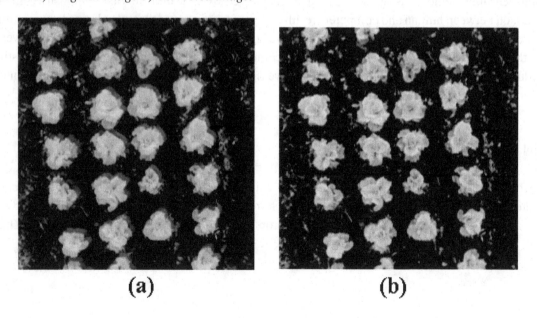

(a)　　　　　　　　　(b)

represents a threshold value is selected manually automatically, where each pixel greater than such value will be converted into a maximum value, while lower values will be transformed into zero as in the following expression:

$$g(x,y) = \begin{cases} 1 & f(x,y) \geq T \\ 0 & f(x,y) < T \end{cases} \qquad (2)$$

T value choice was based on the Otsu's, this method aims at identifying a threshold that maximizes the variance between the classes of an image that will be binarized thresholding (Gonzalez, 2009; Otsu, 1979; Xu, Xu, Jin, & Song, 2011).

In this phase, besides binarization, there are two types of morphological operations commonly used in image processing (Sujaritha et al., 2017; Szeliski, 2010). Normally the connection between weed and lettuce occurs in each crop grid, which might generate problems in the detection phase since two objects merge into a single one. Consequently, it was necessary to conduct erosion operations to disconnect these two links between weed and lettuce. Erosion is an operation that shrinks the objects in a binary image under the domain of a structuring element (Fig 4) (Gonzalez, 2009), whose function is to rule how the image gets weaker throughout its domain (equation 9). Through this operation weeds beside lettuce crops are disconnected trying to avoid false negatives during detection.

Figure 3. a) Original image b) NDVI c) Background segmentation using the Otsu thresholding

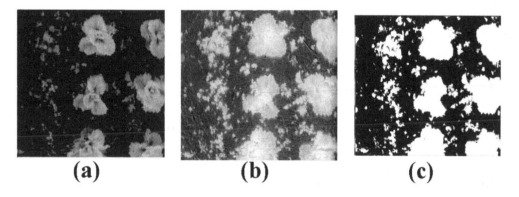

(a)   **(b)**   **(c)**

Figure 4. a) Original image b) Eroded image c)Dilated image

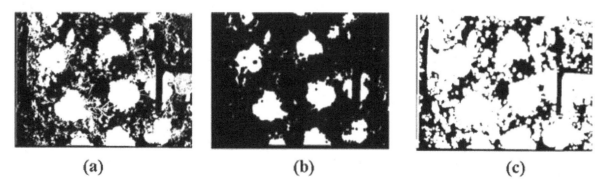

(a)   (b)   (c)

$$A \ominus B = \left\{ z \mid (B)_z \cup A^c \neq \varnothing \right\}$$ (3)

Contrary to erosion, the dilation operation Gonzalez, (2009) was also applied (Fig. 4 c), which enables to recover eroded spaces, expanding the objects using an ellipse as structuring element in order to recover certain zones that require a morphological expansion (equation 10). At this moment, some weakened lettuces during erosion stage should be recovered in certain way and this mask will be applied to original image eliminating background completely, finally this image will be ready for detection under SVM without undesirable objects, except for a little bit of salt and pepper noise that will be cleaned at the final stage (Fig 5).

$$A \oplus B = \left\{ z \mid (\hat{B})_z \cup A \neq \varnothing \right\}$$ (4)

Description and Interpretation

After segmentation, shape features of each object in the image are extracted once NDVI mask is applied. These features will determine whether a segmented object is weed or lettuce. For this phase, individual lettuce and weed images were collected from a set of multispectral images manually acquired by the research team the photographic base acquired. Then, the features were extracted using histograms of oriented gradients (HOG) (Xia, Zhu, Gan, & Shang, 2014). HOG are a representation of an image based on the distribution of the gradients' direction and intensity in a specific region. These histograms are useful, since they enable to characterize the shape of a specific object.

Generally, the representation obtained is a vector with certain amount of features that depend on the parameters set for their recollection. These parameters are classified according to the size of a cell that will divide the image (Fig 5) and the number of bins on each histogram; generally, there are 9 bins that will range between 0 and 180 grades.

The next step is to estimate the gradient of pixels on each cell and gather in the histogram the contributions given by the magnitude of each gradient (Xia et al., 2014). Subsequently, the histograms are normalized in blocks twice the size of the cell, sliding them across the image by intervals of the same size as the cell (Fig. 5). Finally, they are concatenated generating a final vector of features.

According to Tsolakidis, *et al.*, 2014, Xia *et al.*, 2014, and Zhao *et al.*, 2015, the gradient magnitude (equation 5) and its direction (equation 6) are defined, respectively.

$$g_p = g(x, y) = \sqrt{\Delta x^2 + \Delta y^2}$$ (5)

$$\theta_p = \theta(x, y) = \arctan \frac{\Delta x}{\Delta y}$$ (6)

where Δx and Δy are the derivatives in the corresponding axis.

The contribution of the gradients direction is given by the equation 7

$$\delta_p = \frac{\theta_p m}{180} - 0.5f \tag{7}$$

where m is the number of bins specified for the histogram. Finally, the magnitude contribution is given by the equations 9 and 10

$$w_p = \delta p - FLOOR(\delta p) \tag{8}$$

$$v1 = gp_{(}1 - wp_{)} \tag{9}$$

$$v_2 = = g_p w_p \tag{10}$$

Once the features have been collected individually for a set of training data, they are used as an input in order to train a SVM (Mountrakis, Im, & Ogole, 2011). Is important to remark that 3 different kernels were used to train this model (Linear, polynomial and radial basis function). Subsequently, sets of test and validation data were used to verify whether the model was able to make generalizations (Bakhshipour & Jafari, 2018). Once the test step was concluded, the SVM is applied to the image of a lettuce crop grid, where each contour of the image is classified to determine whether it is lettuce or weed. Once the objects belonging to the crop class are detected, an inverted binary image is generated showing the lettuce detected. This image is multiplied by the NDVI image; since the lettuce image is binary, the multiplication will eliminate the lettuces from the NDVI image, thus obtaining an image with only weeds. Finally, this result is thresholded to estimate the weed coverage in that grid and is transformed in a percentage of the total area covered in the image. The final illustration of the process is shown in Fig 6.

Figure 5. a) Image divided in 8x8 cells b) 16x16 block (blue) c) Visualization of the features vector d) Hog vector for weed

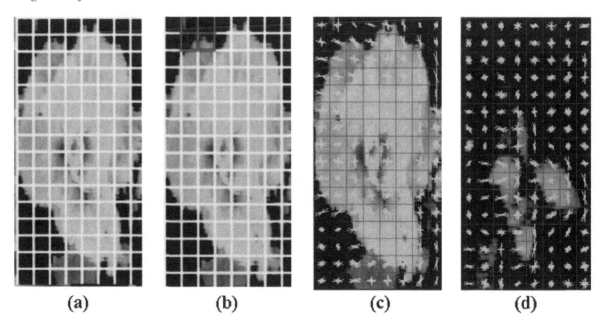

(a)   (b)   (c)   (d)

Figure 6. Detection process diagram

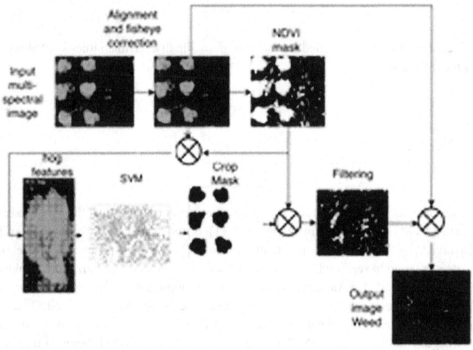

Once a weed detection on a specific area was established, a set of tests are performed using the criteria of a group of weed experts taking a set of images and each one will calculate their respective coverage. Then, each image will be processed by this algorithm giving its respective estimation by each image. Finally, by a non-parametric test that determinate significant statistical differences between experts estimations and computer's estimation. On that way, this test can detect overestimations and underestimations on weed coverage calculation given by the experts.

EXPERIMENTS

The method designed was validated through the detection of weeds in a lettuce crop of $1000m^2$ in the Marengo research center of the Universidad Nacional of Colombia. The Images were taken at two different heights (80cm and 2m) with a parrot sequoia multispectral camera. These heights were selected to get more detailed images, because in this way it will be easier to obtain an appropriate shape descriptor given that the multispectral sensors have a 1 Mpx resolution. 1127 photos were taken for the 80cm images, while 592 photos were taken for the 2m images. According to the camera resolution, the grid covered using the 80cm images was 94x71 cm, while the grid covered with the 2m images was 211x281 cm.

The photographs taken at 80cm height had 4 lettuces in average, while the images at 2m had 20 lettuces in average, for a total of 16348 lettuces aprox. In weed case, counting individual plants is a difficult task due to overlapping between them. Finally, for the SVM training process, only 703 random individual images of lettuce and 1000 of random sized weed patches were collected.

Figure 7. a) Lettuce crop example b) individual lettuce image c) individual weed patch image

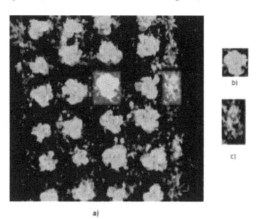

The SVM was trained using a laptop with 6 GB RAM and 2.4 GHz processor with Linux OS trough OpencV C++ library

Training

In the training phase, three sets of 1362 mixed images from 1703 were chosen to perform tests with three types of kernel (Linear, 2nd order polynomial and rbf). The openCV machine learning package provides a tool for parameter auto adjusting in order to keep optimum values (Table 1).

Individual Validation

341 images from each group were used for validation purposes in order to compare performance between each training model (Figures 8 and 9).

The following table (Table 2) shows the average value of each validation test.

Test Phase

Models were tested with 406 images of individual lettuces and weed (167 and 239 respectively), In this case. The dataset was distributed in the following way, 1362 images for training who corresponds a 65%, 341 for validation who corresponds a 16% and 406 for test who corresponds a 19%. Results of every model are included in Tables 3,4 and 5.

As a last test, same three Kernells were chosen in order to train models using 1703 images directly. Finally 406 images were used for test purposes(Table 6).

Despite the good results regarding individual classification, it is important to test weed detection using the image of an entire crop grid. Clearly, the response to detection could deliver interesting results, but not perfect, because there is a chance of finding weed connected to lettuces. Furthermore, it might be necessary to implement erosion before detecting weed. This operation will certainly reduce small weed and disconnect the weed linked to lettuce within the same object. However, some information may get lost, so the final detection must be combined with the mask generated from NDVI as seen in

Table 1. Parameters obtained from the training groups

Kernel	Group 1			Group 2			Group 3		
	C	gamma	coef0	C	gamma	coef0	C	gamma	coef0
Linear	1,00E-01			1,00E-01			5,00E-01		
Poly(2nd order)	1,00E-01	3,38E-02	1,00E-01	1,00E-01	3,38E-02	1,40E+00	1,00E-01	3,38E-02	1,40E+00
Rbf	1,25E+01	2,25E-03		6,25E+01	2,25E-03		6,25E+01	2,25E-03	

Figure 8. Model performance for class lettuce   *Figure 9. Model performance for class weed*

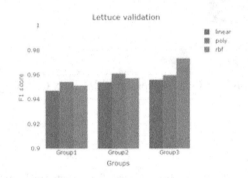   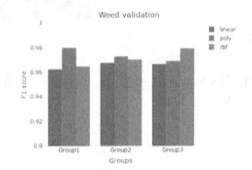

Table 2. Averages values of validation tests

Kernel	Class lettuce F1 Score	Class weed F1 Score
Linear	0.95	0.96
Polynomial	0.96	0.97
Rbf	0.96	0.97

Table 3. Test results for linear kernell

Linear Kernel	Lettuce [x100%]			Weed [x100%]		
	Precision	Recall	F1 score	Precision	Recall	F1 score
model 1	0,953	0,964	0,958	0,975	0,967	0,970
model 2	0,953	0,970	0,961	0,979	0,967	0,973
model 3	0,970	0,958	0,964	0,971	0,979	0,975

Table 4. Test results for polynomial kernell

Polynomial Kernel	Lettuce [x100%]			Weed [x100%]		
	Precision	Recall	F1 score	Precision	Recall	F1 score
model 1	0,976	0,970	0,973	0,979	0,983	0,981
model 2	0,988	0,964	0,976	0,975	0,992	0,983
model 3	0,988	0,964	0,976	0,975	0,992	0,983

Table 5. Test results for rbf kernell

	Lettuce [x100%]			Weed [x100%]		
Rbf Kernel	**Precision**	**Recall**	**F1 score**	**Precision**	**Recall**	**F1 score**
model 1	0,964	0,964	0,964	0,975	0,975	0,975
model 2	0,976	0,964	0,970	0,975	0,983	0,979
model 3	0,970	0,970	0,970	0,979	0,979	0,979

Table 6. Test results with 1703 images used as training data

Kernel	**Class Lettuce F1 Score**	**Class Weed F1 Score**
Linear	0,964	0,975
Polynomial	0,973	0,981
Rbf	0,979	0,985

Fig 6. Fig 10 shows the process of weed detection in an image taken at 80 cm height. In spite of having detected every lettuce, there are still weeds that were false negatives. It is worth noting the type of object obtained from the NDVI mask, because it is possible to obtain lettuces in the edges but with erroneous classifications. Also, the centroid position on each object affects the bounding box of the object, therefore their classification could be erroneous. Accordingly, the physical parameters of the bounding box can be adjusted to set a better detection as showed in Fig 20: a photograph taken at 2m height where the presence of weed hinders the detection.

According to this, it was necessary to test SVM models in an entire image in order to establish a reliable measurement based on accuracy of the detection. Is important to notice that the main interest in this case is a well tuned lettuce detection due to it's needed to create an accurate crop mask. So a test was conducted with ten entire images with a total of 232 lettuces and 23 lettuces as average number per image. Despite results in individual validation Table 7 shows the performance in the new test.

Once SVM was tested, the following step is to prove if computer weed calculations follows the same path for a typical aggregated distribution. Then there is a comparison between weed experts estimation and a computer's estimation following by a set of non-parametric test, the first test was made with Shapiro Wilk to check if the variable coverage with human estimation distributes normal. In most of the cases patched variables does not have normal distribution and this was confirmed by the test. The same test was performed on the algorithm developed whose estimation giving same results. As a consequence Kruskall-Wallis test was performed in order to detect significant statistical differences between coverage estimated by experts and software.

Table 7. Perfomance using a set of complete images

	Linear	**Polynomial**	**Rbf**
Precision	0,995	0,975	0,974
Recall	0,801	0,836	0,823
F1 score	0,887	0,900	0,893

Figure 10. a) Histograms of oriented gradients for the objects detected b) Weed detection (blue)

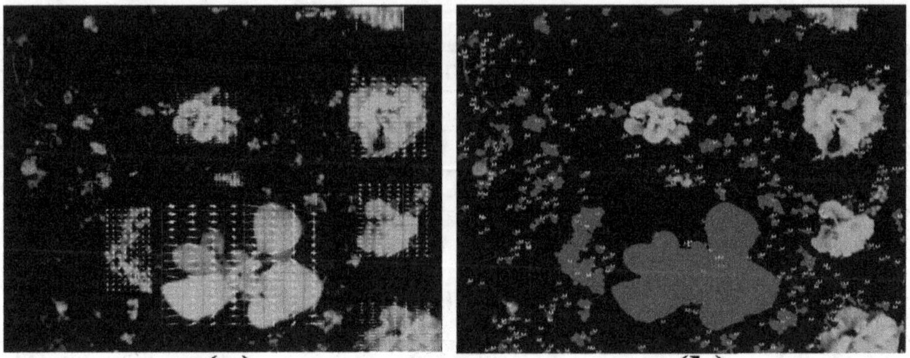

Weed Coverage Distribution

A estimation of coverage by the software conducted with 100 images at 80cm of height from a same ground at Marengo research center, each image has an area of 0.7 m^2 and respective weeds coverage percentage of each one was obtained (Figure 11). In addition, the same experiment was performed with 100 images at 2m of height with an area of 6 m^2 (Figure 13). As it knows weed density is a discrete variable and when has an aggregated distribution usually presents a negative binomial distribution (Gold, Bay, & Wilkerson, 1996), instead, coverage measurement is a continue variable despite of discrete values given by conventional visual inspections (Nkoa, Owen, & Swanton, 2015) but not in the case of computer measurements (Table 8). Taking in account that weeds presents an aggregated distribution the better distribution fitted for coverage measurement was gamma distribution (Figure 12 and 14) with negative binomial as second instance in both cases.

Computer's Results vs Human's Results

To validate the software, it was conducted a test with 19 images evaluated manually by three experts on weed science (Table 9). From these images, ten were taken at 80cm above the ground and nine at 2m above the ground. Subsequently, a variance analysis was conducted for the coverage data of each image analyzed by the experts and the software using the method proposed.

Table 8. Model AIC between 4 probability distributions for weed coverage images

80cm Images		2m Images	
Distribution	**AIC**	**Distribution**	**AIC**
Nbinom	624.85	Nbinom	672.35
Poisson	642.11	Poisson	808.73
Gamma	624.07	Gamma	666.81
Norm	629.83	Norm	716.06
Weibull	629.13	Weibull	693.79

Figure 11. Coverage histogram for 100 images taken at same ground at 80 cm

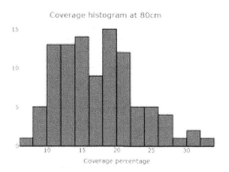

Figure 12. Gamma distribution fitting model for 80cm images

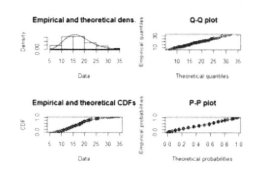

Figure 13. Coverage histogram for 100 images taken at same ground at 2m

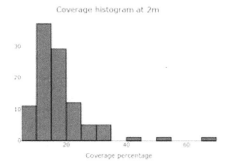

Figure 14. Gamma distribution fitting model for 2m images

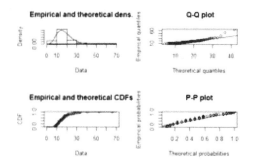

Shapiro Wilk and Homoscedasticity Test

Before to start with the Homoscedasticity test, it was necessary to submit the data to Shapiro Wilk test, this test can determine whether the data has or not a normal distribution, for not normal distributions, Homoscedasticity test and Kruskall wallis test can be applied.

According to data collected shown in table 10, this test was performed to know if each manually weed estimation has the same variance with computer estimation. For this, "data are homoscedastic" null hypothesis was established. The test used in this case was Breusch-Pagan test (Tables 11 and 12).

In this case, only one null hypothesis were rejected with Expert 3 and computer at 80 cm height.

Kruskall Wallis Test

According to table 10, the estimations made by the experts manually might have a significant variance for some images. However, it is necessary to establish a similarity measure between the data to determine whether the methods applied by the experts are similar to the method set by computer. In order to do so, the Kruskall Wallis test was implemented – previously confirming the absence of normality in these values using the Shapiro Wilk test.

Table 9. Coverage estimated by computer and the three experts

Image	Height	Coverage (%)			
		Expert 1	Expert 2	Expert 3	Computer
1	80 cm	6	5	3	9.1
2	80 cm	18	10	20	14
3	80 cm	20	20	15	11.8
4	80 cm	10	7	10	16.5
5	80 cm	9	10	10	9.3
6	80 cm	12	15	15	11.4
7	80 cm	8	10	10	7.15
8	80 cm	20	15	10	10.7
9	80 cm	8	5	8	7.25
10	80 cm	10	20	25	15.6
11	2 m	5	5	15	6.7
12	2 m	15	15	10	13
13	2 m	12	5	10	8.14
14	2 m	10	10	10	12
15	2 m	25	20	25	18
16	2 m	23	7	30	8.5
17	2 m	20	20	20	18
18	2 m	45	70	40	34
19	2 m	15	35	35	21.3

Table 10. Shapiro-Wilk test for experts an computer estimations

Height	Shapiro Wilk Test p-value
80cm	0.02864
2m	9.38×10^{-5}
80cm and 2m	1.87×10^{-9}

Table 11. Breusch-Pagan test for 80cm height

Height		Computer
	Expert 1	p=0,4959
80 cm	Expert 2	p=0,06753
	Expert 3	p=0,03462

Table 12. Breusch-Pagan test for 2m height

Height		Computer
	Expert 1	p=0,9422
2 m	Expert 2	p=0,305
	Expert 3	p=0,2669

Using the set of data, the Shapiro Wilk test had a $p=1.87\times10^{-9}$, validating that the data does not have a normal distribution. Therefore, the Kruskall Wallis test was conducted and its $p=0.6688$ validated there is no significant difference between the methods implemented by the experts and the method implemented by the software using the null hypothesis "localization of distribution parameters are the same in all the estimations." (Fig 8).

The data are divided in two groups to better understand the situation,. One is assessed using only the data obtained for heights of 80cm, where the data did not have a normal distribution since the Shapiro Wilk test had a $p=0.02864$; the Kruskall Wallis did not have significant differences between the methods at such height either (Fig 16). Conversely, in the second group, the data collected at 2m height behaved similarly as the later with a $p=9.38\times10^{-5}$ for Shapiro Wilk and $p=0.6401$ for Kruskall Wallis (Fig 17).

Finally, the experts methods were evaluated with respect to the height of the images previously taken. In this case, the data did not follow a normal distribution ($p=8.492\times10^{-8}$), but in this case the methods are significantly different ($p=0.01567$).

Figure 15. Box plot for the data in table 2.

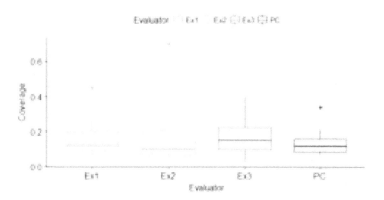

Figure 16. Box plot only evaluated at 80 cm height

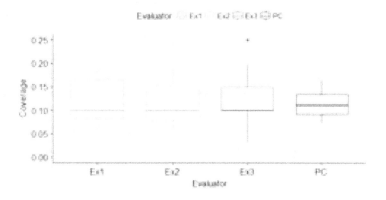

Figure 17. Box plot only evaluated at 2m height

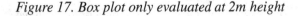
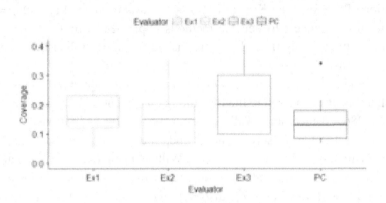

Figure 18. Box plot for evaluators according to the images height

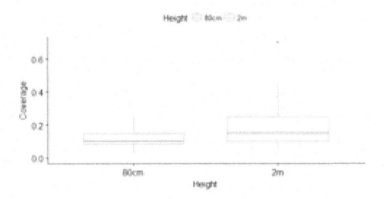

Issues, Controversies, Problems

Notwithstanding, there are some limitations with respect to the detection on crops with overlapping distribution of the weeds at this scale, since these plants connect several objects to each other merging weeds and crops together. This makes it necessary to use a very high erosion parameter, but it increases the possibility of losing information during the detection when using a support vector machine as classifier. However, being able to count with a multi-spectral sensor enables to recover the information lost when combining the NDVI mask with the detection resulted from using the histograms of oriented gradients, also is important to avoid lettuces at image borders because if a lettuce is cropped it may be detected as weed giving false positive during detection stage

Another important factor consists in the centroid of the object detected before the classification, since it generates the bounding box that determines the extraction of the image that will be processed by the support vector machine. As a result, it was necessary to adjust the bounding box (see solutions and recommendations section) to facilitate the correct classification due to false positives.

Figure 19. a) Histogram of oriented gradient for an image without adjusting the bounding box b) Final detection without adjustment

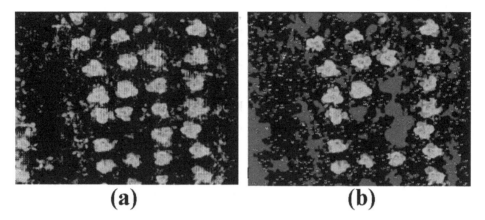

(a)   **(b)**

SOLUTIONS AND RECOMMENDATIONS

Due to some false positives (Fig 19b), was necessary introduce some adjustments on their respective bounding boxes during detection process. Those parameters were based on initial bounding box parameters at detection moment (APPENDIX 1.

Once NDVI mask was obtained and applied to original image to avoid background, the algorithm will detect different kind of objects inside the image by means of the contour and later bounding box, each one will pass through SVM to detect whether that object is a lettuce or weed object. Sometimes due to weed-crop overlapping (Fig 16b) that bounding box will detect something that can be confused and generally will be classified as weed.

To correct that issue, a bounding box adjustment was established during detection stage changing its origin, width height and a small area filter in order to avoid small objects that could be detected as lettuce in cases when some weed have lettuce appearance. Basically this solution was implemented generating some relative origin displacements in bounding boxes and enlarging their respective width and height depending of crop distribution (Fig 17b).

FUTURE RESEARCH DIRECTIONS

The combination of multi-spectral images and histograms of oriented gradients produced accurate results when detecting weed in this specific height and image size. Using the NDVI as a background estimator and describing the shapes allowed to detect weeds in two different types of heights. NDVI brings a good background estimation avoiding background computation algorithms, this advantage falls in infrared sensor that permits calculate this index through a couple of computational operations focusing in elements with photosynthetic activity, by other way this index saves computational costs with the fact that a background subtraction algorithm was avoided. Also an accurate background segmentation brings benefits at the image binarization moment in order to avoid manual thresholding and taking advantage of Otsu thresholding as a adaptative binarization technique. Using only one height for the training model works accurate for the two heights, this confirms that these types of features represents an advantage

Figure 20. a) Histogram of oriented gradient for an image without adjusting the bounding box b) Final detection with adjustment

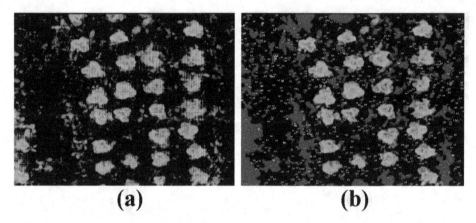

(a)   **(b)**

regarding the variation in scale, meaning that training model with a higher spatial scale could lead to more precise results and it could be used after in a lower spatial scale condition or higher height or bigger area fields. This confirms support vector machines as an applicable technique that can still be used in complex situations due to its low computational cost and capability to adapt no linear models. Even this model was tested on a low cost computational system giving acceptable times of processing, this gives us to think that we can embed this detection model in an autonomous device that can detect and afterwards eliminate weeds inside a given crop.

Additionally, this method may be useful for weed researchers, since it enables them to refine their skills as evaluator. Another advantage of this algorithm is based on automation, not only to assess the weed coverage in lettuce crops, but also to assess the herbicides efficiency or to improve the quality of the data in other fields in weed science. Moreover, it has the possibility of being embedded in an autonomous system with the ability to apply herbicides or foliage fertilizers according to the class of vegetation to be treated, whether it is weed or a crop.

As a future work, this dataset will be tested using convolutional neural networks models with the main purpose of compare with SVM results, Also, different architectures will be evaluated in order to enhance weed estimation computer capabilities specially on images where weed has invaded all the crop and could be hard to get HOG features between weed and lettuces.

CONCLUSION

Based on previous results, SVM based model confirms in its calculations the aggregated behavior of weeds and the possibility of fits their distribution in a more precisely way with fastest data and less bias due to evaluator skills. Furthermore, the statistical analysis proved that a person with sufficient training can make coverage estimates close to those made by the algorithm. Nonetheless, it was also found that as the height increases, people tend to overestimate the amount of weed, departing from the estimations given by the software which are usually more precise.

In this case, histogram of oriented gradients was a shape descriptor who fits according to irregular form given by different kinds of lettuce and weed allowing detections in places when weed can reach same size and even look alike form with lettuces. In this case Rbf and Polynomial kernells proven to be the best choices in order to get a better accuracy when detecting lettuces.

ACKNOWLEDGMENT

The study formed part of electronics and agronomy departments of National University of Colombia, also will include acknowledgements to professors Enrique Darghan and Guido Plaza from agronomy department also including Ph.D. Student Veronica Hoyos and Masters student Diana Zabala. Finally a special acknowledgement to Professor Oscar García from agricultural engineering department of National university of Colombia.

REFERENCES

Alchanatis, V., Ridel, L., Hetzroni, A., & Yaroslavsky, L. (2005). Weed detection in multi-spectral images of cotton fields. *Computers and Electronics in Agriculture*, *47*(3), 243–260. doi:10.1016/j.compag.2004.11.019

Ambrosio, L., Iglesias, L., Marin, C., & Del Monte, J. P. (2004). Evaluation of sampling methods and assessment of the sample size to estimate the weed seedbank in soil, taking into account spatial variability. *Weed Research*, *44*(3), 224–236. doi:10.1111/j.1365-3180.2004.00394.x

Bakhshipour, A., & Jafari, A. (2018). Evaluation of support vector machine and artificial neural networks in weed detection using shape features. *Computers and Electronics in Agriculture*, *145*, 153–160. doi:10.1016/j.compag.2017.12.032

Bossu, J., Gée, C., Jones, G., & Truchetet, F. (2009). Wavelet transform to discriminate between crop and weed in perspective agronomic images. *Computers and Electronics in Agriculture*, *65*(1), 133–143. doi:10.1016/j.compag.2008.08.004

Camps-Valls, G., & Bruzzone, L. (2005). Kernel-based methods for hyperspectral image classification. *IEEE Transactions on Geoscience and Remote Sensing*, *43*(6), 1351–1362. doi:10.1109/TGRS.2005.846154

dos Santos, A., Freitas, D., Silva, G., Pistori, H., & Folhes, M. (2017). Weed detection in soybean crops using ConvNets. *Computers and Electronics in Agriculture*, *143*, 314–324. doi:10.1016/j.compag.2017.10.027

Feyaerts, F., & Van Gool, L. (2001). Multi-spectral vision system for weed detection. *Pattern Recognition Letters*, *22*(6), 667–674. doi:10.1016/S0167-8655(01)00006-X

Gold, H. J., Bay, J., & Wilkerson, G. G. (1996). *Scouting for Weeds, Based on the Negative Binomial Distribution*. *Weed Science*. Cambridge University Press Weed Science Society of America; doi:10.2307/4045627

Gonzalez, R. C. (2009). *Digital Image Processing*. Pearson Education. Retrieved from https://books.google.com.co/books?id=a62xQ2r_f8wC

Irias Tejeda, A. J., & Castro Castro, R. (2019). Algorithm of Weed Detection in Crops by Computational Vision. In *2019 International Conference on Electronics, Communications and Computers (CONIELE-COMP)* (pp. 124–128). IEEE. 10.1109/CONIELECOMP.2019.8673182

Jadhav, B. D., & Patil, P. M. (2014). Hyperspectral Remote Sensing For Agricultural Management: A Survey. *International Journal of Computers and Applications*, *106*(7).

Jamaica Alejandro, D., & ... (2013). *Dinámica espacial y temporal de poblaciones de malezas en cultivos de papa, espinaca y caña de azúcar y su relación con propiedades del suelo en dos localidades de Colombia. Weed Research*. Universidad Nacional de Colombia.

Junguito, R., Perfetti, J. J., & Becerra, A. (2014). *Desarrollo de la agricultura colombiana*. Retrieved from https://www.repository.fedesarrollo.org.co/handle/11445/151

Kim, K. H., Kim, H. J., & Chung, Y. S. (2019). Case Study: Cost-effective Weed Patch Detection by Multi-Spectral Camera Mounted on Unmanned Aerial Vehicle in the Buckwheat Field. *Hangug Jagmul Haghoeji*, *64*(2), 159–164. doi:10.7740/kjcs.2019.64.2.159

Lehoczky, É., Riczu, P., Mazsu, N., Viktória Dellaszéga, L., & Tamás, J. (2018). Applicability of remote sensing in weed detection. *20th EGU General Assembly, EGU2018, Proceedings from the Conference Held 4-13 April, 2018 in Vienna, Austria, p.14644*, *20*, 14644. Retrieved from http://adsabs.harvard.edu/abs/2018EGUGA.2014644L

López-Granados, F. (2011). Weed detection for site-specific weed management: Mapping and real-time approaches. *Weed Research*, *51*(1), 1–11. doi:10.1111/j.1365-3180.2010.00829.x

Lottes, P., Khanna, R., Pfeifer, J., Siegwart, R., & Stachniss, C. (2017). UAV-based crop and weed classification for smart farming. In *Robotics and Automation (ICRA), 2017 IEEE International Conference on* (pp. 3024–3031). 10.1109/ICRA.2017.7989347

Louargant, M., Jones, G., Faroux, R., Paoli, J.-N., Maillot, T., Gée, C., & Villette, S. (2018). Unsupervised Classification Algorithm for Early Weed Detection in Row-Crops by Combining Spatial and Spectral Information. *Remote Sensing*, *10*(5), 761. doi:10.3390/rs10050761

Mountrakis, G., Im, J., & Ogole, C. (2011). Support vector machines in remote sensing: A review. *ISPRS Journal of Photogrammetry and Remote Sensing*, *66*(3), 247–259. doi:10.1016/j.isprsjprs.2010.11.001

Murawwat, S., Qureshi, A., Ahmad, S., & Shahid, Y. (2018). *Weed Detection Using SVMs. Technology & Applied Science Research* (Vol. 8). Retrieved from www.etasr.com

Nkoa, R., Owen, M. D. K., & Swanton, C. J. (2015). Weed Abundance, Distribution, Diversity, and Community Analyses. *Weed Science*, *63*(SP1), 64–90. doi:10.1614/WS-D-13-00075.1

Oerke, E. (2006). Crop losses to pests. *The Journal of Agricultural Science*, *144*(1), 31–43. doi:10.1017/S0021859605005708

Otsu, N. (1979). A threshold selection method from gray-level hostgrams. *IEEE Transactions on Systems, Man, and Cybernetics*, *9*(1), 62–66. doi:10.1109/TSMC.1979.4310076

Perez, A. J., Lopez, F., Benlloch, J. V., & Christensen, S. (2000). Colour and shape analysis techniques for weed detection in cereal fields. *Computers and Electronics in Agriculture, 25*(3), 197–212. doi:10.1016/S0168-1699(99)00068-X

Piron, A., Leemans, V., Lebeau, F., & Destain, M.-F. (2009). Improving in-row weed detection in multispectral stereoscopic images. *Computers and Electronics in Agriculture, 69*(1), 73–79. doi:10.1016/j.compag.2009.07.001

Rew, L. J., & Cousens, R. D. (2001). Spatial distribution of weeds in arable crops: Are current sampling and analytical methods appropriate? *Weed Research, 41*(1), 1–18. doi:10.1046/j.1365-3180.2001.00215.x

Rodríguez, M., Plaza, G., & Gil, R. (2008). Reconocimiento y fluctuación poblacional arvense en el cultivo de espinaca (Spinacea oleracea L.) para el municipio de Cota, Cundinamarca. *Agronomia Colombiana, 26*(1), 87–96.

Sakthi, P., & Yuvarani, P. (2018). Detection and Removal of Weed between Crops in Agricultural Field using Image Processing. *International Journal of Pure and Applied Mathematics, 118*, 201–206. Retrieved from http://www.ijpam.eu

Serrato, N., & Castillo, C. (2018). Colombia Land of Opportunities to Apply Precision Agriculture: An Overview. *International Journal of Agricultural Sciences, 3*. Retrieved from http://iaras.org/iaras/journals/ijas

Sujaritha, M., Annadurai, S., Satheeshkumar, J., Sharan, S. K., & Mahesh, L. (2017). Weed detecting robot in sugarcane fields using fuzzy real time classifier. *Computers and Electronics in Agriculture, 134*, 160–171. doi:10.1016/j.compag.2017.01.008

Szeliski, R. (2010). *Computer vision: algorithms and applications*. Springer Science & Business Media.

Thorp, K. R., & Tian, L. F. (2004). A review on remote sensing of weeds in agriculture. *Precision Agriculture, 5*(5), 477–508. doi:10.100711119-004-5321-1

Tsolakidis, D. G., Kosmopoulos, D. I., & Papadourakis, G. (2014). Plant leaf recognition using Zernike moments and histogram of oriented gradients. In *Hellenic Conference on Artificial Intelligence* (pp. 406–417). 10.1007/978-3-319-07064-3_33

Umamaheswari, S., Arjun, R., & Meganathan, D. (2018). Weed Detection in Farm Crops using Parallel Image Processing. In *2018 Conference on Information and Communication Technology (CICT)* (pp. 1–4). IEEE. 10.1109/INFOCOMTECH.2018.8722369

Wang, A., Zhang, W., & Wei, X. (2019). A review on weed detection using ground-based machine vision and image processing techniques. *Computers and Electronics in Agriculture, 158*, 226–240. doi:10.1016/j.compag.2019.02.005

Westwood, J. H., Charudattan, R., Duke, S. O., Fennimore, S. A., Marrone, P., Slaughter, D. C., ... Zollinger, R. (2018). Weed Management in 2050: Perspectives on the Future of Weed Science. *Weed Science, 66*(3), 275–285. doi:10.1017/wsc.2017.78

Whiting, M. L., Ustin, S. L., Zarco-Tejada, P., Palacios-Orueta, A., & Vanderbilt, V. C. (2006). Hyperspectral mapping of crop and soils for precision agriculture. In *SPIE Optics* (pp. 62980B–62980B). Photonics. doi:10.1117/12.681289

Xia, Q., Zhu, H.-D., Gan, Y., & Shang, L. (2014). Plant leaf recognition using histograms of oriented gradients. In *International Conference on Intelligent Computing* (pp. 369–374). 10.1007/978-3-319-09339-0_38

Xu, X., Xu, S., Jin, L., & Song, E. (2011). Characteristic analysis of Otsu threshold and its applications. *Pattern Recognition Letters*, *32*(7), 956–961. doi:10.1016/j.patrec.2011.01.021

Zhao, Z.-Q., Ma, L.-H., Cheung, Y., Wu, X., Tang, Y., & Chen, C. L. P. (2015). ApLeaf: An efficient android-based plant leaf identification system. *Neurocomputing*, *151*, 1112–1119. doi:10.1016/j.neucom.2014.02.077

Zimdahl, R. (2007). *Fundamentals of weed science*. Elsevier. doi:10.1016/0378-4290(95)90065-9

ADDITIONAL READING

Anderson, W. P. (1977). *Weed science: principles*. West Publishing Co.

Bradski, G., & Kaehler, A. (2008). *Learning OpenCV: Computer vision with the OpenCV library*. O'Reilly Media, Inc.

Jensen, J. R., & Lulla, K. (1987). Introductory digital image processing: a remote sensing perspective.

Klingman, G. C., Ashton, F. M., & Noordhoff, L. J. (1975). *Weed science: principles and practice*. John Wiley & Sons, Inc.

Ng, A. (2017). Machine learning yearning. URL: http://www. mlyearning. org/(96)

Russ, J. C. (2016). *The image processing handbook*. CRC Press. doi:10.1201/b10720

Smola, A. J., & Schölkopf, B. (2004). A tutorial on support vector regression. *Statistics and Computing*, *14*(3), 199–222. doi:10.1023/B:STCO.0000035301.49549.88

Sonka, M., Hlavac, V., & Boyle, R. (2014). *Image processing, analysis, and machine vision*. Cengage Learning.

KEY TERMS AND DEFINITIONS

AIC: Akaike information criterion is an estimator used for obtain relative quality in a certain statistical model.

Homoscedasticity: A random sequence presents homoscedasticity when all of its random variables have the same finite variance.

Kruskal-Wallis Test: Non-parametric statistical method used to compare whether a set of datasets follows the same distribution, to determinate significative statistical differences between them.

NDVI: Normalized difference vegetation index is a simple indicator used to analyze remote sensing based on objects with photosynthetic activity.

Shapiro-Wilk Test: Statistical test to determine whether a data set follows normal probabilistic distribution.

SVM: Support-vector machine is a machine learning model based on non-probabilistic results used for linear or nonlinear regressions also as linear or nonlinear data classifications.

Weed Coverage: Percentage of weed coverage inside of a determined crop area.

Weed Density: Amount of weed plants inside of a determined crop area.

APPENDIX

Table 13. First 20 images at 80cm of a set of 100 with processing results including bounding box adjustments.

Image	Coverage (%)	Bounding box adjustments				
1	16,526	9	1	10	7	7
2	20,9769	11	5	5	0	0
3	12,3262	7	0	0	0	0
4	9,60092	7	0	0	0	0
5	8,57783	7	0	0	0	0
6	13,0107	7	0	0	0	0
7	13,6144	7	0	0	0	0
8	9,7261	25	35	35	15	15
9	11,5459	25	35	35	15	15
10	9,29296	7	0	0	0	0
11	9,44001	7	0	0	0	0
12	9,57714	7	0	0	0	0
13	10,7914	11	0	0	0	0
14	14,42	11	0	0	0	0
15	11,7023	25	35	35	15	15
16	12,8159	7	0	0	0	0
17	12,947	7	0	0	0	0
18	10,2486	25	35	35	15	15
19	8,50721	25	35	35	15	15
20	10,892	7	0	0	0	0

Table 14. First 20 images at 2m of a set of 100 with processing results including bounding box adjustments.

Image	Coverage (%)	Bounding box adjustments				
1	16,526	9	1	10	7	7
2	20,9769	11	5	5	0	0
3	12,3262	7	0	0	0	0
4	9,60092	7	0	0	0	0
5	8,57783	7	0	0	0	0
6	13,0107	7	0	0	0	0
7	13,6144	7	0	0	0	0
8	9,7261	25	35	35	15	15
9	11,5459	25	35	35	15	15
10	9,29296	7	0	0	0	0
11	9,44001	7	0	0	0	0
12	9,57714	7	0	0	0	0
13	10,7914	11	0	0	0	0
14	14,42	11	0	0	0	0
15	11,7023	25	35	35	15	15
16	12,8159	7	0	0	0	0
17	12,947	7	0	0	0	0
18	10,2486	25	35	35	15	15
19	8,50721	25	35	35	15	15
20	10,892	7	0	0	0	0

Chapter 10
Analytical Observations Between Subjects' Medications Movement and Medication Scores Correlation Based on Their Gender and Age Using GSR Biofeedback:
Intelligent Application in Healthcare

Rohit Rastogi

https://orcid.org/0000-0002-6402-7638

ABES Engineerig College, Ghaziabad, India

Devendra Kumar Chaturvedi

https://orcid.org/0000-0002-4837-2570

Dayalbagh Educational Institute, Agra, India

Mayank Gupta

Tata Consultancy Services, Noida, India

Heena Sirohi

ABES Engineering College, Ghaziabad, India

Muskan Gulati

ABES Engineering College, Ghaziabad, India

Pratyusha

ABES Engineerig College, Ghaziabad, India

ABSTRACT

Increasing stress levels in people is creating higher tension levels that ultimately result in chronic headaches. To get the best result, the subjects are divided into two groups. One group will be introduced under EMG, and the other will be handled under GSR. The change in the behaviour of subject (i.e., the change in the stress level) is measured at the intervals of one month, three months, six months, and twelve months. The main aim of the research is to study the effects of tension type headache using bio-

DOI: 10.4018/978-1-7998-1839-7.ch010

feedback therapies on various modes such as audio modes, visual modes, and audio-visual modes. The groups were randomly allocated for galvanic skin resistance (GSR) therapies, and the other one was control group (the group that was not under any type of allopathic or other medications). Except for the control group, the groups were treated in a session for 20 minutes in isolated chambers. The results were recorded over a specific period of time.

INTRODUCTION

Biofeedback

In Introduction, we shall discuss different medication domains in which a subject can lie, social annals of TTH as per gender and age and some process of biofeedback which are used to control TTH and important basic technical terminology used for research work like Bigdata, IoT, TTH, Machine Vision, Biofeedback etc.

The main three professional organizations of biofeedback, (AAPB), (BCIA), and (ISNR) come to a common definition for biofeedback as " the process which enables the individual for learning to change the routine and method of physiological activities so that the health of individual could get improve" (Fumal & Scohnen, 2008).

- **Electromyograpgh:** Earlier the device for biofeedback was developed by Dr.Harry Garlan and Dr.Roger Melenin, 1971. The muscle whistler shown in this with EMG electrode.
- **Feedback Thermometer**: This device detects the skin temperature and is usually attached to a finger and is measured in degree Celsius. The hand-warming and hand-cooling signs are marked by another mechanism.
- **Electrodermograph**: In this process this device detectsskin electrical activities indirectly and directly by the electrode which is placed above the hand.This device detects all the electrical sensation from the brain when it is placed over the scalp item which lies above the human cortex.
- **Photoplethysmograph:** In this process the device measure the blood flow through a particular body part when an infrared light is passes through the body part and the intensity of light transmit through the tissue.
- **Electrocardiogram:** In this process the electrodes are placed over torso, wrist or legs and records the electrical sensation of heart and records the multiple beat differences.
- **Pneumograph:** This device uses a gauge which have a stretchable sensor band which is placed across above the stomach,abdomen and measure the respiration rate.
- **Capnometer:** This system record the information by using as infrared detector which measure the pressure partially when carbon dioxide is excreted through the latex tube.
- **Rheoencephalograph:** This is also known as HEG biofeedback, and it is clear from the name that this records the difference in the color produce by light which is transmitted back by hair scalp based on oxygenated and un oxygenated blood in the brain
- **Pressure:** It can be measured when the patient perform any exercise while continuously resting along an air-filled cushion.

AC (ANALGESIC CONSUMPTION)

Analgesics can be defined as drugs that produce diminished sensations to the severe pain without losing the actual consciousness of mind.In daily language, they are termed as "pain killers" as they kill the pain to a extent but the continuous use of these analgesic drugs cause damage to mind and body in various ways .They reduce the sensations or reflex actions of nerves. Analgesics can be classified into many ways depending on their actions: Paracetamols (acetaminophen), nonsteroidal anti-inflammatory drugs(NSAIDs) such as the salicylates, and opioid drugs such as morphine and oxycodone(Chaturvedi et al., 2018b).

The analgesic is the synonym of painkiller or we could say that it is the other name of painkillers .It is that type of drug that control the pain without interrupting the nerve movement and it doesn't affect to us consciously. There is a difference between Analgesics and Anesthesia. The analgesic may work as they reduce the sensation of pain by local inflammatory methods, as they sends signals to brain to reduce the sensation of pain. In today's time everyone wants to remain fit and healthy but the world is becoming faster and faster and in the race of becoming best, we neglect our health and because of this if any pain occur we treat it as very underrated and try to consume any analgesic drug that could relieve pain, but these drugs could have heinous disadvantages. The studies reveals that we should not be always depend upon these drugs. Some drugs that under analgesic are:

- **Aspirin:** It is commonly used in tooth decay problems, there are various confusions about its consumption that whether to use it in fever for youngsters among 16 years could create the risk of developing Reye's syndrome which results in level and brain malfunctioning. Because of this it advised to keep children away from products which could contain aspirin in any forms.
- **Ibuprofen:** This drug has milder affects than the aspirin. This drug is used in the pain which is mild to moderate and used in pains such as headaches and muscular joints.This drug also reduces the levels of prostaglandins which will lead in increase in the secretion of gastric acid due to which ulceration might be possible. And it is also listed as the drug whose consumption could lead to heart disease and stroke.
- **Opioids:** The drug which is highly used in illegal consumptions and for the production of morphine which is then used as painkiller to reduce the severe amount of pain. One of the name of this drug is Codeine which is somehow similar to morphine and reduce to treat mild to moderate amount of pain. This drug should be kept out away from the reach of children as it could cause them asthmatic problems. The main side-effect of this drug is loose motions.(Yadav V. et al., 2018j)

PM(PROPHYLACTICMEDICATIONS)

A prophylactic is a treatment used to prevent diseases from occurring. "Prevention is better than cure"; these treatments works on this principle. They are used to prevent something from happening, for example, a prophylactic hepatitis vaccine prevents the patient from getting hepatitis. These are also called anti-depressants as they help to reduce headaches. The word prophylactic comes from Greek whose actual meaning is 'to prevent'in this sort of medications some methods are designed so that could not fall for any disease because of the prevention. The preventive measure are not used to treat the acute attacks. The prophylactic medications depend upon various factors like blood pressure,asthma, diabetes, or pregnancy therefore we should choose right type of measure for ourselves. Before starting the medications the points to be noted are the measure that had been opted should not provide any discomfort to the patient and the physician should prescribe such a method so it will not affect other ongoing disease of the patient. The best application of Prophylactic medications we could see in the case of migraine, as no is likely to have the attack of migraine frequently and for this the people suffering from migraine start taking preventive measure(Boureau, Luu & Doubrere, 2001).

Goals of Prophylactic Medications

The preventions that are taken in migraine is to reduce the frequency of attacks, the problems that are occur during the period of attack. In the studies and experiments it is found that the preventive methods are not really successful in reducing the problems associated during the attack but it helps in reducing the frequency and level of pain.

The major goals are:

- To reduce the frequency and the time of attack.
- To get better with the acute attacks.
- To reduce the overall cost of the whole treatment.
- To reduce the adverse use of acute therapies.

It is very important to decide before starting any treatment that which treatment would suits you best, because you any wrong treatment could affect you adversely.

And for this the first thing is to review your doctor about the particular treatment and should questions the doctor that what could be the various negative and positive result of the treatment other than the efficiency of treatment.

You should be very careful from the starting that mild doses are to be taken in initial days and if they doesn't produce any positive result then only switch to higher doses.

It may take a significant amount of time for you to notice any change in the health.

The medication may have any side effect so if you feel any kind of abnormality then you should report to your doctor immediately (Gupta et al., 2019a).

The Preventive medication include some measure which are listed below:

1. **Tricyclic Antidepressants:** This drug include some components such as amitriptyline and protriptyline, these are the most common medicines for the prevention of tension type headache, the major side of this drug is constipation and sore mouth.
2. **Other Antidepressants**: There are some experiments which support the use of the antidepressants such as venlafaxine (Effexor XR) and mirtazapine (Remeron).
3. **Anticonvulsants and Muscle Relaxants:** Other medications that could prevent tension type headaches include anticonvulsants, such as topiramate (Topamax).

OM(Other Medications)

This is a group of medications which include muscle relaxants and Triptans. Muscle relaxants are the drugs which affect the skeletal muscle system and used to control muscle pains. Triptans are the tryptamine based drugs used to treat migraines and severe chronic headaches like hypertension, etc. Some different methods to deal with are listed below (Arora et al., 2017a):

MANAGE LIFESTYLE

- **Control the Stress Level:** The only and best possible solution to reduce tension type headache is by not thinking about the problem much and for this you must have to plan things accordingly from previous itself, start organizing yourself in proper manner, take proper time to take mental as well as physical rest and if you ever find yourself stuck in any difficult or stressful situation always to keep back from it as soon as possible.
- **Go Hot or Cold:** Another effective way to reduce tension type headache is by relaxing the muscles and for this try to apply hot or cold pack (any one which you would prefer), this process sore the muscles and reduce extent of tension type headache.

If you want to apply hot pack then you can use a heating pas which is set on a hot table and start compressing on the affected area by slowly padding. And if you want to apply cold one then for that wrap the ice in a cloth and apply it on the effective area and try to avoid rashes.

- **Perfect Your Posture:** If you feel lazy to perform the methods to reduce the level of tension type headache then this the simplest and easiest way to reduce it.

As correct posture doesn't allow muscles to get tense. Some tips thatyoushould need to follow are when you are s standing, hold your shoulders back upto your head level. Pull in your abdomen and buttocks. When sitting, make sure your thighs are parallel to the ground and your head isn't slumped forward. (Singhal P. et al., 2019b)

ALTERNATIVE MEDICINES

There are some non-conventional therapies are listed below-

- **Acupuncture:** This is one the easiest way to get rid of tension type headache. It provides relief in the chronic headache. The practitioners of thus treatment treat the patient by using very thin needles which produces some amount of pain and discomfort but relief the headache for very long period(Haynes, Griffin, Mooney & Parise, 2005).
- **Massage:** It helps to reduce the stress level and relieve tension. It is mainly used to give relaxation to tightened and sensitive portion in the back part of head, some part of neck and shoulder.

Deep breathing, biofeedback and behavior therapies. Various types of relaxation therapies are helpful in dealing with tension headaches, which include deep breath and biofeedback.

Coping and Support

It is very difficult to live with chronic pain. It could make you very anxious or depressed and affect your relationships, your productivity and the quality of your life(Rastogi, R. et al., 2018c).

Here are some suggestions are listed below to avoid all this:

- **Talk to a Counselor or Therapist:** Talk therapy helps you in a significant way to cope with the effect of chronic pain.
- **Join a Support Group**: Support groups could be good option for dealing with this problem. Group members usually know about the various latest treatments. And you also feel better after having a good company of people (Singh A. et al., 2019d).

SOCIAL ANNALS FOR TTH ALONG WITH AGE

Tension type headache increases with increase in age. It was found that TTH increases with higher frequency and longer duration in adult age. It was seen that use of analgesics was more in adults as compared to children and adolescents, but the rates of TTH decline in older ages (>50). Tension type is headache is now becoming a very common and severe disease among everyone. During studies it had been found that around 60% population is the victim of TTH. There are many who don't even know that they are suffering from tension type headache. They took it as migraine and don't get the right treatment which is required for it. It is clinically proven that it is difficult to difference between migraine and tension type headache, while there are some researches shows that both are not different they just have different levels of pain (Saini et al., 2019c).

By the clinic records and some studies it had been found that the most people will get affected by tension type headache is around 20 years of age and there is a severe increment in pain in between 30-39 years of age. Studies reveal that the around 34.5% under age group of 7-13 years are experiencing severe headaches even from that age onwards, and from these who continues to have tension type headache is something around 9% . During experiments on various age group it has been found that the people of around 26 years of age and when they reach at 32 years there is a increment of 17.6% of patients of tension type headache and amongst them 10.4% are having episodic TTH, 87.3% have frequent TTH and around 2.7% have chronic TTH. The overall fluctuation of TTH varies in the population from 31% to 77% and on increasing the age the TTH gets reduce. Infected rates of TTH differ widely across the whole world .In research it has been found that in just one year the people who are suffering from TTH are increased by 33% under which 31% are for episodic TTH, 2.5% are for chronic TTH. The people whoare affected from Episodic TTH range from 10.9% to 37.4% and people suffering from chronic TTH varies from 0.7% to 3.4% (Cassel, 1997).

GENDER BASED STUDIES ON CHRONIC TTH AT A GLANCE

Emotional and psychological factors such as stress, tension, anxiety, depression due to work load, failing relationships, etc. lead to chronic tension type headache. Females are seen to be more emotional than men and also the reports show that females have higher rates of TTH than men. The ratio of female to male is 1.6:1. Every person has a different lifestyle of living and this is largely seen on gender basis. Females are known be more sensitive and tender than males and because of this any problem affect females more, which increase stress level in them more as compared to males. A research was conducted amongst 1973 participants whose age group varies from 17-64 years. Around 445 participants that is approx. 22.4% are found to infected with tension type headache and out of these infected 26%(273) are females and 18%(168) are males. And after the complete research it had been found that around 62.5% females are suffering from tension type headache and 37.5% of males are suffering from tension type headache. And if we compare the result we come to conclusion that the tension type headache male to female ratio is 1:6 and it also concluded that the mean age of females who are affected to tension type headache is around 42 years while that of males is 39 years. The females which are affected from TTH out of that number almost 7% are illiterate, 26% are primarily educated. 5% are in middle of their education.15% are studying in high school. 9% of them are college going young girls and if we talk about economy then from affected female population, 21.5% has low financial status, 38% has medium financial status and 11% of them have high financial status. On the other hand if we talk about males then we find that, the males which are affected from TTH out of that number almost 3% are illiterate, 9% are primarily educated. 5.5% are in middle of their education.8% are studying in high school. 8.5% of them are college going young boys. And if we talk about economy then from affected female population, 17% has low financial status, 20% has medium financial status and only 2% of them have high financial status (Satya et al. 2019a; Wenk-Sormaz, 2005).

Machine Vision

In today's world machine vision can be seen as an important tool to cure the disease along with the medical science and in future machine vision could be seen as very important technology which can combine the hard work or doctors and surgeon and the scientific technology which could improve the efficiency of treating any disease and reduce the effective cost of any treatment. From the feedback of various patients from past 2 years it has been observed that medical technology's importance and need is increasing day by day. And from the world wide it can be seen as the fifth industry which the most number of visitors.

This technology includes works as use of cameras during endoscopy during surgery to have a crisp and clear image during operating, use of scanner which could improve the quality of dentures and scanners are also used to detect the skin cancer. It is fast as well as accurate as it takes around 100 images in just 1 second which reduces a maximum amount of time during operating which is consumed in scanning. In future we would be able to see more camera based medical technologies due to which the patients need not to go to the doctor everytime, he/she could get prescribed even at home, there would be a physical distance between the patient and doctors and if we are talking about machine vision then we shouldn't forget about artificial intelligence and machine learning. These two are very essential in today's medical science. AI has become so smart that you don't need to code to searchany specific image, because of its knowledge and the ability to identifying patterns the AI find the images when any similar image is presented(Carlson & Neil, 2013).

Medical Images

It is a way through which the images of human body part for any clinical purpose is required for the study of anatomy and physiology of humans or operating upon it. Its main purpose is to get the information of the part which is under the skull and bones for diagnosing and treating disease.The medical imaging uses technologies like X-ray radiography, magneticresonance imaging, medical ultrasonography or ultrasound, endoscopy, elastography, tactile imaging, thermography, medical photography. It helps the doctor to understand the problems and the damage of internal organs so that they could operate well and could provide a better and efficient treatment. The process of medical images is hustle free and it doesn't require any special preparations and in cases like cancers medical imaging can serve as a boon for the patient. One of the technology is ultrasound which helps the doctor to look internal body structure clearly for example tendons, muscles and other internal organs. The medical images are used in various surgeries. The high resolution image allow the doctor to have a look at real time progress of surgery and work according to it. It may look like it is very costly and the surgery which involve medical images are little bit costly, but the situation is just opposite as the doctor just take the image of the affected region and plans the treatment for it. This process reduce the efforts of doctor due to which the overall cost of the surgery gets reduced. And it is consider to be the safer technology because there are chances that human could interpret the wrong disease but because of the medical imaging all the work become safer and smooth. In previous time it is difficult to share the medical files over a long distance and because of this there were chances of the patient of not getting the proper treatment or could had died, but after medical imaging it become very easy to share the reports from one place to another and get the best treatment.(Gulati et al., 2019e)

ANALYSIS OF MEDICAL IMAGES

In the previous segment we have discussed about medical imaging, machine vision and how much they are useful for treating the disease effectively, efficiently and accusatively. In this section we will talk about the analysis of medical image and particular for headache and for which medical imaging is known as Neuroimaging.

So for Neuroimaging the technologies which are used are Computed tomography also known as CT and Magnetic resonance imaging which is generally known as MRI.

We have already discussed about TTH and migraine that they have frequent and severe attacks and now we have two technologies through which we could analysis their medical images.

1.   Computed Tomography(CT)

This process provides the help to diagnostic or the therapist to look at the internal changes of the brain or the positioning of muscles when the attack occurs. There are various 3D high resolution images which make it easy for the doctor to look at the issue clearly, the doctor have the clear view of the person's anatomy, but there are some risk in CT scan such as radiation exposure which could lead to cancer. As the radiation which is emitted from the CT scan emits the effective amount of 3-5mSv to take the CT of a normal head(Lee & Yoon, 2017).

2.   Magnetic Resonance Imaging(MRI)

This medical image process helps the diagnosis but it is slightly differ from CT as there is no emission of radiations in it. It uses a strong magnetic field for the imaging of human anatomy. If we compare MRI from CT we come to know that it provide better result and there will be increase in the contrast in between the soft tissues which are present in the body. And this is because of the high magnetic field. The environment which is magnetic in nature because of this special care is provided to the patient and is it recommended not to wear anything which could be attracted by the magnetic field as if it does then it harm the machine as well as the patient(Chauhan, 2018d; Chaturvedi, 2017b; Satya et al., 2018e).

BIG DATA AND IoT APPLICATIONS

As we are growing with technologies we experience that the different technologies are making out tasks easier and cheaper. And some of those technologies are reducing the overall cost of treatments and surgeries. These technologies include the devices which can have continuously monitoring system for patients, system that provides automatic therapies and for that patient need not go to the doctor, he just have to use these technologies and will be prescribed according to his problem. These technologies are faster as they have internet access and could have the real time status of the health of the person For this IoT and Big Data are using fast.

IoT(Internet of Things) is the network of physically connected devices and various other devices which are embedded with different software, GPS connectivity, which enables to record the real time data. Its impact on medical science will be huge and will be very important. The speed by which this technology is growing we could make an estimate that large number of tasks of medical science will be done by IoT which will create a billion market and reduce the effective cost of various treatments, and saving more lives(Binde & Blettner, 2015).

Let's take an example and understand that how this will change the procedure in medical science. Suppose a person is a diabetic patient, he have his id card, and when it is scanned, it will link to the cloud server which is storing his previous details such as lab records, medical and prescription history.

It might look really simple and easy but this task is little bit difficult and a game changer in the field of medicine. After implementing of this in avery less time all the records which are hundreds and thousand years old will get digitized,and the information will be easily sharable. The challenge that come across implementing these technologies is communication,as there are so many devices which are enabled with sensors which will record data, and sometimes they will talk to server in some language. While the each manufacturer have their own protocol due to which sensors with different manufacturer can't speak to each other much. The environment thus created which is coupled with private data, there might be possibility that the data might get steal and IoT could get failed(Costa, 2014).

The data which will be inbuilt into the sensors will be done by the drivers and that's how Big Data comes into the picture in medical science. The Big Data works mainly on 3 V's:

- **Volume**: Data in large volume.
- **Variety**: The variation in data that had been recorded.
- **Velocity**: The speed by which the data will get analyzed.

As defined the medical science is becoming the emerging user of big data. Some most difficult high dimensional documents data sets include X-rays, MRIs, some wave analysis for example EEG and ECG. The thing that data analysis should be given to its users is the constantly update based on the knowledge that had been gained, while keeping all the data at one place.

Therefore these two technologies IoT and Big Data will change the medical science in coming years. And it will reduce the treatment duration and effective cost and the medical science will be the combination of business methods and real time decisionas per(Sharma, 2018g; Scott & Lundeen, 2004).

PREVIOUS STUDIES (LITERATURE REVIEW)

Tention Type Headache and Stress

There are various researches carried out on stress and illness. Design of EMG model for biofeedbackis efficient and it decides some protocol that would be for TTH. The main mode to treat people who are affected from TTH is Pharmacotherapy((Saini H., 2018j), it is very effective in decreasing the extent and duration of TTH. But it also have drawback as it is all over used as antidepressant medicine((Bansal I. et al., 2018k) and those have risk of having adverse effect. Other than that it has potential risk for analgesic medication because of the overuse(Rastogi, 2018f). In 2008 there was an article which state the efficiency of biofeedback in treating TTH(Rastogi, 2018f). Galvanic skin resistance(GSR) therapies are used to treat these headaches and are found very effective. At times,it can be very severe that it becomes difficult to differentiate among tension type headache and migraine attack. The most commonly used preventive measures include medications such as allopathic treatments and non-medication treatments such as biofeedback therapies. The term "tension type headache" (TTH) has been declared through the International Classification Headache Diagnosis I (ICHD I). The terms "tension" and "type" represents its different meanings and reflect that there are sort of mental or physical tension could cause an impact. However,studies at large extent at this topic shows that there somehow a doubt about its neurobiological nature(McCrory, 2001).

TTH is the most basic form of headache occurring in about three quarters of the world. Tension type headache can usually last from 31 minutes to 8 days .The pain is generally mild and moderate in most of the times and sometimes it is worst. There are no diagnosis tests to conform Tension type headache(TTH). Galvanic skin resistance(GSR) therapies are used to treat these headaches and are found very effective. At times,it can be very severe that it becomes difficult to distinguish between tension type headache and migraine attack. The most commonly used preventive measures include medications such as allopathic treatments and non-medication treatments such as biofeedback therapies. The term "tension type headache" (TTH) has been declared by the International Classification Headache Diagnosis I (ICHD I). The terms "tension" and "type" represents its different meanings and reflect that some sort of mental or muscular tension could cause a impact. However, studies at large extent at this topic shows that there somehow a doubt about its neurobiological nature(Rubin, A., 1999). It is one of the common form for headache and in normal cases people took it lightly and they don't consult to the doctor and they treat themselves with various drugs in most cases people got cured but sometime the drugs have chronic impact on the health which could be severe.

- **Causes:** The main reasons due to which TTH happen could be the contraction between the neck and scalp. The victim of this in majorities are females. In many researches it is also concluded that if you are sitting in only one position for a longer period of time and if there is strain in neck and head then TTH could occur. The people who suffer from this problem are mainly from IT industries because of working on the computer for longer period of time .Other circumstances may include consumption of heavy caffeine, alcohol, taking any amount of physical or mental pressure,chronic fever, and intake of cigarettes.
- **Symptoms:** We have discussed about the causes that how TTH could may occur, but in this section we will discuss that what are the possible ways or signs which confirms that one is suffering from TTH.
  - The first sign could be if you are feeling constantly dull and in pain.
  - If you wrap a tight band around your head.
  - If you are feeling pain all over the head not only on particular points.

In these conditions the problem one may feel could be as difficulty in sleeping. There is no problem such as nausea and vomiting by TTH.

- **Tests:** If the pain is mild or moderate and after applying the home remedies the pain gets settle down then there is no need to see any doctor.
- **Treatments:** The treatment of this disease could be easily done by oneself by keeping track of the amount and duration of pain on a daily basis and if you see serious issue in those record then reach to the doctor immediately. Its treatment also includes of some drugs such as aspirin, acetaminophen to prevent pain. In this disease narcotic pain relievers are not prescribed.

Just make sure to have a proper treatment because if you don't do it properly then there are chances that you will be again affected by it very soon (Rastogi et al., 2018a).

BACKGROUND AND PURPOSE OF STUDY WITH EXPERIMENTAL SETUP

All the experiments are done in randomized, single blinded and protected environment so that the results wouldn't deviate much. Topics are taken from various neurology clinics and subjects referred by neurologists to the OPD of Hardwar Physiotherapy University.

Study Duration and Approval from Candidates:

The candidates were taken from Middle of January '17 tillMay'18 and uptoJuly 2018. Candidates were taken in the experiment only after when we got a consent approval signed from them. Another information-Consent Form was signed by the ethical committee was got completed by the subjects and clinical verification for the experiment was recorded and clearance was given by the ethical committee formed by the University.

Intervention

When all the candidates are divided into three groups, all the candidates are informed about the treatment that is going to be tested upon them. They were given biofeedback training in a separated room of, Hardwar research laboratory, which had very low lighting and negligible external noise, so that they could remain in relaxation state. All candidates underwent respective (EMG/GSR) BF training for 20 minutes per session for 07 sessions(Turk et al., 2008; Kropotov, 2009).

METHODOLOGY and COLLECTION OF DATA

1. **Study Design:** This study will be a randomized, single blinded, prospective controlled trial.
2. **Source of Data:** Data will be obtained from the subjects recruited from various neurology clinics and subjects referred by neurologists to the outpatient department of Department of psychology, DevSanskritiVishwavidyalaya, Hardwar, Uttarakhand and ABESEC, Ghaziabad for biofeedback therapy.
3. **Study Duration:** Subjects will be recruited from June-2019 up to June-2020 and followed up till October-2020.
4. **Informed Consent**: Subjects will be recruited in the trial only after obtaining informed consent from them. (Informed Consent Form approved by the ethical committee attached).
5. **Sampling Design**: Simple random sampling will be used with lottery method for allocation of subjects to seven groups. Subjects with stress and TTH(Tension Type Headache) will be enrolled in the study. Subjects who will not give consent and who will not meet the eligibility criteria will be excluded from the study. The rest of the subjects will be randomized using the lottery method for allocation. They will also be scrutinized after 30 days, 60 days.
6. **Allocation Procedure**: Chits numbered one to seven will be placed in a bowl and the subjects will be asked to pick the chits. Subjects with the following chit numbers will be allocated to the corresponding groups:
   a. EMG auditory biofeedback (EMGa) group.
   b. EMG visual biofeedback (EMGv) group.
   c. EMG auditory +visual (EMGav) group.
   d. GSR auditory (GSRa) group.
   e. GSR visual (GSRv) group.
   f. GSR auditory +visual (GSRav) group.
   g. Control group.
7. **Sample size**: Sample size will be calculated using the following formula:
   a. Probability of Type I error will be set at 0.05 Power of the study is expected to set at 80% (0.8).
8. Let p1= 1.0 and p2=0.75 are the mean differences of pre and post (baseline to one year) average frequency of headache per month in the EMG biofeedback training group and pain management group respectively from a study by (Mullay et al., 2009).
9. Let p=0.875 was calculated as (p1+p2)/2 and q=0.125 was calculated as 1-p. The sample size thus calculated was 26.6 per group. To accommodate for drop outs the sample size was chosen as 30 per group.

10. Universe and Sample

    a. This study will be a randomized single blinded controlled prospective study. We will select a no. of recruited subjects n, h (f and m males) will be randomly assigned to seven groups receiving electromyography feedback auditory (EMGa) (let n =27), visual (EMGv) (let n=28), combined audio-visual (EMGav) (let n=27), galvanic skin resistance biofeedback auditory (GSRa) (let n =26), visual (GSRv) (let n=29) and combined audio-visual (GSRav) (let n=28) and a control group (let n = 27). Each subject (except the control group) will receive 10 sessions of respective biofeedback for 15 minutes each in an isolated room. The control group will receive only medication prescribed by their treating doctor. Each patient will be kept blinded to the type of biofeedback (EMG or GSR) being given. Pain variables (average frequency, duration and intensity of headache per week), SF-36 quality of life scores will also be measured from survey to all n subjects.

All the psycho challenged cases living in Uttarakhand and NCR region (Delhi, Meerut, Ghaziabad, Faridabad, Gurugram, Modinagar and Muradnagar) will be the universe of study.

The college going students and Technocrats of different giant MNCs will be under study.

A control group of around 100 to 150 persons will be chosen and for carrying out this work the methodology employed shall be as follows:

a. Literature survey.

b. Identification of the location (Cluster near the region of NCR-national capital region Zone) and perform the study of different given parameters of the psychosomatic disorders due to life deregulation which disturb one's complete health.

c. Development of a model of Biofeedback based experiments which will be performed at Research Labs and Scientific Spirituality Centers of DevSanskritiVishwaVidyalaya, Hardwar and Patanjali Research Foundations, Uttarakhand.

d. Experimental investigation of Mental and spiritual health will be on various medical parameters.

e. We will Apply some spiritual techniques as per the symptoms observed, suitable to the patient as per his/ her age, diet, culture and habits.

f. Data Analysis of the Comparative study of both EMG and GSR machines with various spiritual techniques (Guided meditations) will be applied over the patients and their performance.

g. Analysis of the results obtained and verification of the efficiency of the technique and suggesting appropriate one will be done in repeated process in case the method doesn't work.

h. Impact of the result obtained on the society, the employee, company/ college and environment.

i. 25% area (approximately 25 wards) out of 103 wards will be selected as sample purposively.

GRAPHICAL RESULTS and INTERPRETATION

About the Study and Analysis

The Whole Analysis is being done in Tableau software which is a data Analytics tool and gives data analysis in different visualizations and was designed by 3 Professors of Standford university in their research work in 1996-97 and was launched in 2003.

The Microsoft had also various data visualization tools but Tableau is more powerful among all. Now MS has launched Power BI as latest data visualization tool which has more features and very less cost than Tableau (one seventh), but the data integrity and facility in Tableau are more popular with 10 + years of research experience.

The Current Study has been done with 95% Confidence interval and on different trend models. Where m is the slope of data pattern and c is intercept on Y-axis.

Linear y= mx+c
Logarithmic y=m*ln x + c
Exponential y=m*ex+ c
Polynomial y=ax³+bx²+cx+d – Cubic polynomial was used here but the degree may vary from 2 to 8
    as per need.
(Here the Cubic Polynomial is most suitable)
Power y=m*aˣ+c
In whole study the t-value and R² value of analysis model has been evaluated.
The T-value tells the significance of the model when the T-value<0.05
The R-value tells the Correlation and its value is between -1 and +1.
where the two values are positively correlated when the R-value is between 0 to 1.
where the two values are negatively correlated when the R-value is between -1 to 0.
where the two values are not correlated or less correlated when the R-value is near to 0 in both sides.

CORRELATION OF TTH START AGE, CHRONOCITY AND GENDER FOR GSR THERAPY

GSRa Mode

Trend/ Analysis of Correlation of TTH Start Age and Chronicity of TTH

The figure 1 is depicting two cluster in GSRa group has 4 possible Clusters
    The first cluster shows the Average Intensity of TTH chronicity in Later Age of 26 to 40.
    The Second and third clusters Show the Average intensity of TTH chronicity in later stage of life of 10 to 32 years.
    The fourth Cluster is fixed on TTH intensity of 12 in average age of life for 4 subjects.
    This has generally applied in the GSRa group.(By Fig. 1)

GSRa and Gender

The figure 2 depicts the Analysis Gender wise in GSRa group in which the 5 males and 21 females were selected in experiment and it was observed that the Male were facing the Chronic TTH in their middle age (# 27 to 39 Years) of life.
    Majority of the Females were spanned towards the later stage and their Chronic TTH span was from (#23 Years to 50 Years) of life. The 2 females suffered from it until the age of 50 years.
    The Rate of Chronic TTH in females is faster than male as per age factor.

Figure 1. The correlation of TTH start age and chronicity in GSRa mode

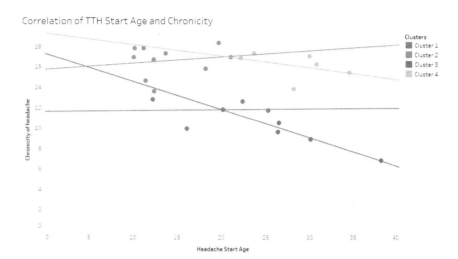

Figure 2. The correlation of age vs TTH start age based on gender GSRa mode

It shows the analysis that generally Females suffer from chronic TTH in later stage of their lives than Men. (By Fig. 2)

GSRv and Age

Trend/ Analysis of Correlation of TTH Start Age and Chronicity of TTH

The figure 3 is depicting two cluster in GSRv group has 4 possible Clusters

The first cluster shows the Average Intensity of TTH chronicity in Average Age of 10 to 24.

The Second and third clusters Show the High and Very high intensity of TTH chronicity in respectively average age and later age of life.

Figure 3. The correlation of TTH start age and chronicity in GSRv mode

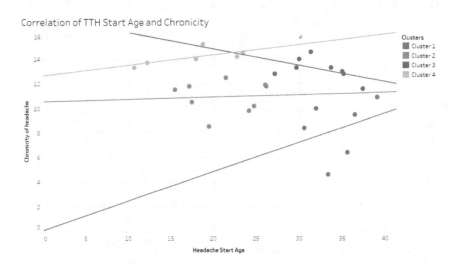

The fourth Cluster is average of TTH intensity of 12 in average age of life of 15 tom 36 years. This has generally applied in the GSRv group.(By Fig. 3)

GSRv and Gender

The figure 4 depicts the Analysis Gender wise in GSRa group in which the 9 males and 19 females were selected in experiment and it was observed that the Male were facing the Chronic TTH in their middle age (# 24 to 50 Years) of life.

Majority of the Females were spanned towards the later stage and their Chronic TTH span was from (#27 Years to 48 Years) of life.

Figure 4. The correlation of age vs TTH start age based on gender GSRv mode

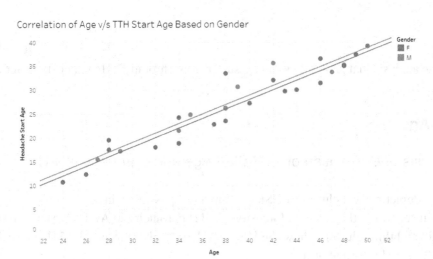

The Rate of Chronic TTH in females is faster than male as per age factor.

It shows the analysis that generally Females suffer from chronic TTH in later stage of their lives than Men.(By Fig. 4)

GSRav Mode and Age

Trend/ Analysis of Correlation of TTH Start Age and Chronicity of TTH

The figure 5 is depicting two cluster in GSRav group has 3 possible Clusters

The first cluster Shows the Average Intensity of TTH chronicity in Average Age of 13 to 31.

The Second and third clusters Show the low and Medium intensity of TTH chronicity in respectively average age and later age of life.

This has generally applied in the GSRav group.(By Fig. 5)

GSRav and Gender

The above figure 6depicts the Analysis Gender wise in GSRav group in which the 9 males and 19 females were selected in experiment and it was observed that the Male were facing the Chronic TTH in their middle age (# 24 to 50 Years) of life.

Majority of the Females were spanned towards the later stage and their Chronic TTH span was from (#27 Years to 48 Years) of life.

The Rate of Chronic TTH in females is faster than male as per age factor.

It shows the analysis that generally Females suffer from chronic TTH in later stage of their lives than Men.(By Fig. 6)

Figure 5. The correlation of TTH start age and chronicity in GSRav mode

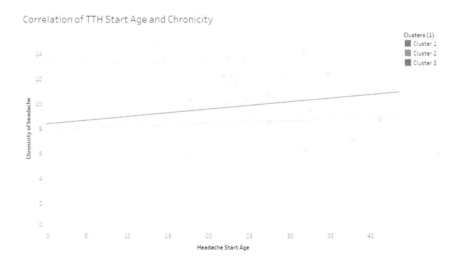

Figure 6. The correlation of age vs TTH start age based on gender GSRav mode

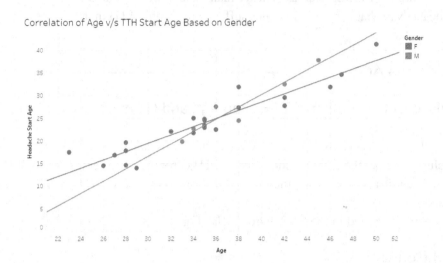

TREND LINE MODELS

Trend Lines Model (Fig.1) GSRa and Age

A linear trend model is computed for sum of Chronicity of headache (actual & forecast) given sum of Headache Start Age (actual & forecast). The model may be significant at p <= 0.05. The factor Clusters may be significant at p <= 0.05.\

Trend Lines Model (Figure 2) GSRa and Gender

A linear trend model is computed for sum of Headache Start Age given sum of Age. The model may be significant at p <= 0.05.

Table 1. It shows that on 26 subjects Under Study, all were under study and no one was filtered with 1.34 standard error and showing highly correlation with the subject health

Formula used to calculate model:	Clusters*(Headache Start Age + intercept)
Total observations that are modeled:	26
Total observations that are filtered:	0
Degrees of freedom for Model:	8
Residual degrees of freedom (DF):	18
SSE (sum squared error):	32.3744
MSE (mean squared error):	1.79858
R-Squared:	0.877614
Standard error:	1.34111
t-value (significance):	< 0.0001

Table 2. Analysis of Variance:

Field	DF	SSE	MSE	F	t-value
Clusters	6	182.7492	30.4582	16.9346	< 0.0001

Table 3. Individual Trend Lines:

Panes		Color	Line		Coefficients				
Row	Column	Clusters	t-value	DF	Term	Value	StdErr	t-value	t-value
Chronicity of headache	Headache Start Age	Cluster 3	0.030326	2	Headache Start Age	-0.275787	0.0491516	-5.61095	0.030326
					Intercept	17.3064	1.50785	11.4775	0.0075058
Chronicity of headache	Headache Start Age	Cluster 1	0.944562	4	Headache Start Age	0.0081618	0.110291	0.0740019	0.944562
					Intercept	11.6083	2.19013	5.30028	0.006086
Chronicity of headache	Headache Start Age	Cluster 2	0.651164	8	Headache Start Age	0.0598652	0.127477	0.469615	0.651164
					Intercept	15.7301	1.85554	8.47736	< 0.0001
Chronicity of headache	Headache Start Age	Cluster 4	0.427809	4	Headache Start Age	-0.113765	0.129051	-0.881548	0.427809
					Intercept	19.2972	3.68627	5.23488	0.0063625

Table 4. It shows that on 26 subjects Under Study, all were under study and no one was filtered with 3.41 standard error and showing highly correlation with the subject health(as R square vale is higher to 0.5)

Formula used to calculate Model:	Gender*(Age + intercept)
Total observations that are modeled:	26
Total observations that are filtered:	0
Degrees of freedom forModel:	4
Residual degrees of freedom (DF):	22
SSE (sum squared error):	256.354
MSE (mean squared error):	11.6524
R-Squared:	0.841744
Standard error:	3.41357
t-value (significance):	< 0.0001

Table 5. Variance Analysis:

Field	DF	SSE	MSE	F	t-value
Gender	2	7.9168459	3.95842	0.339708	0.715647

Table 6. Individual Trend Lines:

Panes		Color	Line		Coefficients				
Row	**Column**	**Gender**	**t-value**	**DF**	**Term**	**Value**	**StdErr**	**t-value**	**t-value**
Headache Start Age	Age	M	0.0601019	3	Age	1.26688	0.429683	2.9484	0.0601019
					intercept	-22.8272	14.8138	-1.54095	0.220982
Headache Start Age	Age	F	< 0.0001	19	Age	0.9968	0.0938828	10.6175	< 0.0001
					intercept	-14.177	3.40644	-4.16181	0.0005295

Trend Lines Model (Figure 3) GSRv and Age

To calculate the sum of Chronicity of headache either it is actual or forecasted alinear trend model could be used which give the result accordingly. The model value should lies as p<=0.05. The factor Clusters may be significant at p <= 0.05.

Trend Lines Model (Figure 4) GSRv and Gender

To compute the sum of Headache starting age when the start age is given is calculated by Trend line model. The model may be significant at p <= 0.05.

Trend Lines Model (Figure 5) GSRav and Age

To calculate the sum of Chronicity of headache either it is actual or forecasted alinear trend model could be used which give the result accordingly. The model value should lies as p<=0.05. The factor Clusters may be significant at p <= 0.05.

Table 7. It shows that on 29 subjects Under Study, all were under study and no one was filtered with 1.44 standard error and showing highly correlation with the subject health(as R square vale is higher to 0.5)

Formula used for Model:	Clusters*(Headache Start Age + intercept)
Total Observations that are Modeled:	29
Observations that are filtered:	0
Degrees of freedom for Model:	8
Residual degrees of freedom (DF):	21
SSE (sum squared error):	43.945
MSE (mean squared error):	2.09262
R-Squared:	0.751938
Standard error:	1.44659
t-value (significance):	< 0.0001

Table 8. Variance Analysis:

Field	DF	SSE	MSE	F	t-value
Clusters	6	116.25648	19.3761	9.25925	< 0.0001

Table 9. Individual Trend Lines:

Panes		Color	Line		Coefficients				
Row	Column	Clusters	t-value	DF	Term	Value	StdErr	t-value	t-value
Chronicity of headache	Headache Start Age	Cluster 2	0.878632	7	Headache Start Age	0.0185882	0.117367	0.158377	0.878632
					intercept	10.5469	2.55141	4.13374	0.0043834
Chronicity of headache	Headache Start Age	Cluster 3	0.190853	6	Headache Start Age	-0.133681	0.0906773	-1.47425	0.190853
					intercept	17.513	2.95683	5.92289	0.0010322
Chronicity of headache	Headache Start Age	Cluster 1	0.545113	4	Headache Start Age	0.238613	0.361338	0.660359	0.545113
					intercept	0.0598912	12.5209	0.0047833	0.996413
Chronicity of headache	Headache Start Age	Cluster 4	0.139687	4	Headache Start Age	0.0846463	0.0460171	1.83945	0.139687
					intercept	12.6532	0.843364	15.0032	0.000115

Table 10. It shows that on 29 subjects Under Study, all were under study and no one was filtered with 2.61 standard error and showing highly correlation with the subject health(as R square vale is higher to 0.5).

Formula to calculate Model:	Gender*(Age + intercept)
Total observations that are modeled:	29
Total observations that are filtered:	0
Degrees of freedom for model:	4
Residual degrees of freedom (DF):	25
SSE (sum squared error):	171.356
MSE (mean squared error):	6.85425
R-Squared:	0.90362
Standard error:	2.61806
t-value (significance):	< 0.0001

Table 11. Variance Analysis:

Field	DF	SSE	MSE	F	t-value
Gender	2	5.7890634	2.89453	0.422297	0.660129

Table 12. Individual Trend Lines:

Panes		Color	Line		Coefficients				
Row	Column	Gender	t-value	DF	Term	Value	StdErr	t-value	t-value
Headache Start Age	Age	M	< 0.0001	8	Age	0.996274	0.110781	8.99317	< 0.0001
					Intercept	-10.9195	4.23831	-2.57639	0.0327998
Headache Start Age	Age	F	< 0.0001	17	Age	1.00127	0.0803573	12.4602	< 0.0001
					Intercept	-12.0486	3.13935	-3.83792	0.0013178

Table 13. It shows that on 28 subjects Under Study, all were under study and no one was filtered with 1.67 standard error and showing highly correlation with the subject health(as R square vale is higher to 0.5)

Formula used to calculate model:	Clusters (1)*(Headache Start Age + intercept)
Total observations that are modeled:	28
Total observations that are filtered:	0
Degrees of freedom of Model:	6
Residual degrees of freedom (DF):	22
SSE (sum squared error):	61.4584
MSE (mean squared error):	2.79356
R-Squared:	0.625675
Standard error:	1.6714
t-value (significance):	0.0003472

Table 14. Variance Analysis:

Field	DF	SSE	MSE	F	t-value
Clusters (1)	4	92.369212	23.0923	8.26626	0.0003149

Trend Lines Model (Figure 6) GSRav and Gender

A linear trend model is computed for sum of Headache Start Age given sum of Age. The model may be significant at $p <= 0.05$.

STRESS ANALYSIS AND EFFECTS

Effect on Analgesic Consumption

Analgesic is also known as a pain killer it is basically a type of drug which is used to achieve pain relief (analgesia). The word analgesia itself means "the absence of the sense of pain while remaining con-

Table 15. Individual trend lines:

Panes		Color	Line		Coefficients				
Row	Column	Clusters (1)	t-value	DF	Term	Value	StdErr	t-value	t-value
Chronicity of headache	Headache Start Age	Cluster 3	0.919937	4	Headache Start Age	0.0237491	0.221945	0.107005	0.919937
					intercept	7.95365	7.66014	1.03832	0.357764
Chronicity of headache	Headache Start Age	Cluster 1	0.74668	10	Headache Start Age	-0.0221214	0.0666134	-0.332086	0.74668
					intercept	13.6147	1.56177	8.71752	< 0.0001
Chronicity of headache	Headache Start Age	Cluster 2	0.675655	8	Headache Start Age	0.0594091	0.136842	0.434144	0.675655
					intercept	8.38419	2.92354	2.86782	0.0208989

Table 16. It shows that on 28 subjects Under Study, all were under study and no one was filtered with 2.41 standard error and showing highly correlation with the subject health(as R square vale is higher to 0.5)

Formula used to calculated model:	Gender*(Age + intercept)
Total observations that are modeled:	28
Total observations that are filtered:	0
Degrees of freedom for Model:	4
Residual degrees of freedom (DF):	24
SSE (sum squared error):	140.246
MSE (mean squared error):	5.84359
R-Squared:	0.893267
Standard error:	2.41735
t-value (significance):	< 0.0001

Table 17. Variance Analysis:

Field	DF	SSE	MSE	F	t-value
Gender	2	22.123552	11.0618	1.89298	0.172437

Table 18. Individual Trend Lines:

Panes		Color	Line		Coefficients				
Row	Column	Gender	t-value	DF	Term	Value	StdErr	t-value	t-value
Headache Start Age	Age	M	0.0027439	4	Age	1.36545	0.207156	6.59142	0.0027439
					intercept	-24.4539	7.92176	-3.08693	0.0366834
Headache Start Age	Age	F	< 0.0001	20	Age	0.913222	0.0746196	12.2384	< 0.0001
					intercept	-8.00701	2.65732	-3.01319	0.0068688

scious". Analgesic drug affects our mind and nervous system in different ways. They are different from anesthetics. Which either temporarily and in some cases completely vanish the sensation. Analgesics include paracetamol, NSAIDs such as the salicylates, and opioid drugs. Nonsteroidal anti-inflammatory drugs are kinds of drug which lower the fever, reduces the pain and in case of Higher dose decrease inflammation. Some of the common drugs are aspirin, naproxen, ibuprofen. These all are available across all the countries easily. So basically, it was the group that consumed analgesic medicines to get relief from the stress and headache.

The Trend Lines show that there is significant reduction in analgesic consumption after the GSR therapy.

Effect of GSR on PM (Prophylactic Medication)

Along with traditional headache medication there are other drugs that help to reduce headache these are antidepressants.in medical this approach is referred as prophylaxis. An analysis published in 2014 proposed that antidepressants also reduce the depression in the people suffering from hepatitis c by 40 percent. Some studies even depict that is the person is pretreated with anti-depressants there are likely less chance to have depression symptom. Even the person which take depressants after them stroke is less likely to get depression.

Effect of Therapy on AM (Alternative Medicines)

Alternative medicine is also known as pseudo medicine or questionable medicine. Alternative medicine is claimed to have the healing effect like that of medicine which is unproven or impossible to prove and are likely to be harmful. According to scientist a therapy is said to be unproven and not working when it does not follow the natural laws and violate it. Alternative medicine is not same as traditional medicine. alternative medicine is a bit dangerous as it does not give proper results .it is used my significant number of people, but still large amount of funding is raised by united state government. The industry of alternative medicine is highly profitable. This fact is overlooked, or we can say that hidden by media, with alternative medicine portrayed positively when compared to big pharma. The subjects involved under this studies were provided with this Prophylactic medication and the anti-depressant to reduce their pain caused due to TTH (Fumal et al., 2008).

Effect on OM (Other Medication)

This was the group which was called as OM (Other medication) which include muscle relaxants and use of triptans. Triptan is a drug of the family tryptamine which is used as abortive medication in the treatment of headache and migraines. This drug is effective in treating individual headache, but it is not a cure for tension type headache especially for a person suffering from migraine. They are sometime effective in disabling the tension type headache. Triptans do not relieve other kinds of pain.

While a muscle relaxant is a drug that decrease muscle tone by affecting skeletal muscle. They are used to decrease muscle pain and hyperflexes.it may have following effects on our body.Dependency, Allergic Reactions, Sleepiness, Tapering offfor the treatment purpose, the subject were provided with various other treatment facilities like physical exercises which reduces their body stress to a large extent. These exercise consist of yoga, meditation and various other therapies which provide internal peace to the subjects so that they can feel stress free all the time without any side effects. It also include treatment like body massage which results in muscle relaxation and finally works in the direction of reduction of stress pain as stated by (Bouraue et al., 1991). Use of triptans are also the best ways for its cure, as triptans are basically the drugs which are used for the treatment of cluster headaches and migraines.

The Trend Lines show that there is significant reduction in PM, AM and OM after the GSR therapy.

NOVELTY IN OUR WORK

Nowadays, stress is associated with everyone's life and every person of any society at some levels is affected by stress problems and thus suffer TTH. The work in this research is unique as it aimsat reducing the headache level and to control the analgesic and prophylactic medications. Based on results it was seenthe transfer of people from one group to other. As the number of sufferings are increasing in the society and the world,these type of researches provide a means to treat various disorders according to (Millea & Brodie, 2002; Mullaly, 2009).

FUTURE SCOPE, LIMITATIONS & POSSIBLE APPLICATIONS

In our research paper, the experiment was conducted on a limited and fixed number of patients. And depending on the needs, it may be experimented on various patients based on different type of factors like age, gender, climate, etc. Biofeedback therapies provide a way to treat mental disorders with minimal use of medications and these therapies conducted on different age groups, genders of various countries. Biofeedback therapies help individuals to improve their physical and mental health by gaining control on their bodies and minds. These therapies increase awareness of these functions,so patients can control stress and physical ailments. Although it is bit costly for common people and efforts have been made to make this therapy accessible to every group of society. These therapies are also useful in treating:

1. Alcohol use disorder.
2. Drug addiction.
3. Insomnia.
4. Anxiety.
5. Chronic pains.
6. Migraines.
7. Nausea.

Various scientists believe that these therapies will bring a revolution in the treatment of TTH.The patients will be able to control their own pains and can record their real time reactionsas per(Crystal & Robbins, 2010; Zanella et al., 2004).

CONCLUSION

At starting of the experiment, we had 96 peoples in AC (Analgesic Consumption) group in the 1st month, which got reduced to 69 in the 3rd month,57 in 6th month and 37 in the 12th month. On seeing the results of AM group there was 1 single person in the 1st month,4 in the 3rd month,4 in 6th month and 7 in the 12th month. The number of persons kept on increasing in this group. In no medication group there were 14 people in the 1st month,17 in the 3rd month 22 in the month,31 in the 12th month. In PM group there were 22 people in the 1st month,14 in the 3rd month,8 in the 6th month, 4 in the 12th month. Here the people decreased with increased in time.

While in NA group there were 3people in the 1st month,17 in 3rd month,27 in 6th month and 48 in the 12th month.

A prophylactic is a treatment used to prevent diseases from occurring. "Prevention is better than cure"; these treatments works on this principle. They are used to prevent something from happening, for example, a prophylactic hepatitis vaccine prevents the patient from getting hepatitis. These are also called anti-depressants as they help to reduce headaches. Results show that there is significant effect of the GSR therapy on the prophylactic treatment.

The license for practicing varies based on different health care provider and different countries. Though the practice of alternative medicine is illegal, yet it is promoted by various practitioner in cancer treatment. Alternative medicine is criticized for taking advantage of the weakest members of society. PM- Prophylactic Medication, other Anti-Depressants and AM- Alternative Medicines. Prophylactic id basically a Greek word which means "an advanced guard". A prophylactic medication of a treatment or a medication which is designed to prevent a disease which a subject is suffering from.

ACKNOWLEDGMENT

We are extremely grateful to the researchers in the fields of scientifically spirituality for their guidance in the studies and our paperwork. Authors convey their esteem gratitude to Management of Dev Sanskrit VishwavidyalayaHaridwar and ShantiKunj where the Biofeedback experiments and study were conducted. Also, the infrastructure, staff support and students of ABESEC, Ghaziabad are specially thanked for contributing their time and to be an important part of this study.

REFERENCES

Arora, N., Rastogi, R., Chaturvedi, D. K., Satya, S., Gupta, M., Yadav, V., . . . Sharma, P. (2019b). Chronic TTH Analysis by EMG & GSR Biofeedback on Various Modes and Various Medical Symptoms Using IoT. In Big Data Analytics for Intelligent Healthcare Management. doi:10.1016/B978-0-12-818146-1.00005-2

Arora, N., Trivedi, P., Chauhan, S., Rastogi, R., & Chaturvedi, D. K. (2017a). Framework for Use of Machine Intelligence on Clinical Psychology to study the effects of Spiritual tools on Human Behavior and Psychic Challenges. *Proceedings of NSC-2017*.

Bansal, I., Rastogi, R., Chaturvedi, D. K., Satya, S., Arora, N., & Yadav, V. (2018k). Intelligent Analysis for Detection of Complex Human Personality by Clinical Reliable Psychological Surveys on Various Indicators. In *National Conference on 3rd MDNCPDR-2018 at DEI*, Agra.

Binder, H., & Blettner, M. (2015). Big Data in Medical Science—a Biostatistical View. *Deutsches Ärzteblatt International, 112*(9), 137.

Boureau, F., Luu, M., & Doubrere, J. F. (2001). Study of experimental pain measures and nociceptive reflex in Chronic pain patients and normal subjects. *Pain, 44*(3), 131–138. PubMed

Carlson, N. (2013). *Physiology of Behavior*. Pearson Education, Inc.

Cassel, R. N. (1997). Biofeedback for developing self-control of tension and stress in one's hierarchy of psychological states. Psychology: A Journal of Human Behavior, 22(2), 50-57.

Chaturvedi, D. K., Rastogi, R., Arora, N., Trivedi, P., & Mishra, V. (2017b). Swarm Intelligent Optimized Method of Development of Noble Life in the perspective of Indian Scientific Philosophy and Psychology. *Proceedings of NSC-2017*.

Chaturvedi, D. K., Rastogi, R., Satya, S., Arora, N., Saini, H., & Verma, H., … Varshney, Y. (2018b). Statistical Analysis of EMG and GSR Therapy on Visual Mode and SF-36 Scores for Chronic TTH. Proceedings of UPCON-2018.

Chauhan, S., Rastogi, R., Chaturvedi, D. K., Satya, S., Arora, N., Yadav, V., & Sharma, P. (2018d). Analytical Comparison of Efficacy for Electromyography and Galvanic Skin Resistance Biofeedback on Audio-Visual Mode for Chronic TTH on Various Attributes. *Proceedings of the ICCIDA-2018*.

Costa, F. F. (2014). Big Data in Biomedicine. *Drug Discovery Today, 19*(4), 433–440. doi:10.1016/j.drudis.2013.10.012 PubMed

Crystal, S. C., & Robbins, M. S. (2010). Epidemiology of tension-type headache. *Current Pain and Headache Reports, 14*(6), 449–45. doi:10.1007/s11916-010-0146-2 PubMed

Fumal, A., & Scohnen, J. (2008). Tension-type headache: Current research and clinical management. *Lancet Neurology, 7*(2), 70–83. doi:10.1016/S1474-4422(07)70325-3 PubMed

Gulati, M., Rastogi, R., Chaturvedi, D. K., Sharma, P., Yadav, V., Chauhan, S., & Singhal, P. (2019e). Statistical Resultant Analysis of Psychosomatic Survey on Various Human Personality Indicators: Statistical Survey to Map Stress and Mental Health. In *Handbook of Research on Learning in the Age of Transhumanism*. Hershey, PA: IGI Global; doi:10.4018/978-1-5225-8431-5.ch022

Gupta, M., Rastogi, R., Chaturvedi, D. K., Satya, S. A., Verma, H., Singhal, P., & Singh, A. (2019a). Comparative Study of Trends Observed During Different Medications by Subjects under EMG & GSR Biofeedback. IJITEE, 8(6S), 748-756. Retrieved from https://www.ijitee.org/download/volume-8-issue-6S/

Haddock, C. K., Rowan, A. B., Andrasik, F., Wilson, P. G., Talcott, G. W., & Stein, R. J. (1997). Home-based behavioral Treatments for chronic benign headache: A meta-analysis of controlled trials. *Cephalalgia, 17*(1), 113–118. doi:10.1046/j.1468-2982.1997.1702113.x PubMed

Haynes, S. N., Griffin, P., Mooney, D., & Parise, M. (2005). Electromyographic BF and Relaxation Instructions in the Treatment of Muscle Contraction Headaches. *Behavior Therapy*, *6*(1), 672–678.

Kropotov, J. D. (2009). *Quantitative EMG, event-related potentials and neurotherapy*. San Diego, CA: Academic Press.

Lee, C. H., & Yoon, H. J. (2017). Medical big data: Promise and challenges. *Kidney Research and Clinical Practice*, *36*(1), 461–475. doi:10.23876/j.krcp.2017.36.1.3

McCrory, D., Penzien, D. B., Hasselblad, V., & Gray, R. (2001). *Behavioral and physical treatments for tension Type and cervocogenic headaches*. Des Moines, IA: Foundation for Chiropractic Education and Research.

Millea, J. P., & Brodie, J. J. (2002). Tension type headache. *American Family Physician*, *66*(5), 797–803.

Mullaly, J. W., Hall, K., & Goldstein, R. (2009). Efficacy of BF in the Treatment of Migraine and Tension Type Headaches. *Pain Physician*, *12*(1), 1005–1011. PubMed

Rastogi, R., Chaturvedi, D. K., Satya, S., Arora, N., & Chauhan, S. (2018a). An Optimized Biofeedback Therapy for Chronic TTH between Electromyography and Galvanic Skin Resistance Biofeedback on Audio, Visual and Audio Visual Modes on Various Medical Symptoms. In the National Conference on 3rd MDNCPDR-2018 at DEI, Agra.

Rastogi, R., Chaturvedi, D. K., Satya, S., Arora, N., Singhal, P., & Gulati, M. (2018f). Statistical Resultant Analysis of Spiritual & Psychosomatic Stress Survey on Various Human Personality Indicators. *The International Conference Proceedings of ICCI 2018*. Doi:10.1007/978-981-13-8222-2_25

Rastogi, R., Chaturvedi, D. K., Satya, S., Arora, N., Sirohi, H., Singh, M., . . . Singh, V. (2018i). Which One is Best: Electromyography Biofeedback Efficacy Analysis on Audio, Visual and Audio-Visual Modes for Chronic TTH on Different Characteristics. Proceedings of ICCIIoT- 2018. Retrieved from https://ssrn.com/abstract=3354375

Rastogi, R., Chaturvedi, D. K., Satya, S., Arora, N., Yadav, V., Chauhan, S., & Sharma, P. (2018c). 'F-36 Scores Analysis for EMG and GSR Therapy on Audio, Visual and Audio Visual Modes for Chronic TTH. *Proceedings of the ICCIDA-2018*.

Rubin, A. (1999). Biofeedback and binocular vision. *Journal of Behavioral Optometry*, *3*(4), 95–98.

Saini, H., Rastogi, R., Chaturvedi, D. K., Satya, S., Arora, N., Gupta, M., & Verma, H. (2019c). An Optimized Biofeedback EMG and GSR Biofeedback Therapy for Chronic TTH on SF-36 Scores of Different MMBD Modes on Various Medical Symptoms. In *Hybrid Machine Intelligence for Medical Image Analysis*. Springer Nature Singapore Pte Ltd.; doi:10.1007/978-981-13-8930-6_8

Saini, H., Rastogi, R., Chaturvedi, D. K., Satya, S., Arora, N., Verma, H., & Mehlyan, K. (2018j). Comparative Efficacy Analysis of Electromyography and Galvanic Skin Resistance Biofeedback on Audio Mode for Chronic TTH on Various Indicators. Proceedings of ICCIIoT- 2018. Retrieved from https://ssrn.com/abstract=3354371

Satya, S., Arora, N., Trivedi, P., Singh, A., Sharma, A., Singh, A., ... Chaturvedi, D. K. (2019a). Intelligent Analysis for Personality Detection on Various Indicators by Clinical Reliable Psychological TTH and Stress Surveys. *Proceedings of CIPR 2019 at Indian Institute of Engineering Science and Technology.*

Satya, S., Rastogi, R., Chaturvedi, D. K., Arora, N., Singh, P., & Vyas, P. (2018e). Statistical Analysis for Effect of Positive Thinking on Stress Management and Creative Problem Solving for Adolescents. *Proceedings of the 12th INDIA-Com*, 245-251.

Scott, D. S., & Lundeen, T. F. (2004). Myofascial pain involving the masticatory muscles: An experimental model. *Pain, 8*(2), 207–215. doi:10.1016/0304-3959(88)90008-5 PubMed

Sharma, S., Rastogi, R., Chaturvedi, D. K., Bansal, A., & Agrawal, A. (2018g). Audio Visual EMG & GSR Biofeedback Analysis for Effect of Spiritual Techniques on Human Behavior and Psychic Challenges. *Proceedings of the 12th INDIACom*, 252-258.

Singh, A., Rastogi, R., Chaturvedi, D. K., Satya, S., Arora, N., Sharma, A., & Singh, A. (2019d). Intelligent Personality Analysis on Indicators in IoT-MMBD Enabled Environment. In *Multimedia Big Data Computing for IoT Applications: Concepts, Paradigms, and Solutions.* Springer; doi:10.1007/978-981-13-8759-3_7

Singhal, P., Rastogi, R., Chaturvedi, D. K., Satya, S., Arora, N., Gupta, M., . . . Gulati, M. (2019b). Statistical Analysis of Exponential and Polynomial Models of EMG & GSR Biofeedback for Correlation between Subjects Medications Movement & Medication Scores. IJITEE, 8(6S), 625-635. Retrieved from https://www.ijitee.org/download/volume-8-issue-6S/

Turk, D. C., Swanson, K. S., & Tunks, E. R. (2008). Psychological approaches in the treatment of chronic pain patients- -When pills, scalpels, and needles are not enough. *Canadian Journal of Psychiatry, 53*(4), 213–223. doi:10.1177/070674370805300402 PubMed

Vyas, P., Rastogi, R., Chaturvedi, D. K., Arora, N., Trivedi, P., & Singh, P. (2018h). Study on Efficacy of Electromyography and Electroencephalography Biofeedback with Mindful Meditation on Mental health of Youths. *Proceedings of the 12th INDIA-Com*, 84-89.

Wenk-Sormaz, H. (2005). Meditation can reduce habitual responding. *Advances in Mind-Body, 3*(4), 34–39.

Yadav, V., Rastogi, R., Chaturvedi, D. K., Satya, S., Arora, N., Yadav, V., ... Chauhan, S. (2018j). Statistical Analysis of EMG & GSR Biofeedback Efficacy on Different Modes for Chronic TTH on Various Indicators. *Int. J. Advanced Intelligence Paradigms, 13*(1), 251–275. doi:10.1504/IJAIP.2019.10021825

Zanella, A., Bui, N., Castellani, A., Vangelista, L., & Zorzi, M. (2004). Internet of things for smart cities. IEEE Internet Things J., 1(1), 22–32. doi:10.1109/JIOT.2014.2306328

Chapter 11
HAAR Characteristics– Based Traffic Volume Method Measurement for Street Intersections

Santiago Morales
Universidad Nacional de Colombia, Colombia

César Pedraza Bonilla
Universidad Nacional de Colombia, Colombia

Felix Vega
Universidad Nacional de Colombia, Colombia

ABSTRACT

Traffic volume is an important measurement to design mobility strategies in cities such as traffic light configuration, civil engineering works, and others. This variable can be determined through different manual and automatic strategies. However, some street intersections, such as traffic circles, are difficult to determine their traffic volume and origin-destination matrices. In the case of manual strategies, it is difficult to count every single car in a mid to large-size traffic circle. On the other hand, automatic strategies can be difficult to develop because it is necessary to detect, track, and count vehicles that change position inside an intersection. This chapter presents a vehicle counting method to determine traffic volume and origin-destination matrix for traffic circle intersections using two main algorithms, Viola-Jones for detection and on-line boosting for tracking. The method is validated with an implementation applied to a top view video of a large-size traffic circle. The video is processed manually, and a comparison is presented.

DOI: 10.4018/978-1-7998-1839-7.ch011

INTRODUCTION

As cities increase in size, vehicular traffic becomes increasingly dense and mobility more complex, thus affecting the lives of millions of people. For this reason it is necessary to measure and model traffic through studies that collect primary information directly from the roads. This is critical when deciding how to design intersections, roads, traffic lights, crosswalks, bridges, among others, since without prior and accurate knowledge of the vehicular flow, such designs may not be optimal.

Vehicle counting is a tool that allows to know in detail how mobility behaves in cities. Knowing how many cars pass through a certain point, as well as their average speed, helps to update vehicle traffic models. The most commonly used method to obtain this information is manual vehicle counting (Federal Highway Administration, 2013). However, this method has fundamental problems, such as significant measurement errors (due to human intervention), lack of traceability, low frequency in taking measurements and high costs. Additionally, the problem is exacerbated depending on where the measurements are to be obtained – for example, it is more difficult to measure the amount of vehicles at an intersection than in a straight segment of the road (Tuydes-Yaman, Altintasi, & Sendil, 2015).

As technology advances, alternative technological methods have begun to be implemented, allowing to solve some of the problems explained. Among all emerging technologies, image processing and computer vision stand out, as they are alternatives that are gaining acceptance due to their reduced errors, the validation and traceability scope, and the ability to take measurements more frequently and in more locations within a city.

This paper introduces a solution that seeks to measure vehicular flow at intersections (particularly large intersections) through image processing algorithms, both for detection and tracking. This article is organized as follows: section 2 presents the background of the problem, section 3 sets out in detail the methodology used, section 4 describes the design of the experiment that will validate the results, which are presented in section 5. Sections 6 and 7 present the results and conclusions, respectively.

BACKGROUND

Below is the background of the problem from the point of view of the measurement and modeling of vehicular traffic in cities, and the tools currently used to collect primary information that allows designing such models. The second point of view illustrates the most common current technologies used to detect and track objects through image processing. Finally, the central problem of the article is presented, i.e. the difficulty of obtaining the origin-destination matrix (known as OD Matrix) at intersections.

Traffic Volume Measurement

As the volume of vehicles increases in large cities, it becomes necessary to measure, analyze, model and evaluate vehicular traffic, in order to more accurately determine their behavior on the roads of modern cities. This allows making wise decisions that contribute to the improvement of mobility. For example, traffic models influence the design of roads, vehicular intersections, or the optimization of traffic signal timing (de la Rocha, 2010).

The traffic problem began to be analyzed in the 50s, when mathematical models began to compare vehicular flow with the movement of particles in fluids (Hoogendoorn & Bovy, 2001). Since then, the subject has been thoroughly studied and debated. However, it is necessary to collect information in real environments to evaluate and contrast it with reality.

Manual counting is within the most commonly used methods, in which trained personnel are located on roads or intersections to count vehicles (Instituto Nacional de Vías (INVIAS), 2013). However, this method is expensive, is not performed regularly, and manual counts are error-prone. Other alternatives replace the human being with automatic devices such as pneumatic sensors, infrared sensors, or inductive loops, all installed on the road (Federal Highway Administration, 2013). These alternatives reduce human-induced error but are expensive and invasive with the infrastructure.

Automated Traffic Measurement

Some more sophisticated alternatives for vehicle counting include the use of platforms based on location systems such as GPS (Suda, 2017) or VANETS (Chen et al., 2016), which allow to know the precise location of vehicles in the roadway and develop very precise traffic models from this information. These options are still far from being a real alternative to measure the total vehicle flow, since nowadays they are not yet available in a significant number of vehicles (Saini, Alelaiwi, & Saddik, 2015).

Another increasingly accepted technological alternative is video-based vehicle counting, for which there are two approaches. One is to record a video at the desired location and then analyze it manually (Federal Highway Administration, 2013). This approach has similar drawbacks to the case of on-site manual counting, but allows the results to be traced and validated. The other alternative is automatic counting using image processing, which, as technology advances, becomes more common every day. The most relevant methods in the field of image processing and computer vision, which can be used for vehicle counting, are summarized below.

Object Detection Using Image Processing

In this section some of the relevant algorithms related to object detection using image processing are highlighted. For example, the algorithm proposed by Viola-Jones (Viola & Jones, 2004) was the first to provide competitive real-time detection rates. It consists of applying cascade HAAR feature-based (Haar, 1910), such as primary weak classifiers that assess the differences in light intensity between different regions of the image. The best HAAR features and the best weak classifiers are obtained through training using the adaboost algorithm (Freund & Schapire, 1997). With an approach similar to that used by Viola-Jones with HAAR features, other authors have proposed extracting other properties from the image. Some relevant examples are the Local Binary Patterns, abbreviated LBP features (Bouwmans et al., 2018) and Histogram of Oriented Gradients, abbreviated HOG (Silva, Bouwmans, & Frélicot, 2015), which have been used successfully to detect pedestrians (X. Wang, Han, & Yan, 2009). The Integral Channel Features, abbreviated ICF (Dollar, Tu, Perona, & Belongie, 2012) has also been used successfully with regard to this problem. Another example is the Scale-Invariant Feature Transform, abbreviated SIFT (Lowe, 1999), used to detect faces (Luo et al., 2007).

On the other hand, ANN (Artificial Neural Networks) are computer systems that have gained much attention and acceptance in recent years thanks to their performance in a wide variety of problems, ranging from classification or prediction, to speech recognition, or medical diagnosis (Liu et al., 2017). Of course, image processing has also been influenced by neural networks, as they have shown versatility in the detection and classification of objects in digital images. For example, OverFeat (Sermanet et al., 2013) is a framework that allows training and adapting Convolutional Neural Networks (CNN) for object detection. Another example is the Fast Region-Based Convolutional Neural Network or Fast R-CNN method (Girshick, 2015). Finally, the progress achieved by architectures such as RetinaNet (Lin, Goyal, Girshick, He, & Dollar, 2017), YOLO (Redmon, Divvala, Girshick, & Farhadi, 2016) and YOLOv3 (Redmon & Farhadi, 2018) stands out, since they not only detect and classify objects with high success rates, but also—in some cases—allow real-time detections.

Object Tracking Using Image Processing

Below are some of the most relevant algorithms for object tracking. It is worth noting that all the mentioned detection algorithms have two phases. The first one is initialization, where it is required to know in advance the initial position of the object to be tracked. Subsequently, during the tracking phase the algorithm updates the position of the object, frame by frame. There are some well known examples of tracking algorithms like on-line boosting or simply boosting (Grabner, Grabner, & Bischof, 2006). During initialization, this algorithm creates a model from a positive sample (the object in its initial position) and several negative samples (the surrounding environment) using Adaboost. Then, the model adds more positive and negative samples of the object and its environment to adapt to changes and optimize the model.

Another example is the Kernelized Correlation Filter algorithm, known simply as KCF (Henriques, Caseiro, Martins, & Batista, 2015). The basic idea is to estimate an optimal image filter (correlation filter) so that the input image produces the desired response; i.e. a Gaussian distribution focused on the target. During initialization, the model is trained with a set created from spatial movements of the target. The model is updated with each frame to adapt to changes in the object and the environment.

Median Flow (Kalal, Mikolajczyk, Matas, & Republic, 2010) is an algorithm that works well when the movement is slow and predictable. It is based on the idea of predicting the future movement, when compared with the previously recorded trajectory. With each frame, the error between prediction and past records is reduced. This algorithm should not be used over long tracking distances, or with objects with changing trajectories.

Another tracking algorithm called Multiple Instance Learning or MIL (Z. Wang, Yoon, Xie, Lu, & Park, 2016) is similar to on-line boosting, but instead of a positive sample and several negative samples to obtain the model, MIL uses "bags" of samples. Only one of these bags contains several positive samples of the object. According to the authors, thanks to this there is a greater variability to choose the features that best describe the object.

Some other examples of relevant algorithms for object tracking are: Channel and Spatial Reliability Tracker or CSRT (Lukežič, Vojíř, Čehovin Zajc, Matas, & Kristan, 2018), Tracking, Learning and Detection or TLD (Kalal, Mikolajczyk, & Matas, 2010), Minimum Output Sum of Squared Error or MOSSE (Bolme, Beveridge, Draper, & Lui, 2010) and Generic Object Tracking Using Regression Networks or GOTURN (Held, Thrun, & Savarese, 2016).

The Particular Problem of Intersections

Many of the technologies mentioned in Section 2.1 are useful for counting vehicles and obtaining average speeds on straight roads. However, analyzing an intersection often requires the OD Matrix (Tuydes-

Yaman et al., 2015). This matrix corresponds to the vehicle count for each possible entry-exit at the intersection. For example, figure 1 shows a four-way intersection, for a total of sixteen individual vehicle counts. It is very difficult to measure this matrix using conventional indicators (Al-Sobky & Hashim, 2014). In roundabouts and large intersections, error rate increases and it is even more difficult to know that matrix accurately. This article focused on overcoming these limitations in order to measure the OD Matrix of a roundabout of significant size.

Proposed Method

Figure 2 shows the method designed to address the problem of detecting and tracking vehicles in roundabouts. A video was recorded from a top view with a drone, so that all junctions were fully visible (more details about the video's characteristics and the recording conditions in Section 4).

In this method, it is necessary to first establish some initial conditions (such as defining the origin and destination zones where vehicles enter and leave the intersection, or the average size of the vehicles to be analyzed), then it goes into a loop in which each frame of the input video is captured and analyzed sequentially. This analysis consists of two main steps: detection and tracking. The detection focuses only on the areas defined by the origin zones. As new vehicles are detected, a specialized tracking algorithm updates the positions of each vehicle in the frame, so that when one reaches a destination zone, the OD matrix is updated.

1:   procedure: main
2:   initialize **originZones, destinyZones**
3:   load **inVideo**
4:   initialize **vehicles, OD_matrix**
5:   while True:
6:   get new **frame** from **inVideo**
7:   if **frame** is Null:
8:   break
9:   for **originZone** in **originZones**:
10:   detect **vehicles** on **originZone** on **frame**
11:   for **vehicle** in **vehicles**:
12:   track **vehicle** and update its center
13:   for **destinyZone** in **destinyZones**:
14:   if **destinyZone** contains **vehicle**:
15:   delete **vehicles** from **vehicles**
16:   update **OD_Matrix**
17:   exit

Algorithm 1. Detailed version of the flow chart illustrated in figure 2

The algorithm 1 explains the previous process in greater detail and can be divided into the following parts, which will be described in the following subsections:

- Initialization and frame capture: From lines 1 to 8 (Section 3.1).
- Detection: Lines 9 and 10 (Section 3.2).

Figure 1. Example of a four-way intersection and its corresponding OD Matrix

Figure 2. Illustrates the flow chart of the proposed method. The two most relevant steps (detection and tracking), for which independent specialized algorithms were used, are highlighted in red

- Tracking: Lines 11 and 12 (Section 3.3).
- Update of OD Matrix and end of application: Lines 13 to 17 (Section 3.4).

Initialization and Frame Capture

At initialization, the user sets the initial conditions of the application. The most relevant information to be provided is described below:

- **Input and Output Videos**: The users must define the path of the video file they want to analyze. They can also decide if they want to obtain an output video along with the analysis result.
- **Origin Zones**: An origin zone is an area that denotes an entry point to the intersection and is where vehicles are detected. They must be parallel to the track, so they must have a user-defined angle of rotation as illustrated in figure 3.a. It is worth noting that the area of an origin zone (in the conditions of the video) is between 7k and 17k pixels, while the area of the whole frame is 4.1M pixels. This significant reduction of the area facilitates vehicular detection. The user can define an origin zone with several detection areas to avoid occlusion as shown in figure 3.b.

Figure 3. Origin zones defined by the user. Figure 3.a shows the point of origin that is always the relative upper left corner; the width (W) and the height (H); and the angle of rotation parallel to the direction of the road. Figure 3.b illustrates an origin zone partitioned in several detection areas, so the occlusion caused by the tree can be minimized.

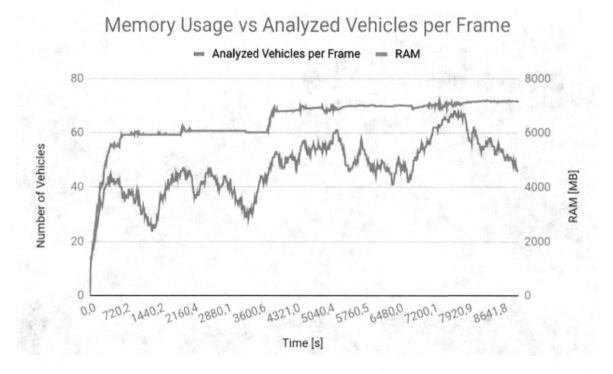

- **Destination Zones**: A destination zone is an un-rotated area, which must be located at the exit paths of the intersection. When a vehicle crosses it, It helps to determine the vehicle's destination to update the OD Matrix accordingly.
- **Standard Vehicle Size (in Pixels)**: The apparent size in pixels of the vehicles to be detected and tracked is a sensitive parameter that the user must define.

The user must set all possible destination zones, but the number of origin zones is optional. In the OD Matrix, the columns correspond to the origin zones and the rows correspond to the destination zones, at each position inside this matrix, a counter registers how many vehicles entered and leaved the intersection at the corresponding origin and destination zones.

Then, the main loop starts. The frames of the source video are captured and analyzed sequentially. This is necessary, since the information collected from one frame will be required in the next cycle. This process is repeated until the origin video ends, or until the user determines it (interruption).

Detection

In this step, new vehicles are detected on each origin zone. To successfully achieve this task, a previous conditioning of the origin zone and posterior validations of the detections are required. The whole process is explained below in algorithm 2.

Algorithm 2.
1:    procedure: *detectNewVehicles* (**originZone**, **frame**, **vehicles**)
2:    obtain **rotatedROI** from **originZone** on **frame**
3:    use Viola-Jones cascade classifier to get **detections** on **rotatedROI**
4:    depure **detections**
5:    if **detections** not Null:
6:    for **detection** in **detections**:
7:    if **detection** not contains **vehicle** in **vehicles**:
8:    create new **vehicle** in **vehicles**
9:    set new **vehicle** validity to False
10:   else if **detection** contains **vehicle** in **vehicles** and **vehicle** is not valid:
11:   validate **vehicle** in **vehicles**
12:   else:
13:   depure **vehicles**

The steps of this algorithm are described below.

- **Line 1**: The procedure receives the following input parameters:
  - Origin zone: Entrance to the intersection to be analyzed.
  - Frame: Current video frame.
  - Vehicles: List containing the vehicles detected so far.
- **Line 2**: The ROI is obtained after rotating and trimming the current frame with respect to the origin zone (figure 4). This implies that the training set only requires images of vehicles facing one direction.

Figure 4. a) shows a user-defined origin zone with a 200° rotation angle, with respect to the X axis. Figure 4.b illustrates the ROI obtained after trimming and rotating the frame with respect to the origin zone.

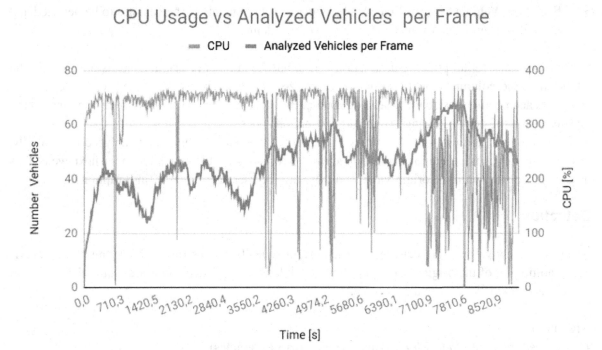

- **Line 3**: In this line, the classifier from the Viola-Jones algorithm performs the detection (more details below). The result is a list of detections (rectangles referenced with respect to ROI).
- **Line 4**: The previous line does not necessarily produce one vehicle detection. The same vehicle may have two or more detections. Therefore, the results are refined to ensure that there is only one detection per vehicle.
- **Lines 5 and 6**: Lines 7 to 11 are processed if at least one detection exist. The procces is executed once per detection.
- **Lines 7 to 9**: Each detection is compared with the vehicles in the list. Those that do not match any, are added as a new invalidated vehicle.
- **Lines 10 and 11**: On the other hand, detections that match invalidated vehicles increase an internal validation counter of the vehicle. In order to be validated, each vehicle must be detected in 5 frames. This reduces the false positive rate. Detections that match validated vehicles have no effect.
- **Lines 12 and 13**: If there are no detections for 10 frames, invalidated vehicles are removed. This eliminates false positives that have been validated and release computational resources.

Algorithm 2 is used once per frame, per detection area in each origin zone.

Viola-Jones for Vehicular Detection

In order to use the algorithm proposed by Viola-Jones (Viola & Jones, 2004), it was necessary to train a classifier to detect vehicles passing through the origin zones. To train the classifier, a data set with two classes (vehicles and not vehicles) was created. Also, several suggestions, concluded from the training processes can be noted. For example, it is recommended to increase the amount and heterogeneity of positive images. It is convenient to have a variety of backgrounds, colors and shapes in the set of positive samples. Likewise, the amount of positive samples is relevant, more robust classifiers can be obtained by increasing the quantity of vehicles in the data set. On the other hand, the width and height of the images are sensitive parameters. On the processed video, the analyzed vehicles (sedans, vans or SUVs) fitted on an area of approximately 100 x 50 pixels, this area was reduced to 50 x 25 pixels, within which 1'176'509 possible HAAR features exist and should be filtered. It is recommended to use real adaboost (RAB). There are several variations of "Boosted" classifiers (Friedman, Hastie, & Tibshirani, 2000), among which the following were evaluated: DAB (Discrete Adaboost), RAB (Real Adaboost) and GAB (Gentle Adaboost), in section 5 will be shown that the best performance was obtained with RAB. As noted above (algorithm 2), the detection areas were rotated, so positive images contain vehicles facing only one direction (figure 5). This reduces significantly the quantity of positive samples. Negatives should include images of the usual environment. It is preferable to include in the set of negative samples, images of roads with crosswalks, asphalt and traffic signs, but always without vehicles.

500 images of vehicles were used. The samples were heterogeneous and all were oriented at an angle of 0° to the X axis of the frame. Figure 5 illustrates some positive samples used in training.

For negative samples, 1600 images with no vehicles were used. Distinctive backgrounds of roads and highways (with crosswalks or traffic signs) were included.

Figure 6 illustrates some of the stages of the final classifier obtained during the training process. It also shows the HAAR features that make up those stages.

A fourteen stage cascading classifier was obtained, each with 3 to 7 HAAR features. This is the classifier that was obtained using the architecture that offered the best results, "Real Adaboost". Section 5.1 shows the comparison with other classifiers.

Tracking

Each vehicle that has already been properly detected and validated is tracked until it reaches any of the user-defined destination zones. The position of the centroid of the vehicle updates frame by frame, so that it can be determined where it leaves the intersection and next the OD Matrix can be updated. In the context of the presented problem, tracking should ideally meet the following requirements:

- It must be able to perform multiple tracking.
- It must be very accurate; under the conditions of the video, 1 meter equals approximately 8 pixels, so maximum tolerances of $\pm$ 4 pixels are expected with respect to the true centroid.
- It must have low execution times (1 millisecond or better per centroid update).
- It must be tolerant to disturbances and changes in drone movement and changes in light conditions.
- It must be tolerant of high vehicle flows.
- It must be able to track the vehicle over long distances.
- It must be able to track vehicles with low speeds that may even be static.

Figure 5. Some positive samples used in training. All images had a resolution of 50 x 25 pixels and were right-oriented. The variation in background, color and shapes is relevant.

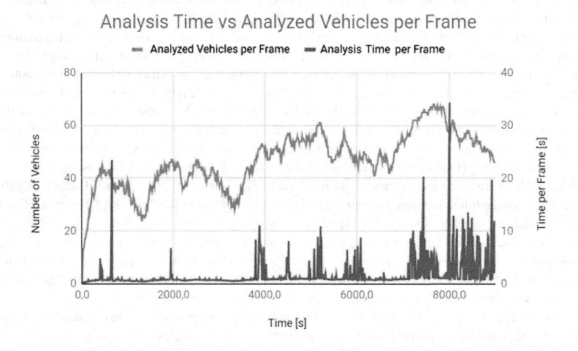

Figure 6. Some stages of the obtained classifier, each with its corresponding HAAR features.

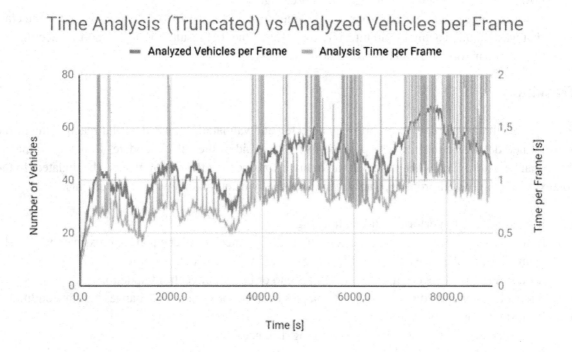

- Preferably, it should be resistant to occlusion.

In the application, the following five algorithms were tested to determine which one meets most of the above conditions:

- Kernelized Correlation Filters (KCF).
- Multiple Instance Learning (MIL).
- Median Flow.
- Tracking-Learning-Detection (TLD).
- On-line Boosting.

On section 5 will be shown that the best results were obtained with on-line boosting, therefore the following explanation will focus on this algorithm.

Tracking Initialization

In general, tracking algorithms need an first reference of the object to be tracked in order to initialize parameters. In the case of on-line boosting, a model is created from an initial position (obtained from the detection process) and the current frame. A minimal data set is defined, one positive sample is obtained from the reference and several negative samples are taken from the surroundings to model the object background. With this information, the problem acquires a binary classification approach in which it seeks to differentiate between the object and the background. Figure 7 illustrates the above. The initial model is trained at runtime, identifying through Adaboost the HAAR and LBP (Local Binary Patterns) features that allow to differentiate more precisely between the object and the background.

The on-line boosting initialization process is more expensive than its execution. The results obtained showed that initialization may take significantly longer than position updating.

Figure 7. The initial position (from the detection stage) helps to form the first positive sample of the binary classification model (Grabner et al., 2006). The background is modeled using negative samples obtained from areas of the same size surrounding the positive sample.

Tracking Procedure

After initialization, as new frames arrive, the on-line boosting process repeats cyclically, but instead of training a completely new model, an update is executed. To do so, the current classifier is first assessed in order to obtain a new vehicle position. Using this position, a positive sample and several surrounding negative samples are created again, thus updating the model. This allows the algorithm to adapt to changes in the background and the object. It is worth noting that square areas were used instead of rectangular areas to generate the positive and negative samples on which the detection is based. This is because the vehicles are rotating around their Z axis (perpendicular to the ground). If the tracking process had used a rectangular areas, eventually, much of the vehicle would be out of the rectangle, and in the long term this may cause tracking errors. On the contrary, this does not happen with square areas. This is illustrated in figure 8.

OD Matrix Update

Once vehicles have been detected, validated and tracked, it is evaluated whether the updated centroid of each vehicle at the intersection is within a destination zone. If so, the OD Matrix is updated by increasing the corresponding counter according to the origin and destination of the vehicle. Other metrics, such as average speed, can be calculated at this point. Finally, the vehicle is removed from the analysis list. Figure 9 illustrates the real OD Matrix, with the vehicle counting and the average speed, updated to the frame 6503 in the output video.

Figure 8. Illustrates the difference between a rectangular area in figure 8.a. and a square area in figure 8.b. The vehicle in the rectangular area initially fits perfectly, but it does not do so when it turns at the intersection, which causes tracking errors. In the case of the square area, this does not happen regardless of the angle of rotation of the vehicle.

Figure 9. The 4x4 OD Matrix, up to frame 6503 of the output video obtained from the application. It shows the vehicle counting and the average speed of each possible combination between the four origin zones versus the four destination zones.

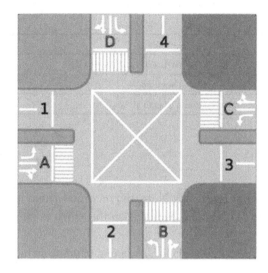

EXPERIMENT DESIGN

A video of about 44 minutes was recorded in a roundabout (divided into 3 sections of 12 minutes and one of 8 minutes), of which 5 minutes were used to generate positive samples for the classifier training (section 3.2) and 6 minutes and 15 seconds for the analysis and validation. The most relevant conditions and characteristics of the sample video and the chosen location are the following:

- **Time:** Between 8:00 and 9:00 (GMT -5).
- **Location:** 4.6594225, -74.0886375 (Roundabout of Carrera 60 - Calle 63, Bogotá - Colombia).
- **Light Conditions:** Daylight, partly cloudy.
- **Vehicle Flow:** High.
- **Dimensions and Registered Area:** Approximately 340 x 191 meters, equivalent to 64940 m².
- **Video resolution:** 2720 x 1530 pixels.
- **Pixel-meter Ratio:** 8 pixels / m.
- **Number of Origin Zones:** 4.
- **Number of Detection Areas:** 6.
- **Number of Destination Zones:** 4.
- **Vehicle Size:** 100 x 50 pixels (reduced to 50 x 25 pixels in training).

Table 1 shows the characteristics of the hardware used to perform the analysis.

Table 1. Characteristics of the hardware where the application was run

Feature	Value
CPU	Intel(R) Core(TM) i7-4500U CPU
Cores / Logical CPUs	2 / 4
RAM	8 GB
Disk	HDD 1 TB
Graphics Card	Not used
O.S	Linux Ubuntu 16.04

Experiment Settings

First of all, the initial conditions were defined as shown in figure 10. Next, the recorded video is analized by the developed application, and an output video is obtained, each car on the video is tagged as shown in figure 11.

To validate detection, the results in the output video are compared with a manual validation, in which the exact amount of vehicles that entered the intersection is determined, both detectable (sedans, vans or SUVs), and not detectable, (buses, trucks and motorcycles).

Regarding tacking, it is suggested to use evaluation strategies where the position of the tracked object is verified frame by frame against a known reference (Lehtola, Huttunen, Christophe, & Mikkonen, 2017) . However, measuring each vehicle frame-by-frame would have been too demanding (considering that around 500 vehicles were tracked with approximately 700 frames for each one), so it is proposed to use the labels on the vehicles of the output video as a visual aid, to account for those that could not be tracked successfully throughout their trajectory at the intersection (see figure 17 in section 6).

Finally, the performance of the application is estimated using external tools[1], and the time intervals are measured using libraries within the application itself[2].

The above procedure can be repeated with the same input video but with different tracking algorithms (TDL, KCF, Boosting, etc.) or with classifiers trained with different algorithms (RAB, GAB, etc.), in such a way that the different output videos can be used to compare the results obtained in each case.

Definitions

Confusion Matrix for Detection

Since detection is a classification problem, it is convenient to record the results in a confusion matrix (Table 2.

Detection Error

From the confusion matrix described above, several metrics can be extracted, especially the TPR (True Positive Rate) and the FNR (False Negative Rate), which can be determined according to Equations 1 and 2.

Figure 10. Status of the video prior to analysis and validation. It shows the origin zones denoted as Z1, Z2, Z3 and Z4; as well as the destination zones denoted as A, B, C and D. At the bottom, it shows the global count of "Frames", vehicles detected "Cars" and fully-tracked vehicles "Counted". Finally, the OD Matrix is shown at the top-left corner, including the vehicle counting and the average speed, with each position set to zero.

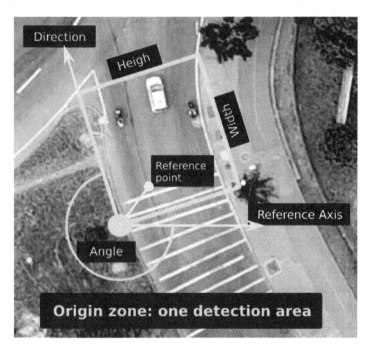

Figure 11. It illustrates how the identifier "Z4 8" is assigned to the tracked vehicle, i.e. it is the eighth vehicle identified during the analysis and belongs to the origin zone "Z4"

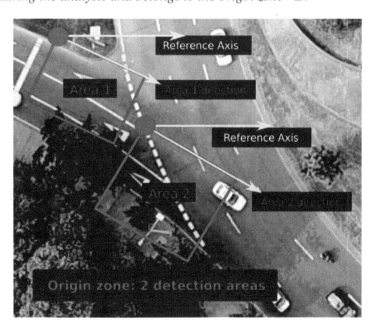

Table 2. Confusion Matrix for the detection algorithm.

		Real vehicles		Total of Evaluation
		Detectables	*Not Detectables*	
Evaluation of Detection	**Detected**	V_{TP}	V_{FP}	$V_{TP}+V_{FP}$
	Not Detected	V_{FN}	V_{TN}	$V_{FN}+V_{TN}$
Total Real Vehicles		$V_P=V_{TP}+V_{FN}$	$V_N=V_{FP}+V_{TN}$	

$$FNR = \frac{|V_P - V_{TP}|}{V_P} \times 100\% \qquad\qquad [1]$$

$$TPR = \frac{V_{TP}}{V_P} \times 100\% = 100 - FNR \qquad\qquad [2]$$

Where:

    FNR: False negative rate

    TPR: True positive rate

    V_P: Number of detectable vehicles that went through the origin zone

    V_{TP}: Number of vehicles that were correctly detected

    FNR is equivalent to the detection error.

Tracking Error

The tracking error is given by Equation 3, and the successful tracking rate is given by Equation XX.

$$E_{TR} = \frac{V_{NTR}}{V_{TP}} \times 100\% \qquad\qquad [3]$$

$$STR = 100 - E_{TR} \qquad\qquad [4]$$

Where:

    E_{TR}: Tracking error

    STR: Successful tracking rate

    V_{NTR}: Number of vehicles in which tracking was not successful

    V_{TP}: Number of vehicles that were correctly detected

Combined Error

In each origin zone, the combined error of the two algorithms can be determined as the rate of vehicles that were not properly included in the OD Matrix. This involves computing the error produced by both the detection algorithm and the tracking algorithm. This error is defined in Equation 5.

$$E_C = \frac{TPR \times E_{TR}}{100} + FNR \qquad\qquad [5]$$

Where:

E_C: Combined error of detection and tracking per origin zone

E_{TR}: Tracking error

TPR: True positive rate

FNR: False negative rate

Given the above, the detection, tracking and combined error can now be defined throughout the intersection. These are defined in Equations 6 to 7.

$$V_{TotalP} = \sum_{i=1}^{n} V_{Pi} \qquad\qquad [6]$$

$$V_{TotalTP} = \sum_{i=1}^{n} V_{TPi} \qquad\qquad [7]$$

$$FNR_{TOTAL} = \frac{1}{V_{TotalP}} \sum_{i=1}^{n} FNR_i \times V_{Pi} \qquad\qquad [8]$$

$$E_{TR-TOTAL} = \frac{1}{V_{TotalTP}} \sum_{i=1}^{n} E_{TRi} \times V_{TPi} \qquad\qquad [9]$$

$$E_{TOTAL} = \frac{1}{V_{PT}} \sum_{i=1}^{n} E_{Ci} \times V_{Pi} \qquad\qquad [10]$$

$$ASR = 100 - E_{TOTAL} \qquad\qquad [11]$$

Where:

n: Number of origin zones defined by the user

V_{Pi}: Number of detectable vehicles that went through the i-th origin zone

V_{TotalP}: Total number of detectable vehicles that went into the intersection

V_{TPi}: Number of vehicles correctly detected in the i-th origin zone

$V_{TotalTP}$: Total number of vehicles correctly detected

FNR_i: False negative rate of the i-th origin zone

FNR_{TOTAL}: Total false negative rate in the whole intersection

E_{TRi}: Tracking error of the i-th origin zone

$E_{TR+TOTAL}$: Total tracking error in the whole intersection

E_{Ci}: Combined error of the i-th origin zone

E_{TOTAL}: Total error of the application

ASR: Application success rate

RESULTS

TPR of Different Adaboost Trainers

Table 3 shows the True Positive Rate (TPR) of different types of Adaboost trainers, as mentioned in Section 3.2.2.

Because RAB had the smallest error, it was chosen as the application's default classifier.

Error and Time Intervals of Different Tracking Algorithms

Table 4 shows the average time it takes for various tracking algorithms to update the position of a single vehicle, as well as their error rate, as mentioned in Section 3.3.

Since the algorithm that has the best compromise between error and speed of execution is on-line boosting, it was chosen as the application's default tracker.

Application Assessment

Below are the results obtained from three different approaches: error of the detection algorithm, error of the tracking algorithm and finally the success rate of the application. The results begin with the assessment of each origin zone and then the total of the intersection.

Table 3. True positive rate of different types of Adaboost trainers

Adaboost Trainer	TPR [%]
Discrete Adaboost	68.7%
Gentle AdaBoost	70.9%
Real AdaBoost	80.7%

Table 4. Results obtained for different tracking algorithms

Tracking Algorithm	Time [ms]	Error [%]
Kernelized Correlation Filters (KCF)	~15	11.3
On-line Boosting	~50	2.0
Median Flow	~67	68.0
Multiple Instance Learning (MIL)	~182	8.2
Tracking-Learning-Detection (TLD)	>500	Very slow, not evaluated

Detection Assessment

Below are the tables with the confusion matrices obtained for each origin zone.

Table 9 shows a confusion matrix that includes the sum of all zones.

Table 10 shows the detection error for each zone and the total error as defined in equations XX and XX.

Tracking Error

Table 11 shows the tracking error for vehicles that departed from each origin zone, as well as the total error at the intersection.

Table 5. Confusion matrix of origin zone 1

	Detectable	Not Detectable	Total
Detected	109	0	109
Not Detected	15	33	48
Total	124	33	

Table 6. Confusion matrix of origin zone 2

	Detectable	Not detectable	Total
Detected	228	0	228
Not Detected	50	78	128
Total	278	78	

Table 7. Confusion matrix of origin zone 3

	Detectable	Not Detectable	Total
Detected	59	0	59
Not Detected	17	10	27
Total	76	10	

Table 8. Confusion matrix of origin zone 4

	Detectable	Not Detectable	Total
Detected	125	0	125
Not Detected	71	42	113
Total	196	42	

Table 9. Totalized confusion matrix

	Detectable	Not Detectable	Total
Detected	521	0	521
Not Detected	153	163	316
Total	674	163	

Table 10. Detection errors of each origin zone and of the entire intersection

Error	Valor [%]
FNR_1	12.1
FNR_2	18.0
FNR_3	22.4
FNR_4	36.2
FNR_{TOTAL}	**22.7**

Table 11. Tracking errors for vehicles that departed from each origin zone, as well as the total tracking error throughout the intersection

Error	Value [%]
E_{TR1}	2.0
E_{TR2}	6.1
E_{TR3}	2.0
E_{TR4}	4.1
$E_{TR\text{-}TOTAL}$	**4.3**

Application Success Rate

Table 12 shows the combined error of the detection and tracking algorithms in each origin zone and finally the total combined error of all the zones, which is equivalent to the total error of the application.

Performance

It is worth mentioning that to evaluate the performance, the CPU was intensively used with the parallelization illustrated in figure 12 (which in turn is a summary version of the flow chart of figure 2).

Table 13 shows the minimum, average and maximum times that each relevant application routine took to execute its task.

In addition, three different variables were observed in the analysis time: memory usage (actual RAM consumption), CPU usage, and analysis time per frame. Each of these measurements was compared with the number of vehicles that the application had to analyze at each given moment (frame). Figures 13 to 16 illustrate the application performance.

Finally, it was empirically established that in the hardware used to run the application, the total analysis time is 24 times greater than the playback time, or what is the same, the analysis speed is approximately 1 fps. Of course, this relationship depends on factors such as the density of vehicular traffic or the distances between the origin and destination zones, as well as the very nature of the intersection to be analyzed (in general, small intersections imply faster analyzes).

Table 12. Combined errors in each of the origin zones, as well as the total error of the application

Error	Value [%]	Detection Contribution [%]	Tracking Contribution [%]
E_{C1}	13.9	12.1	1.8
E_{C2}	23.0	18.0	5.0
E_{C3}	23.9	22.4	1.5
E_{C4}	38.8	36.2	2.6
E_{TOTAL}	**26.0**	**22.7**	**3.3**

Figure 12. Flow chart illustrating the division of the tracking process so that it can be processed in parallel by several threads.

Table 13. Time intervals in microseconds measured for each relevant application routine

Application Routine	Time [Microseconds]		
	Minimum	Average	Maximum
Negative Detection	272.7	402.4	1783.8
Positive Detection	285.3	447.1	1994.3
Tracking Initialization	263365.0	283192.3	341027.0
Tracking Update	22382.2	50267.9	106262.0

Figure 13. RAM memory usage versus vehicle counting, in time

Figure 14. CPU usage compared to vehicle counting per frame

Figure 15. Analysis time per frame versus vehicle counting, in time

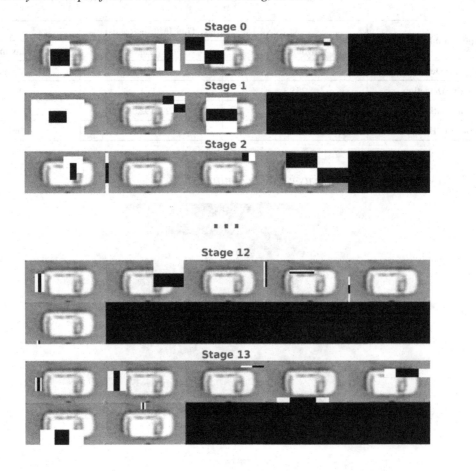

Figure 16. Similar to figure 15, it illustrates vehicle counting versus the analysis time, but the latter with a truncated vertical axis. This allows to see (contrary to figure 15) the similarity between both variables.

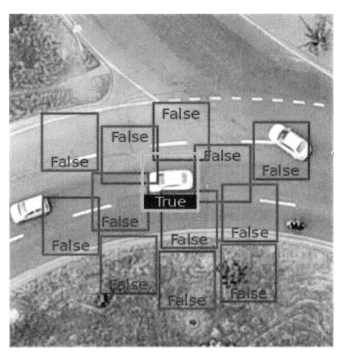

Discussion

Several observations and analysis can be made from the tables and graphs presented in section 5. First, regarding the total error of the application (26.0%), it can be seen that the main contributor is the detection (22.7%), while tracking contribution is significantly smaller (3.3%). However, the opposite happens when the processing time is examined, on average tracking takes much longer than detection (50 ms versus 447 us). This is understandable, since vehicles have similar regular forms, which allows a fast detection method (like Viola-Jones) to be successful, instead tracking should be done over long distances, in changing conditions (drone movement, or light changes) and with dense vehicular traffic. Thus, tracking took longer times, but had smaller error rates, while the detection being much faster, had higher error rates. This, however, is not of great concern, since the detection error can be reduced by increasing the amount of positive samples in the training set.

Another problem that was faced was occlusion. None of the algorithms presented, proved to be resistant to this obstacle. This is clearly reflected in the higher error rates in the "Z4" zone, where a tree was located and standed out over the road (figure 17).

On the other hand in figure 13, in the first 190 seconds, it can be seen that as the number of vehicles analyzed in each frame increases, the RAM consumption increases proportionally. This increase is sustained until the system limit is reached, a value close to 7.1 GB. The above indicates that tracking consumed most of the system resources. As for the CPU (figure 14), its intensive use can be appreciated during the entire analysis process (350% where the maximum is 400% for the 4 available logical CPUs). The lower peaks of the CPU usage correspond to high peaks of the processing time per frame (figure

Figure 17. A tracking error can be seen in the vehicle labeled "S4 3", caused by the occlusion generated by the tree that stands out over the road

15), this can be attributed to the need of the system to migrate memory blocks to the swap space. This process is slowed considering that the disk was HDD and not SSD. Despite the presence of peaks, figure 16 illustrates the direct relationship between the analysis times and the number of vehicles. If hardware had more available RAM, these variables would be identical. This indicates that tracking not only takes the longest, but also consumes the most resources, which again leads us to reflect on the difficulty of tracking vehicles at this type of intersection.

CONCLUSION AND FUTURE WORK

Once the results have been presented and discussed, the following conclusions can be drawn.

An application was designed based on the method illustrated in figure 2 and explained in detail in section 3. It made use of two specialized algorithms to obtain the OD matrix of an intersection, one for detection and the other for tracking. The detection was carried out with the algorithm proposed by Viola-Jones (Viola & Jones, 2004), which is based on HAAR features and cascade classifiers filtered and trained through adaboost. On the other hand, the tracking algorithm used was on-line boosting (Grabner et al., 2006).

The Viola-Jones detection algorithm showed a TPR of 77.3%, obtained with a classifier trained with 500 positive samples. However, this detection rate can be improved by training a classifier with a bigger data set.

The vehicular tracking based on on-line boosting showed a success rate of 95.7%. Despite the low error, this algorithm is not robust to occlusion, obstacles such as bridges or trees could significantly increase the error rate.

The total error of the application, that is, the combination of the error contributed by detection and tracking is 26.0%. Of this percentage, 22.7% corresponds to the first one while the other 3.3% corresponds to the second one.

Application performance is impacted mainly by the tracking algorithm, which consumes the most hardware resources and takes most of the processing time.

Due to the above-mentioned conclusion, the processing time depends on the number of vehicles to be analyzed per frame, and this depends on the size of the intersection and the vehicle traffic density.

The processing speed in the specified circumstances was 1fps. In respect of future work, it is proposed to reduce the detection error by implementing more robust classifiers obtained with training data sets with a larger number of positive samples. It is also proposed to improve performance, for this the most viable option is to implement a tracking algorithm that makes use of GPU (GOTURN is an alternative worth trying). Other relevant aspects that should be improved are, to eliminate the problem of occlusion in tracking, to include vehicle classification and to face the problem in nighttime conditions.

REFERENCES

Al-Sobky, A. S. A., & Hashim, I. H. (2014). A generalized mathematical model to determine the turning movement counts at roundabouts. Alexandria Engineering Journal, 53(3), 669–675. doi:10.1016/j.aej.2014.06.012

Bolme, D., Beveridge, J. R., Draper, B. A., & Lui, Y. M. (2010). Visual object tracking using adaptive correlation filters. 2010 IEEE Computer Society Conference on Computer Vision and Pattern Recognition, 2544–2550. doi:10.1109/CVPR.2010.5539960

Bouwmans, T., Silva, C., Marghes, C., Zitouni, M. S., Bhaskar, H., & Frelicot, C. (2018). On the role and the importance of features for background modeling and foreground detection. *Computer Science Review, 28*, 26–91. doi:10.1016/j.cosrev.2018.01.004

Chen, C., Hu, J., Zhang, J., Sun, C., Zhao, L., & Ren, Z. (2016). Information congestion control on intersections in VANETs: A bargaining game approach. IEEE Vehicular Technology Conference, 2016-July, 1–5. doi:10.1109/VTCSpring.2016.7504289

de la Rocha, E. (2010). Image-processing algorithms for detecting and counting vehicles waiting at a traffic light. *Journal of Electronic Imaging, 19*(4), 043025. doi:10.1117/1.3528465

Dollar, P., Tu, Z., Perona, P., & Belongie, S. (2012). Integral Channel Features. *Proceedings of the British Machine Vision Conference 2009*, 91.1–91.11. 10.5244/c.23.91

Federal Highway Administration. (2013). Traffic Monitoring Guide. Retrieved from http://www.fhwa.dot.gov/policyinformation/tmguide/

Freund, Y., & Schapire, R. E. (1997). A Decision-Theoretic Generalization of On-Line Learning and an Application to Boosting. *Journal of Computer and System Sciences, 55*(1), 119–139. doi:10.1006/jcss.1997.1504

Friedman, J., Hastie, T., & Tibshirani, R. (2000). Additive logistic regression: A statistical view of boosting. *Annals of Statistics, 28*(2), 337–407. doi:10.1214/aos/1016218223

Girshick, R. (2015). Fast R-CNN. 2015 IEEE International Conference on Computer Vision (ICCV), 1440–1448. doi:10.1109/ICCV.2015.169

Grabner, H., Grabner, M., & Bischof, H. (2006). Real-Time Tracking via On-line Boosting. Procedings of the British Machine Vision Conference 2006, 6.1–6.10. 10.5244/C.20.6

Haar, A. (1910). Zur Theorie der orthogonalen Funktionensysteme—Erste Mitteilung. *Mathematische Annalen, 69*(3), 331–371. doi:10.1007/BF01456326

Held, D., Thrun, S., & Savarese, S. (2016). Learning to track at 100 FPS with deep regression networks. *Lecture Notes in Computer Science, 9905*, 749–765. doi:10.1007/978-3-319-46448-0_45

Henriques, J. F., Caseiro, R., Martins, P., & Batista, J. (2015). High-speed tracking with kernelized correlation filters. *IEEE Transactions on Pattern Analysis and Machine Intelligence, 37*(3), 583–596. doi:10.1109/TPAMI.2014.2345390 PubMed

Hoogendoorn, S. P., & Bovy, P. H. L. (2001). State-of-the-art of vehicular traffic flow modelling. *Proceedings of the Institution of Mechanical Engineers. Part I, Journal of Systems and Control Engineering, 215*(4), 283–303. doi:10.1177/095965180121500402

Instituto Nacional de Vías (INVIAS). (2013). Volúmenes de Tránsito 2010-2011. Retrieved from https://www.invias.gov.co/index.php/archivo-y-documentos/documentos-tecnicos/volumenes-de-transito-2008/1921-volumenes-de-transito-2010-2011

Kalal, Z., Mikolajczyk, K., & Matas, J. (2010). Tracking-Learning-Detection. *IEEE Transactions on Pattern Analysis and Machine Intelligence, 6*(1). Retrieved from http://info.ee.surrey.ac.uk/Personal/Z.Kalal/

Kalal, Z., Mikolajczyk, K., Matas, J., & Republic, C. (2010). Forward-Backward Error: Autonomous Identification of Tracking Failures. *Proceedings of the 20th International Conference on Pattern Recognition*, 2756–2759. 10.1109/ICPR.2010.675

Lehtola, V., Huttunen, H., Christophe, F., & Mikkonen, T. (2017). Evaluation of visual tracking algorithms for embedded devices. Lecture Notes in Computer Science, 10269, 88–97. doi:10.1007/978-3-319-59126-1_8

Lin, T.-Y., Goyal, P., Girshick, R., He, K., & Dollar, P. (2017). Focal Loss for Dense Object Detection. 2017 IEEE International Conference on Computer Vision (ICCV), 2999–3007. doi:10.1109/ICCV.2017.324

Liu, W., Wang, Z., Liu, X., Zeng, N., Liu, Y., & Alsaadi, F. E. (2017). A survey of deep neural network architectures and their applications. *Neurocomputing, 234*, 11–26. doi:10.1016/j.neucom.2016.12.038

Lowe, D. G. (1999). Object recognition from local scale-invariant features. Proceedings of the Seventh IEEE International Conference on Computer Vision, 1150–1157. doi:10.1109/ICCV.1999.790410

Lukežič, A., Vojíř, T., Čehovin Zajc, L., Matas, J., & Kristan, M. (2018). Discriminative Correlation Filter Tracker with Channel and Spatial Reliability. *International Journal of Computer Vision, 126*(7), 671–688. doi:10.1007/s11263-017-1061-3

Luo, J., Ma, Y., Takikawa, E., Lao, S., Kawade, M., & Lu, B. L. (2007). Person-specific SIFT features for face recognition. ICASSP, IEEE International Conference on Acoustics, Speech and Signal Processing - Proceedings, 2, 593–596. 10.1109/ICASSP.2007.366305

Redmon, J., Divvala, S., Girshick, R., & Farhadi, A. (2016). You Only Look Once: Unified, Real-Time Object Detection. 2016 IEEE Conference on Computer Vision and Pattern Recognition (CVPR), 779–788. doi:10.1109/CVPR.2016.91

Redmon, J., & Farhadi, A. (2018). YOLOv3: An Incremental Improvement. Retrieved from http://arxiv.org/abs/1804.02767

Saini, M., Alelaiwi, A., & Saddik, A. E. (2015). How Close are We to Realizing a Pragmatic VANET Solution? A Meta-Survey. *ACM Computing Surveys, 48*(2), 1–40. doi:10.1145/2817552

Sermanet, P., Eigen, D., Zhang, X., Mathieu, M., Fergus, R., & LeCun, Y. (2013). OverFeat: Integrated Recognition, Localization and Detection using Convolutional Networks. Retrieved from http://arxiv.org/abs/1312.6229

Silva, C., Bouwmans, T., & Frélicot, C. (2015). An eXtended Center-Symmetric Local Binary Pattern for Background Modeling and Subtraction in Videos. Proceedings of the 10th International Conference on Computer Vision Theory and Applications, 395–402. doi:10.5220/0005266303950402

Suda, J. (2017). Misstatements of GPS- location of public transport vehicles in Warsaw. Transportation Overview - Przeglad Komunikacyjny, 2017(2), 37–45. doi:10.35117/A_ENG_17_02_05

Tuydes-Yaman, H., Altintasi, O., & Sendil, N. (2015). Better estimation of origin–destination matrix using automated intersection movement count data. *Canadian Journal of Civil Engineering*, *42*(7), 490–502. doi:10.1139/cjce-2014-0555

Viola, P., & Jones, M. J. (2004). Robust Real-Time Face Detection. *International Journal of Computer Vision*, *57*(2), 137–154. doi:10.1023/B:VISI.0000013087.49260.fb

Wang, X., Han, T. X., & Yan, S. (2009). An HOG-LBP human detector with partial occlusion handling. 2009 IEEE 12th International Conference on Computer Vision, 32–39. 10.1109/ICCV.2009.5459207

Wang, Z., Yoon, S., Xie, S. J., Lu, Y., & Park, D. S. (2016). Visual tracking with semi-supervised online weighted multiple instance learning. *The Visual Computer*, *32*(3), 307–320. doi:10.1007/s00371-015-1067-1

ENDNOTES

[1] Psutil: Librería multiplataforma para monitorización de procesos y del sistema en Python. Url: https://github.com/giampaolo/psutil

[2] Chrono: Librería de C++ para medición de tiempos de ejecución. Url: http://www.cplusplus.com/reference/chrono/

Chapter 12
Detection and Classification of Wear Fault in Axial Piston Pumps:
Using ANNs and Pressure Signals

Jessica Gissella Maradey Lázaro
Universidad Autónoma de Bucaramanga, Colombia

Carlos Borrás Pinilla
Universidad Industrial de Santander, Colombia

ABSTRACT

Variable displacement axial piston hydraulic pumps (VDAP) are the heart of any hydraulic system and are commonly used in the industrial sector for its high load capacity, efficiency, and good performance in the handling of high pressures and speeds. Due to this configuration, the most common faults are related to the wear and tear of internal components, which decrease the operational performance of the hydraulic system and increase maintenance costs. So, through data acquisition such as signals of pressure and the digital processing of them, it is possible to detect, classify, and identify faults or symptoms in hydraulic machinery. These activities form the basis of a condition-based maintenance (CBM) program. This chapter shows the developed methodology to detect and classify a wear fault of valve plate taking into account six conditions and the facilities providing by wavelet analysis and ANNs.

INTRODUCTION

Reliability and safety of industrial processes, with an efficient asset management, have taken a very important role to achieve greater competitiveness and productivity of global companies which working hard to ensure reliability, quality services, safety and maintainability of hydraulic equipment. The manufacturer's usual advice regarding with maintenance is to implement a management strategy of maintenance that includes preventive and predictive activities (Aly, 2015), whose results providing

DOI: 10.4018/978-1-7998-1839-7.ch012

high reduction in maintenance and operational costs, increasing the productivity of the processes, as well as the stability of the system in general. On the other hand, methodologies for on-line and off-line fault diagnosis have been developed for this purpose, allowing to taking actions that avoid either long unscheduled stops or permanent damage of machine's components and as well as preserve their useful life. Hydraulic axial piston pumps are commonly used for high fluid power applications, which are severely affected by valve plate wear failure.

Wavelet analysis has been extensively studied for feature extraction (Guillén, Paredes & Quintero, 2004; Peng & Chu, 2004), which has shown an advantage in the processing of non-linear and non-stationary signals. This allows the management in the time and frequency domain and the decomposition of the series in terms of the approximation coefficients (cA) and detail (cD), separating the information into low and high frequencies respectively. The generalized Wavelet Packet Transform (WPT) algorithm is also a very effective method to apply, which after the decomposition of the signal allows extracting and normalizing the characteristic vectors according to the energy of each component (Tao, Lu, Lu, & Wan, 2013; Tao, Wang, Ma & Fan, 2012).

Furthermore, artificial neural networks (ANNs), has proven their capacity of training - learning, self-organization, accuracy, convergence, performance and fault tolerance (Lázaro, Pinilla & Prada, 2016), as well as their robustness in the solution of complex and nonlinear dynamic systems. ANNs have been especially used in tasks of patterns recognition problems, diagnosis and classification, model predictive faults, to handle analytical and heuristic information, incomplete data among others.

Therefore, combining ANN technique with wavelet data analysis achieve a methodology that enables to detect and classify wear failure in an automatic and effective way. The objectives of this chapter is to show every step followed in the proposed methodology, analyze the results obtained as of pressure signals taken of five wear conditions and verify the performance of test stage.

BACKGROUND

Wear Valve Plate in Variable Displacement Axial Piston Pumps

In the design of the axial piston pump, the piston cylinder is hydrostatically loaded against the valve plate. The valve plate has two ports: The suction port, which is related to the oil tank and the discharge port that is related to the cargo. These two ports exert different pressure against the pistons cylinder (barrel). The slack is a slanting clearance. The distribution of the pressure is very complicated and is not a constant caused by the operating principle of the axial piston pump (i.e. depends on the angle of the swash plate). If the operating pressure exceeds the design limits, the clearance between the valve plate and the barrel-cylinder will be reduced.

When wear occurs, not only increases the clearance between the two surfaces but also the surfaces become increasingly irregular. The valve plate does not make good contact with the pistons cylinder, and the oil film between the surfaces is not stable, which leads to leaks (Chu, Kang, Chang, Liu, & Chang, 2003). Once the oil film is lost (occurs when friction and wear are serious enough), the two surfaces come into contact and abrasion occurs at the ports of entry and exit of the sliding surface of the valve plate (He, Wang, Wang & Li, 2012). Figure 1 shows the structure of the valve plate.

Figure 1. Structure of valve plate
Source: *(Universidad Industrial de Santander, 2016)*

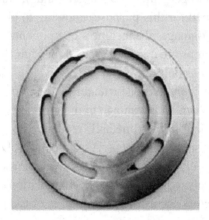

According to statistical data, 75% of the failures in the axial piston pump are due to abrasion between the friction torque, in which the most relevant item is the wear between the valve plate and the pistons cylinder (Hu, Wang, He & Zao, 2010). Figure 2 shows the morphology of the valve plate surface with a high degree of abrasion. This failure occurs because the operating pressure exceeds the limits and the clearance between the valve plate and the piston cylinder decreases. Once the lubricant film is lost, the two surfaces are in contact producing wear on the inlet and outlet edges of the sliding surface of the valve plate (He et al., 2012; Hu et al., 2010; Zhao, Wang & Dou, 2012).

The methodologies used to diagnose faults in axial piston pumps are varied, ranging from dynamic models (Ančić, Šestan & Virag, 2014; Bayer & Enge-Rosenblatt, 2011) to combined models (Hu et al., 2010; Ančić, Šestan & Virag, 2014) or hybrids and some multi-fault classifiers (Du & Wang, 2010; Du, Wang & Zhang, 2013). In general, all the methodologies developed meet the following stages as detailed below:

1. **Signal Processing**: Vibration analysis was generally based on the spectral analysis using the Fourier Transform (FT), Short Time Fourier transform (STFT) and the time sequence analysis. As these methods have been designed to process signals in the domain of frequency or time only. On the

Figure 2. Surface morphology of the valve plate with a high degree of abrasion Source: (ASME 2016 International Mechanical Engineering Congress and Exposition, 2016)

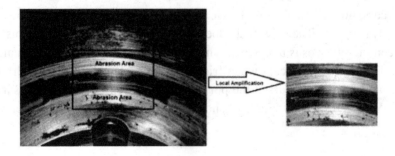

other hand whereas Wavelet Analysis (e.g. used in seismic, monitoring the state of structures and diagnosis of medical equipment, among others) has been widely used, allowing the analysis in an extended time-frequency domain to identify characteristic local signals by decomposition of the signal in the time domain for these frequencies (Hu et al., 2010), achieving the reduction of noise and the effective extraction of the characteristics of the fault. There are more methods to perform during this step, some differences among the methods can be seen in Table 1.

2. Use of statistical tools to measure the reduction and independence of the extracted data. Principal Component Analysis (PCA) is a multivariate statistical technique that has been successfully applied in the detection and diagnosis of faults, as it is a linear dimensionality reduction tool, optimal in terms of capturing the variability of data and applied to perform the stage of extraction of characteristics. PCA determines a set of orthogonal vectors, which are linear combinations of the original variables, these are ordered by the amount of variance explained in the directions of said components (Stalford & Borrás, 2002; Karkoub, Gad & Rabie, 1999).

3. Extraction of the characteristics of the signal (i.e features).

4. Establish the diagnostic rules or classification algorithm to be used, taking into account the supervision and training of the same, in the cases in which it applies.

5. Experimental validations or simulation of faults to verify the validity of the rules and / or algorithm selected in the previous step.

The assessment criteria selected to choose references was fault diagnosis in axial piston pump using neural networks and wavelet analysis such as shown in preliminary section. Lastly, we reported some recent references which the methodology and results obtained are relevant.

Preliminary Literature Review

Ramdén, Krus, & Palmberg (1998) developed a monitoring methodology based on off-line condition of faults related to the wear of the valve plate on a shaft piston pump. Fixed-Displacement Bent-Axis Piston Pump using a Multilayer Perceptron (MLP) network algorithm using Back Propagation (BP) as a training algorithm, obtaining a network of 5 input nodes, 15 hidden nodes and 4 output nodes, which was trained and simulated with Matlab Neural Network Toolbox taking the original valve plate and plates

Table 1. Performance comparison of different methods for signal processing sentence

Methods	Resolution	Interference Term	Speed
CWT	Good frequency resolution and low time resolution for low-frequency components; low frequency resolution and good time resolution for high-frequency components	No	Fast
STFT	Dependent on window function, good time or frequency resolution	No	Slower than CWT
WVD	Good time and frequency resolution	Severe interference terms	Slower than STFT
CWD	Good time and frequency resolution	Less interference terms than WVD	Very slow
CSD	Good time and frequency resolution	Less interference terms than CWD	Very slow

Source: (Mechanical Systems and Signal Processing, 2014)

with wear corresponding to 3%, 6% and 9% loss of volumetric efficiency. As a final recommendation, the authors suggest working with wavelet analysis for the processing of the vibration signal taken from the pump casing.

Karkoub et al., (1999) experimentally tested a model that describes dynamic behavior in steady state and transient state through the prediction of pressures by taking several measurement points in a variable displacement inclined shaft pump. They integrated an MLP network with a Levenberg-Marquardt Optimization technique for network training. As a result, an error of 2% was obtained between the theoretical and experimental results, which allows concluding that neural networks have a very important potential in the modeling of complex systems like this one.

Later, Gao, Zhang, & Kong (2003) proposed a methodology for diagnosis of on-line failure of hydraulic pumps using the pressure signals taken and through of wavelet analysis to obtain the coefficients of detail and approximation to be compared with those obtained from the normal pump state, as a base line reference, having differences in the patterns and amplitudes in the different bands making distinguishable and easy to identify the types of defects analyzed (wear of the oscillating plate and loss of adjustment in the piston shoes). Additionally, on-line validation tests were performed, which allows obtaining a real-time diagnosis of the pump without affecting her normal operation.

In 2012, Tao et al., developed a methodology based on performance evaluation called "SOM Neural Network", designed as a neural network of unsupervised learning to organize herself according to the nature of the input data. The detection of the fault is achieved by calculating the minimum error quantification (MQE), which can be transformed into a standardized reliable value (CV). Then, to verify the method, the experiment was performed for two faults: sliding skid and wear of the valve plate, whose results were satisfactory. As future work, it is proposed that is possible to focus the study of the extraction of characteristics and identification of failure modes taking into account only the information in normal operation available.

Next, Kivela & Mattila (2013), studied the internal leakage for VDAP. They proposed a real-time model structure named Multi-Variable Histogram (MVH) based in condition monitoring purposes. This model is based on detecting internal variables operation point changes and was tested and verified for one condition of leakage. The results obtained shows that the proposal model detects very well the fault and the performance changes in a pump. Future works could be considered other levels of leakage and to detect other faults. They proposed to use a model as Fault Detection and Isolation (FDI).

Also, Plawiak (2014), suggests computational intelligent methods to estimate the state of consumption of a fixed displacement axial piston pump. In addition, it designed and compared methods for signal processing (i.e. elimination of the constant component, normalization, standardization, reduction, Fast Fourier Transform - FFT), as well as the effectiveness of three models of neural networks (i.e. MLP, generalized and probabilistic regression), with Matlab. The analysis performed showed that in terms of accuracy the best result in the frequency domain is achieved with MLP and in the time domain with generalized regression.

In Addition, Bei, Lu, & Wang (2013) proposed a method to evaluate performance and to diagnose faults in hydraulic pumps, by Decomposing Wavelet Pack (WPT), extracting the characteristics based on the concept of energy and carrying out the classification using a Self-Organizing Maps (SOM) network. The flaws studied were the wear of the oscillating plate and the rotor. Additionally, they calculated the mean square error (MQE), optimized the characteristics using the Taguchi method and established a confidence value (CV).

Next, Salimi & Niromandfam (2015), show how using statistical parameters of the vibration spectrum is a simple method to obtain characteristics and through of the data analysis method, the best features can be selected. In addition, the article emphasizes the importance of the use of the wavelet transform for the treatment of unstable signals and different pump conditions, besides, using a neural network with the Levenberg-Marquart training algorithm it was possible to obtain an acknowledgment of defects greater than 95% which shows that the efficiency of the selected network to recognize the defects is good and acceptable.

Some technical papers by SAE of 2017 are revised. First, Schoenau, Stecki & Burton (2000) shows the utilization of ANN's in the control, identification and condition of Hydraulic Systems. Next, Hindman, Burton & Schoenau (2002) make a survey about condition monitoring of fluid power systems, describing some conditions that affect the performance of hydraulic systems and to provide a background of state estimation through Kalman Filter, ANN's and Spectral Analysis. Finally, Gao, Kong & Zhang (2002) presents a hydraulic pump fault diagnosis method based on a discharge pressure pulsation model which was simulated, analyzed and validated against test data. The results conclude that wavelet analysis is capable of detecting pump failure using the outlet pressure pulsating signal.

Recent Works

In the last two years, more sophisticated techniques have been used in fault diagnosis.

Wen, Li & Gao (2017) applied a new convolutional neural network (CNN) based data-driven fault diagnosis method. Firstly, the authors used a signal to image conversion method to extract the 2-D features of raw data with predefined parameters. This model facilitates the feature extraction process and was tested with three cases: Motor bearing, self-priming centrifugal pump and axial piston hydraulic pump. The CNN model was based in LeNet-5, which is effective in the image pattern recognition. Also, zero-padding method was used to control the size of feature dimension. The novelty in this paper is the use of deep learning machine which a better performance was obtained (i.e. 99.79%) vs 87.45% with Support Vector Machine (SVM), and finally 67.70% obtained with ANN. Future works should be focused in modified this method to find an unknown fault conditions and to reduce training time for CNN.

Casoli, Bedotti, Campanini & Pastori (2018) proposed a methodology based in data- acquisition and feature extraction as from acceleration signals through the theory of cyclo-stationarity. The fault studied was related to worn slippers, which are the common faults of the swash–plate type. In this work, the acceleration signal treatment by decomposition in frequency domain using FFT and analysis of every periodic component extracted is relevant. In the future work, they will apply the methodology to detect other incipient faults and suggest to work with features of interest for online applications improving computational cost.

Besides, Lan, Hu, Huang, Niu, Zeng, Xiong & Wu (2018) developed a new intelligent fault diagnosis methodology based in WPT, Local Tangent Space Alignment (LTSA), Empirical Mode Decomposition (EMD), Local Mean Decomposition (LMD) and recognition technique Extreme Learning Machine (ELM) for slipper abrasion of VDAP. The use of ELM is the relevant because provides fastness and good generalization performance on pattern recognition task. For this study, the signals of vibration and flow were taken. The strategy consisted in compare three methods for extraction (i.e FEM1, FEM2 and FEM3) based in WPT and Singular Value Decomposition (SVD). The best performance in extraction stage was obtained with FEM1. In addition, three classifiers were compared: ELM, BP and SVM. The best performance was obtained with ELM (i.e. 84.1%-97.8%, according to the dataset). The authors suggest as future work to modify the approach taking into account imbalanced data and apply online sequential learning algorithm, as well as, apply the modified method in multi-fault diagnosis.

On the other hand, Wang, Xiang, Zhong & Tang (2018) developed a data indicator based deep belief networks to detect multiple faults in VDAP. They used a feature extraction method working in time domain, frequency domain and time-frequency domain, similarly to previous work. Later, using deep belief networks (DBNs) to classify the multiple faults studied (i.e. five faults in rolling bearings). Restricted Boltzmann Machine (RBM) provided an improve in feature learning stage. The performance obtained was 99.17% and 97.4%. This result was compared with other techniques as SVM and ANN, showing better performance respectively. Inside paper, each stage can see followed based on DBN's and the indicators (i.e. features) used and simulations with benchmark data.

In 2019, Ozmen, Simanoglu, Batbat & Guven (2019) studied the pressure distribution and leakage between the slipper and swash plate in VDAP. They used a multi-gene genetic programming (MGGP) and ANN for prediction modeling the mentioned faults. The results were compared with analytic equations. The best performance was obtained with ANN. The future investigations must be targeted on the rotation operational conditions of the slippers, different slipper geometries, load changes, the oil temperature and the operational duration aspects that it will taking into account in a better robust model.

Besides, Wang & Xiang (2019) claimed a novel minimum entropy deconvolution (MED) based convolutional neural network to classify faults in VDAP. The faults studied are worn in three pistons, blocked support hole in static pressure slippers, worn in shaft shoulder and cylinder block with a pitting defect. Simulations were made from benchmark data from Case Western Reserve University bearing data center. The contributions of this work are to solve automatic feature learning problem, to provide stability and classification accuracy and to visualize using the t-distributed stochastic neighbor embedding technique. The classification accuracy obtained were superior to the traditional CNN model.

Finally, Gao, Tang, Xiang, Zhong, Ye & Pang (2019) proposed a hybrid method of Walsh transform denoising and Teager energy operator demodulation (TEO) to solve heavy background noise present in vibration signals. Bearing and cylinder blocks faults were studied. Comparisons were made with other demodulation methods such as wavelet, obtaining that the Walsh transform provides fastness, soft and adaptability way for the raw signals. The future works will be to implement this method in diagnostic of other faults and to test with other classification method with enough data.

MAIN FOCUS OF THE CHAPTER[1]

Methodology for the Detection and Classification of Wear Failure of the Valve Plate in VDAP

Then, to focus the problem of valve plate wear studied in the present chapter, is proposed to use wavelet analysis for signal processing and neural network technique for classification. Wavelet Analysis is recommended for fault diagnosis (Peng & Chu, 2004) especially for the handling of non-linear and non-stationary signals (Tao et al., 2013; Ramdén et al., 1996), being successful to filter signals and to extract characteristics of the signal. The wavelet coefficients provide compact information about a signal in different time and frequency locations, which allows reducing the dimensionality of the data, and obtained the feature vectors (Sorsa & Koivo, 1993). Artificial neural networks have been applied to complex engineering problems, such as fault diagnosis, pattern recognition, signal processing, robotics and control, showing good skills to detect some data without drift, hydraulic process failures (Aly, 2015), and better performance with respect to regression or finite difference methods (Lurette & Lecoeuche, 2003). Also, ANN have been used as a controller to verify the flow and pressure in the VDAP, whose parameters are affected by the change in the angle of the oscillating plate (Sorsa & Koivo, 1993; Karkoub et al., 1999).

Experimental Setup

For the acquisition of the signals, a test bed was built, made up of an Eaton 54 series variable displacement hydraulic axial piston pump, M-21 model hydraulic motor and a 40 hp electric motor. Figure 3. shows the assembly of the test bench.

Furthermore, a DAQ National instruments system were implemented, as well as sensors (a flow meter (manufacturer: Flow Metrics, model FM-8), two pressure transducers (For suction, manufacturer: WIKA, model ECO-1, and for discharge, manufacturer was used: MSI, MSP300 series) and an accelerometer (manufacturer: Piezotronics PCB, model 352C33)), a signal conditioning module and analog-digital converter and software. The hardware and software used are from the National Instruments company, NI USB-9234 card, NI-USB6215 and Measurement & Automation Explorer (MAX), NI-DAQmx and NI LabView 2013 respectively (Bolaño & Niño, 2016; Castillo& Serrano, 2016). The sampling frequency for the pressure signals was 100 KHz.

The pressure signal was taken at the outlet port (i.e point B) and pre-charge pump (i.e point C), as can be seen in the Figure 4 pressure signals were taken with 60,000 samples each one. Also, it should be noted that the samples taken are reliable, since the values obtained were verified with the loss flow data taken. Figure 5 shows the data acquisition process.

In addition, six conditions of wear valve plate were defined as follows bellow:

- Condition 1 corresponding to the normal state (94.9% Volumetric Efficiency).
- Condition 2 corresponding to fault condition (90.1% Volumetric Efficiency).
- Condition 3 corresponding to fault condition (84.6% Volumetric Efficiency).
- Condition 4 corresponding to fault condition (74.8% Volumetric Efficiency).
- Condition 5 corresponding to fault condition (69.5% Volumetric Efficiency).
- Condition 6 corresponding to fault condition (64.9% Volumetric Efficiency).

Then, is necessary to correct the baseline of the signals, to eliminate trends and to make the spectral analysis for each condition (i.e P-Welch), as can be seen in the Appendix 1. In order that, low pass and high pass filters were designed. For practical purposes, the analysis for condition 1 and 6 (i.e more critical) is attached.

The spectra of the pressure signals (i.e condition 1 and condition 6) and the designed filters can be seen in Figures 6 - 11. When analyzing these figures, no appreciable changes are observed in the spectra of each condition; however, is worth highlighting that when filter the signals, high-frequency peaks (i.e in order to 25KHz) are observed and when graphing the power and high-frequency center is not possible to easily distinguish condition 1 and condition 6.

Figure 3. Test Bench
Source: Universidad Industrial de Santander, 2016.

Figure 4. Pressure signals points
Source: Universidad Industrial de Santander, 2016.

Mother Wavelet Selection

A decision has to be taken in order to specify family of mother wavelet to be selected to perform the wavelet analysis, for which there is no closed guideline and given the type of signals to be analyzed, the decision to be taken is not necessarily known. More than a selection method, the procedure usually consists of comparative evaluations of the performance of the different wavelet families, without assuming that this is a universal optimal decision (Arizmendi, 2011).

Figure 5. Data acquisition
Source: *Universidad Industrial de Santander, 2016*

Figure 6. Spectra of the pressure signals condition 1 (94, 9%)
Source: *(Author, 2017)*

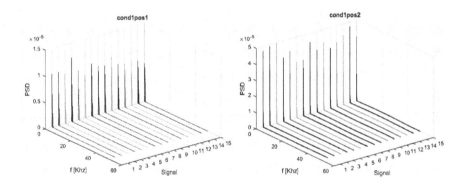

Figure 7. Spectra of the pressure signals condition 6 (64, 9%)
Source: *(Author, 2017)*

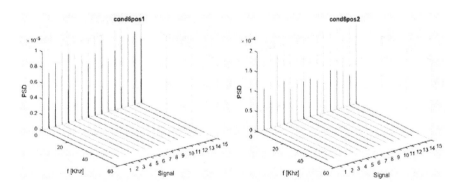

Figure 8. Low-pass filter for pressure signals condition 1 (94, 9%)
Source: (Author, 2017)

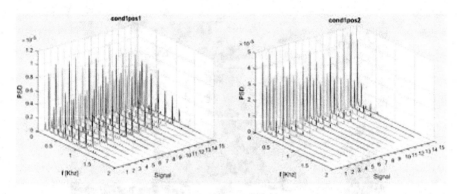

Figure 9. Low-pass filter for pressure signals condition 6 (64, 9%)
Source: (Author, 2017)

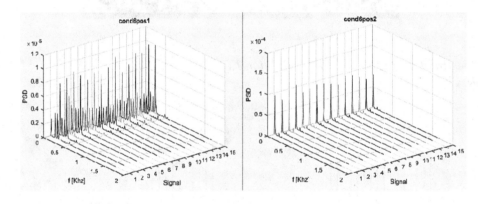

Figure 10. High-pass filter for pressure signals condition 1 (94, 9%)
Source: (Author, 2017)

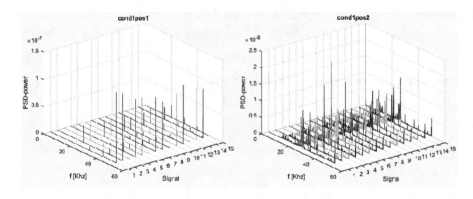

Figure 11. High-pass filter for pressure signals condition 6 (64, 9%)
Source: (Author, 2017)

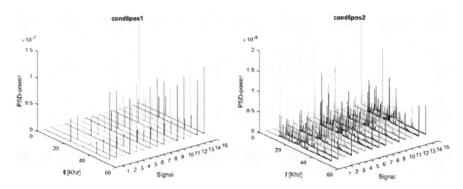

Table 2. Wavelets families and their corresponding orders analyzed

Wavelet Family	Wavelet Degree
Daubechies	1:43
Coiflets	1:5
Symlets	1:25
Biortogonal	1.1,1.3,1.5,2.2,2.4,2.6,2.8,3.3,33.5,3.,3.9,4.4,5.5,6.8

Source: Universitat Politecnica de Catalunya, 2012.

Therefore, a number of different possibilities were analyzed with the limitation that wavelets must be real and must be defined in the discrete domain (Arizmendi, 2011). Matlab® Toolbox has different families of wavelets that can be considered in the analysis, some of them are: Daubechies (Db), Symlets (Sym), Coiflets (Coif) and Biortogonal (Bior). Table 2 shows the different families that were used for analysis in this project and the order of the wavelet implemented.

Discrete Wavelet Analysis was applied to the original signals, first performing the decomposition using the generalized Wavelet Packet Transform (WPT), for which was taken into account that the optimal decomposition with trees with 2^D, where D is the intensity of entropy (i.e. depth) and to select the best level of decomposition is taken the trees with the best intensity of entropy. This means that if the maximum level of decomposition is 16, not all levels are good to get characteristic signal information. Normally, the first ones are the best. In this way, only 5 or 6 of the 16 available could be selected, by heuristic rule can said half of the possible, improving the computational time. According with the Entropian criterion "logenergy" was selected, because of it don't need a parameter and is commonly used among the different options that Matlab® has (i.e shannon, threshold, norm, sure). For each mother wavelet, the absolute values of the decomposition coefficients are sorted in descending order and the spectra of each signal are reconstructed by summing the consecutive coefficients (Arizmendi, 2011). For the decomposition and reconstruction process the functions of Matlab® were used: wavedec, wmaxlev, waverec, wrcoef, wprcoef.

The average mean square error (MSE) and the signal-to-noise ratio (SNR) can be calculated for a complete set of signals for each wavelet order as detailed below:

$$MSE = \frac{1}{N} \sum_{i=1}^{N} \left[x(i) - \hat{x}(i) \right]^2 \tag{1}$$

$$SNR(db) = 10log \left[\frac{\sum_{i=1}^{N} \left[x(i) \right]^2}{\sum_{i=1}^{N} \left[x(i) - \hat{x}(i) \right]^2} \right] \tag{2}$$

where, $\hat{x}$ is the reconstructed signal. Finally, the Q1 index for the order (r) is computed according to the statistics mentioned above as follows:

$$Q1(r) = \frac{SNR(r)}{MSE(r)} \tag{3}$$

The maximum values of Q1 indicate the orders with the best reconstruction error. This procedure is the initial phase for the selection of the mother wavelet [37].

In order to determine the definitive wavelet, the average value of several statistical values can be combined (SNR, Preserved Energy (EP), Distortion Percentage (PRD) and Compression Ratio (CR)) as detailed below (Tao et al., 2013):

$$E_p = \frac{\sum_{i=1}^{N} \left[\hat{x}(i) \right]^2}{\sum_{i=1}^{N} \left[x(i) \right]^2} *100\% \tag{4}$$

$$PRD = \sqrt{\frac{\sum_{i=1}^{N} \left[x(i) - \hat{x}(i) \right]^2}{\sum_{i=1}^{N} \left[x(i) \right]^2}} *100\% \tag{5}$$

$$CR = \frac{L_o}{L_c} \tag{6}$$

where L_o is the cardinality of the decomposition coefficients of the original signal and L_c is the cardinality of the decomposition coefficients other than zero.

These performance measures help to determine the best wavelet for signal reconstruction. Then, an objective criterion to select the optimal wavelet function is through Q2 as shown:

$$Q2 = \left[\frac{SNR + E_p + CR}{MSE + PRD} \right] \tag{7}$$

The maximum value obtained for the Q2 index for the pressure signals was with db2 for the average values and db3 for the values taking into account the standard deviation. In our case, was selected db2. Table 3 and Table 4 show these results.

Feature Extraction

Once the decomposition and reconstruction is performed according to the WPT algorithm, the average values of the approximation coefficients (Ca) are determined, corresponding to the low frequency and detail bands (Cd) corresponding to high frequency bands. The characteristics of the signal that were extracted are: Mean, RMS Value, Standard Deviation, Variance, Skewness, Kurtosis and Interquartile Range (IRS). With these values an increased vector of characteristics is achieved, which can be represented as follows: **[avgA, avgD, rmsA, rmsD, stdA, stdD, varA, varD, skeA, skeD, kurA, kurD, iqrA, iqrD]**. For the pressure signals 15x14 vectors were taken.

Feature Selection

This is a process that identifies the most relevant characteristics that must be extracted to ensure that the classifier performs better. Likewise, aims to prevent oversizing the system allowing to find the most relevant data for it, and to filter those that do not contribute or alter the results of the classification. The selection of characteristics not only implies the reduction of cardinality, that is to say, the imposition

Table 3. Performance indices of different wavelets taking into account their average values

Wavelet	Maxlev	Entrophy	Depth	MSE	SNR	Q1	PRD	CR	EP	Q2
' db2'	14	537621.586	7	0.00	40.10	571730187.35	1.02	48.98	99.99	196.81
'coif1'	13	535626.648	6	0.00	40.09	570255188.82	1.02	48.88	99.99	196.49
' sym3'	13	535657.528	6	0.00	39.96	551998187.45	1.04	50.53	99.99	195.16
'bior1.1'	15	520286.37	3	0.00	40.58	645577392.70	0.96	38.12	99.99	196.00

Source: (Author, 2017).

Table 4. Performance indices of different wavelets taking into account their standard deviation

Wavelet	Entrophy	Depth	MSE	SNR	Q1	PRD	CR	EP	Q2
' db3'	-539027.344	0	0.00	2.20	69133203.58	1.04	3.73	0.01	47.80
'db31'	-539027.344	0	0.00	2.20	68703532.57	1.03	3.59	0.01	47.71
'db34'	-539027.344	0	0.00	2.20	68373996.09	1.03	3.55	0.01	47.70
'db36'	-539027.344	0	0.00	2.20	68332645.89	1.03	3.60	0.01	47.66
' sym3'	-539027.344	0	0.00	2.20	69133203.58	1.04	3.73	0.01	47.80

Source:(Author, 2017).

of an arbitrary or predefined limit on the number of attributes that can be considered when creating a model, but also the choice of attributes, which means that modeling should be select or discard attributes based on their usefulness for analysis (Pinto, 2014).

Although there are several methods such as Bidirectional Search, Genetic Algorithms, Forward Selection and Backward selection. A technique called "Independent Features" was chosen, code developed by Will Dwinnell (predictr@bellatlantic.net) for its ease of implementation and simplicity, whose method consists in calculating the level of significance 'Sig' of the real variables (in columns) of the matrix X, based on their ability to distinguish 2 categories of the column from vector Y (typically, the variables are maintained when the significance is$> = 2.0$). For a continuous output variable, is suggested that the values be divided into 50% lower and higher. Using Weiss / Indurkhya 'Independent Features' (significance testing method). Note that it is not intended to be the final selection of the variables, but only a means of eliminating features that are not of interest. Figures 12 and 13 show the graph of the most significant characteristics by applying the independent features algorithm for the pressure signals respectively, according to this, 19 available can be taken for the study, however, for reasons of convergence of the classifier and computational time, only 12 characteristics are chosen.

Neural networks are a very sophisticated technique capable of modeling complex functions of various kinds. One of the main problems when using neural networks is the correct selection of architecture and network parameters. In order to overcome the problem, it is necessary to have a sufficiently large dataset for training and for validation of the network (Arango, 2007). The number of features that you wish to select from a possible number that is calculated with the code of Independent Features, in which the threshold $s > = 2$ was defined, is defined as input to the network. Subsequently, the data was normalized and the type of network to be trained and tested was selected, in this case, an Adaline network (transfer function: purelin), a Red Multilayer Perceptron (transfer function: softmax) and two non-linear adaptive networks (transfer function: tansig and logsig) were implemented. The 'kfold' command was used which generates a 3-fold stratified partition for the training set, ensuring that 70% of the data will be used for training and 30% for test. The network architecture designed for the Adaline typology and the non-linear adaptive networks consist of an input layer, a hidden layer and an output layer with 6 neurons. The training rule used is Levenverg-Marquart and performance Mean Squared Error. Figure 14 shows the network architecture implemented for the Adaline typology, the non- linear networks (NL1 and NL2), the MLP1 and MLP2 typology (i.e MLP1 one with a hidden layer and MLP2 with two hidden layers).

The training and performance rule were the same as in the case of Adaline and the others. The confusion matrix obtained after the training can be seen can be seen in Figure 15 and 16. A summary of the results according to training stage shows in Figure 17.

As a result for training stage, the networks that best identified the conditions were: NL2, MLP1 and MLP2 achieving 100% in the classification. The network that presented the worst performance was the

Adaline

Test data consists of 30% of the signals taken in the extraction of characteristics according to the k-fold distribution. These were never used for training, so is desirable to check how well the neural network identifies each condition. Next, Figures 17-21 show the detail of the results obtained in the test for each type of network implemented for the pressure signals. A validation is carried out with 5 test cases for each of the conditions in order to evaluate the degree of confusion and the ability to differentiate between conditions. In the case of a linear network (purelin) it is observed that conditions 1,3 and 4 are clearly

Figure 12. Feature selection
Source: (Author, 2017)

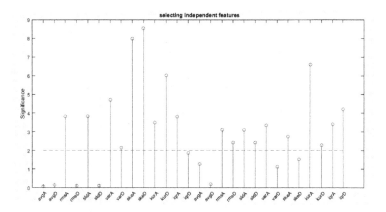

Figure 13. Features of greater significance in 3D
Source: (Author, 2017)

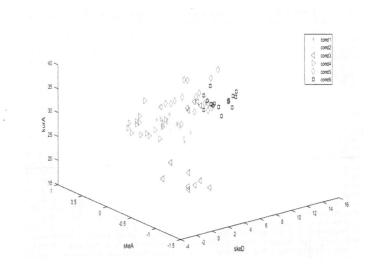

Figure 14. Architecture implemented for a) adaline and non-linear networks, b) MLP1 and c) MLP2 and non-linear networks
Source: (Author, 2017)

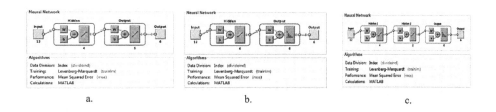

Figure 15. Training confusion matrix for Adaline, NL1 and NL2
Source: (Author, 2017)

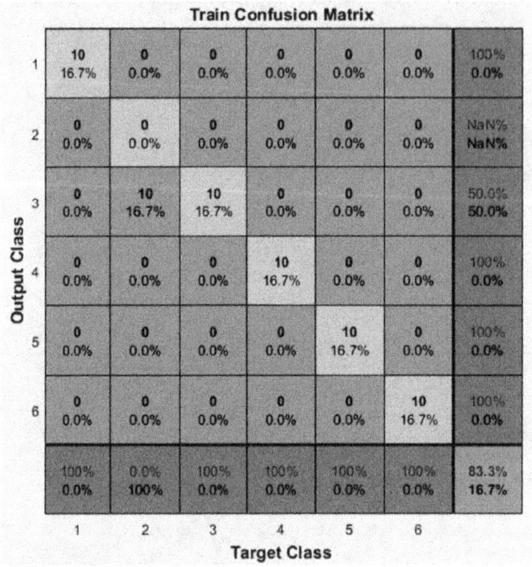

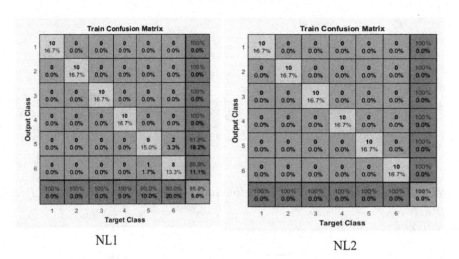

Adaline

Figure 16. Training confusion matrix for MLP1 and MLP2.
Source: (Author, 2017)

NL1

NL2

Figure 17. Summary of results for training stage

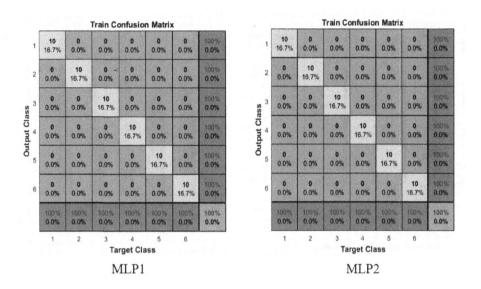

MLP1   MLP2

Table 5. Summary of the results for test stage

ANN	Mean Squared Error Performance MSE		Gradient		Validation Check		% de Classification (Training)
	Value	Epoch	Value	Epoch	Value	Epoch	
Adaline	0,028225	45	0,00722	51	6	51	83,3%
NL1	0,052012	36	0.002211	42	6	42	95%
NL2	0,000020206	15	0,0000000294	15	0	15	100%
MLP1	0,010574	7	0,0000000288	11	4	11	100%
MLP2	0,00000218	24	0,0000000396	24	0	24	100%

Source:(Author, 2017).

differentiated from the other conditions (p> 0.8), although the response of the network for some conditions is negative. However, for cases of validation of condition 2, there is a high degree of confusion with defined decision boundaries for thresholds below 0.4, which does not allow a good classification system. The highest degree of separation is obtained for conditions 5 and 6, which correspond to the response values of the network with greater relative value compared to the other conditions. In addition, NL1 exhibits good results for conditions 1, 2 and 3. For MLP1, MLP2 and NL2 show better results in all conditions. Table 5 shows a summary of the results obtained for test stage and for each network implemented from pressure signals.

In Appendix 1 attached the code using Matlab.

Comparing with other results obtained using traditional ANNs, performances in the lowest classification percentage (Wen et al.,2017; Wang et al., 2018) have been achieved, although they have used vibration signals which are usually noisy (Gao et al.,2019).

Figure 18. Adaline network output (Test)
Source: (Author, 2017)

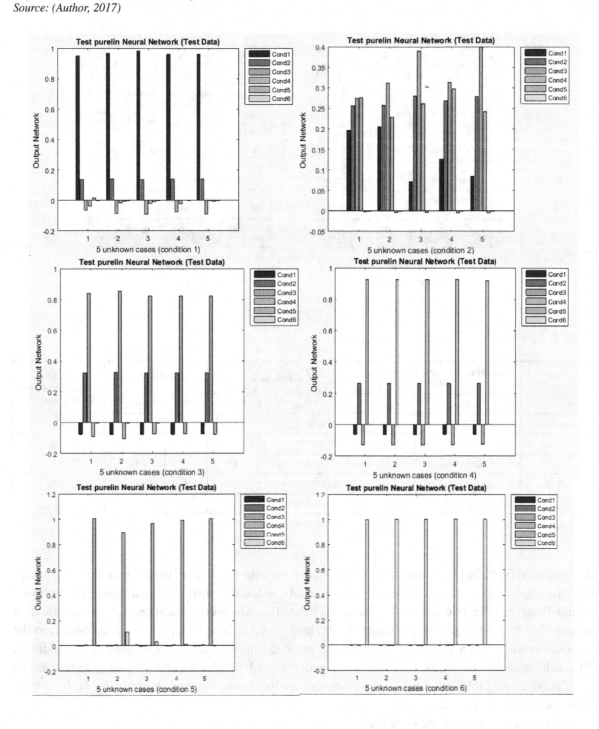

Figure 19. MLP1 network output (Test)
Source: (Author, 2017)

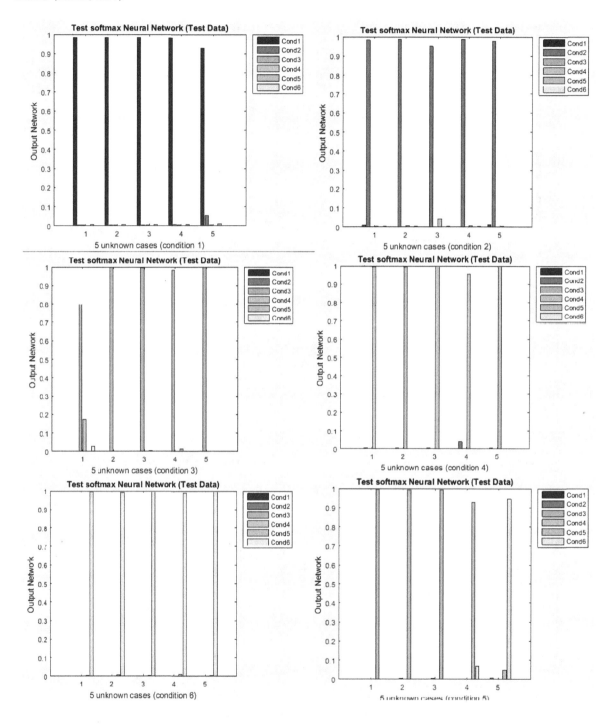

Figure 20. MLP2 network output (Test)
Source: (Author, 2017)

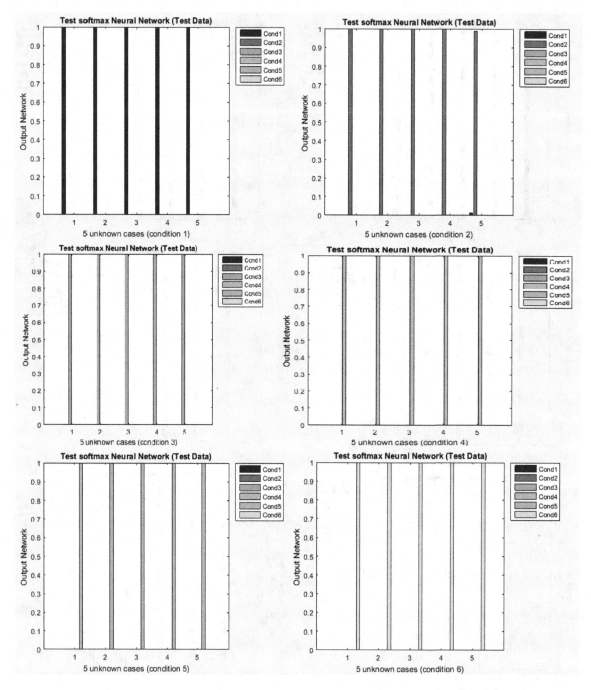

Figure 21. NL1 network output (Test)
Source: (Author, 2017)

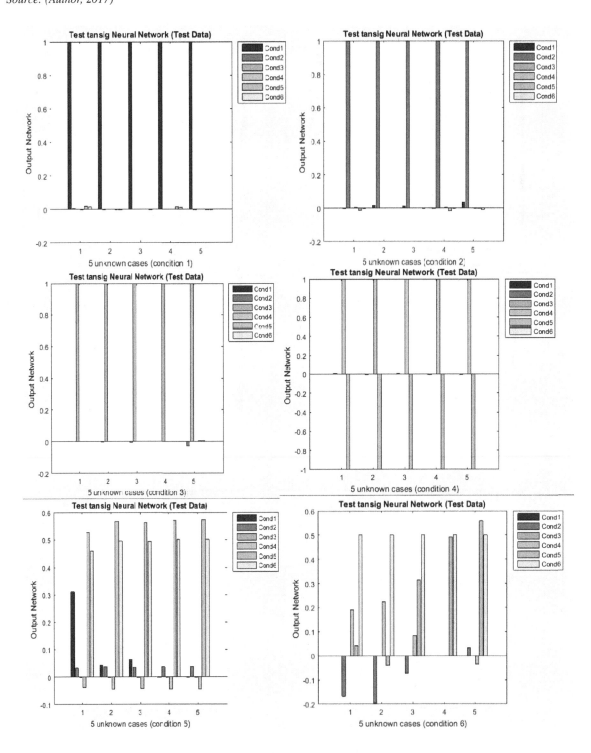

Figure 22. NL2 network output (Test)
Source: (Author, 2017)

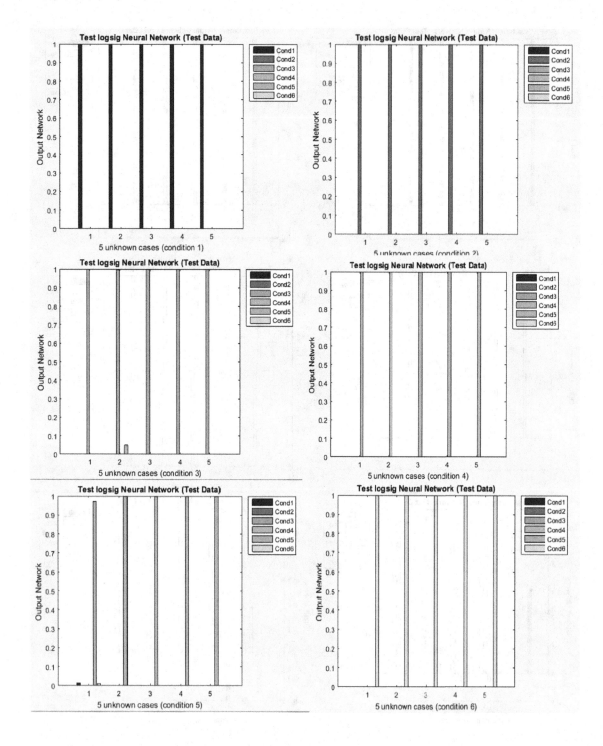

Table 6. Summary results of test stage and for each network for pressure signals.

ANN	% de Clasification (Test)
Adaline	83,3%
NL1	96,7%
NL2	100%
MLP1	96,7%
MLP2	100%

Source: (Author, 2017)

CONCLUSION

An algorithm was developed for the detection, classification and analysis of off-line faults of a hydraulic pump with axial pistons corresponding to different valve plate wear conditions using the pressure signals reaching a precision in the training stage of 100% and in the test stage the obtained results are very similar.

The spectral analysis allows observing differences in the frequency domain allowing comparing and identifying the characteristics of each condition only with simple observation. In addition, the implementation of low pass and high pass filters are of vital importance to verify the frequencies, amplitudes and power of the most relevant peaks. When the high filter is apply to the pressure signals, it can be seen peaks in frequencies of the order of 20Hz, so it is suspected that it can be noise, however not having the frequency curves of the pressure sensor cannot verify whether or not is.

The feature selection method is very helpful in saving computational cost and also improves the performance of the classifier. Likewise, by increasing the number of characteristics, the fault can be better identified. The most significant characteristics are approximation kurtosis, skewness of approximation and detail.

According to the performance of the ANN's, Adaline network is the one that presents the lowest results in both the training and testing stage while NL2 and MLP2 networks providing the best with a 100% classification percentage.

Future works should be focus on to apply other techniques for extraction and classification, specially, deep learning.

ACKNOWLEDGMENT

This work is supported by the Vicerrectoría de Investigación y Extensión (VIE) of the Universidad Industrial de Santander, where the research project has a grant to supports Master´s students in mechanical engineering.

REFERENCES

Aly, A. A. (2015). An Artificial Neural Network Flow Control of Variable Displacement Piston Pump with Pressure Compensation. *International Journal of Control, Automation, and Systems*, *4*(1), 1–7.

Ančić, I., Šestan, A., & Virag, Z. (2014). Determination of Actual Discharge of High-Pressure Low-Discharge Axial Piston Pumps. *Transactions of FAMENA*, *38*(2), 1–10.

Arango, G. (2007). *Clasificación de fallas en motores eléctricos utilizando señales de vibración* (Master's thesis). Universidad Tecnológica de Pereira.

Arizmendi, C. J. (2011). *Signal processing techniques for brain tumour diagnosis from magnetic resonance spectroscopy data* (Doctoral dissertation). Universitat Politecnica de Catalunya.

Bayer, C., & Enge-Rosenblatt, O. (2011, June). Modeling of hydraulic axial piston pumps including specific signs of wear and tear. In *Proceedings of the 8th International Modelica Conference; March 20th-22nd; Technical Univeristy; Dresden; Germany* (pp. 461-466). Linköping University Electronic Press. 10.3384/ecp11063461

Bei, J., Lu, C., & Wang, Z. (2013). Performance Assessment and Fault Diagnosis for Hydraulic Pump Based on WPT and SOM. *Vibroengineering PROCEDIA*, *2*, 23–28.

Bolaño, Y., & Niño, J. (2016). *Rediseño y construcción de un banco de pruebas para diagnóstico de fallas en una hidrotransmisión bomba variable-motor fijo* (Bachelor Thesis). Universidad Industrial de Santander, Bucaramanga.

Casoli, P., Bedotti, A., Campanini, F., & Pastori, M. (2018). A methodology based on cyclostationary analysis for fault detection of hydraulic axial piston pumps. *Energies*, *11*(7), 1874. doi:10.3390/en11071874

Castillo, S., & Serrano, J. (2016). *Metodología de mantenimiento predictivo para diagnóstico de fallas en bombas hidráulicas de pistones axiales asociadas con desgaste de los platos rodantes de presión* (Bachelor Thesis). Universidad Industrial de Santander, Bucaramanga.

Chu, M. H., Kang, Y., Chang, Y. F., Liu, Y. L., & Chang, C. W. (2003). Model-following controller based on neural network for variable displacement pump. *JSME International Journal. Series C, Mechanical Systems, Machine Elements and Manufacturing*, *46*(1), 176–187. doi:10.1299/jsmec.46.176

Du, J., & Wang, S. (2010, January). Hiberarchy clustering fault diagnosis of hydraulic pump. In 2010 Prognostics and System Health Management Conference (pp. 1-7). IEEE.

Du, J., Wang, S., & Zhang, H. (2013). Layered clustering multi-fault diagnosis for hydraulic piston pump. *Mechanical Systems and Signal Processing*, *36*(2), 487–504. doi:10.1016/j.ymssp.2012.10.020

Gao, Q., Tang, H., Xiang, J., Zhong, Y., Ye, S., & Pang, J. (2019). A Walsh transform-based teager energy operator demodulation method to detect faults in axial piston pumps. *Measurement*, *134*, 293–306. doi:10.1016/j.measurement.2018.10.085

Gao, Y., Kong, X., & Zhang, Q. (2002). Wavelet Analysis for Piston Pump Fault Diagnosis. *SAE Transactions*, 131–136.

Gao, Y., Zhang, Q., & Kong, X. (2003). Wavelet-based pressure analysis for hydraulic pump health diagnosis. *Transactions of the ASAE. American Society of Agricultural Engineers*, *46*(4), 969–976.

Guillén, M. L., Paredes, J. L., & Quintero, O. E. C. (2004). Detección y diagnóstico de fallas utilizando la transformada wavelet. *Ciencia e Ingeniería*, *25*(1), 35–42.

He, Z., Wang, S., Wang, K., & Li, K. (2012, July). Prognostic analysis based on hybrid prediction method for axial piston pump. In *IEEE 10th International Conference on Industrial Informatics* (pp. 688-692). IEEE. 10.1109/INDIN.2012.6301185

Hindman, J., Burton, R., & Schoenau, G. (2002). Condition monitoring of fluid power systems: A survey. *SAE Transactions*, 69–75.

Hu, W., Wang, S., He, Z., & Zhao, S. (2010, June). Prognostic analysis based on updated grey model for axial piston pump. In *2010 5th IEEE Conference on Industrial Electronics and Applications* (pp. 571-575). IEEE. 10.1109/ICIEA.2010.5517068

Karkoub, M. A., Gad, O. E., & Rabie, M. G. (1999). Predicting axial piston pump performance using neural networks. *Mechanism and Machine Theory*, *34*(8), 1211–1226. doi:10.1016/S0094-114X(98)00086-X

Kivelä, T., & Mattila, J. (2013, October). Internal leakage fault detection for variable displacement axial piston pump. In *ASME/BATH 2013 Symposium on Fluid Power and Motion Control*. American Society of Mechanical Engineers Digital Collection. 10.1115/FPMC2013-4445

Lan, Y., Hu, J., Huang, J., Niu, L., Zeng, X., Xiong, X., & Wu, B. (2018). Fault diagnosis on slipper abrasion of axial piston pump based on extreme learning machine. *Measurement*, *124*, 378–385. doi:10.1016/j.measurement.2018.03.050

Lázaro, J. G. M., Pinilla, C. B., & Prada, S. R. (2016, November). A Survey of Approaches for Fault Diagnosis in Axial Piston Pumps. In *ASME 2016 International Mechanical Engineering Congress and Exposition* (pp. V04AT05A052-V04AT05A052). American Society of Mechanical Engineers.

Lurette, C., & Lecoeuche, S. (2003). Unsupervised and auto-adaptive neural architecture for on-line monitoring. Application to a hydraulic process. *Engineering Applications of Artificial Intelligence*, *16*(5-6), 441–451. doi:10.1016/S0952-1976(03)00064-2

Özmen, Ö., Sınanoğlu, C., Batbat, T., & Güven, A. (2019). Prediction of Slipper Pressure Distribution and Leakage Behaviour in Axial Piston Pumps Using ANN and MGGP. *Mathematical Problems in Engineering*.

Peng, Z. K., & Chu, F. L. (2004). Application of the wavelet transform in machine condition monitoring and fault diagnostics: A review with bibliography. *Mechanical Systems and Signal Processing*, *18*(2), 199–221. doi:10.1016/S0888-3270(03)00075-X

Pinto, J. (2014). *Análisis e interpretación de señales cardiorrespiratorias para determinar el momento óptimo de desconexión de un paciente asistido mediante ventilación* (Bachelor Thesis). Universidad Autónoma de Bucaramanga, Bucaramanga.

Pławiak, P. (2014). An estimation of the state of consumption of a positive displacement pump based on dynamic pressure or vibrations using neural networks. *Neurocomputing*, *144*, 471–483. doi:10.1016/j.neucom.2014.04.026

Ramdén, T., Krus, P., & Palmberg, J. O. (1996). Reliability and sensitivity analysis of a condition monitoring technique. In *Proceedings of the JFPS International Symposium on Fluid Power* (pp. 567-572). The Japan Fluid Power System Society. 10.5739/isfp.1996.567

Salimi, E., & Niromandfam, B. (2015). *Condition Monitoring of Hydrolic Pump by Wavelet Transform and Artificial Neural Network*. Academic Press.

Schoenau, G. J., Stecki, J. S., & Burton, R. T. (2000). Utilization of Artificial Neural Networks in the Control, Identification and Condition Monitoring of Hydraulic Systems—An Overview. *SAE Transactions*, 205–212.

Siyuan, L., Linlin, D., & Wanlu, J. (2011, August). Study on application of Principal Component Analysis to fault detection in hydraulic pump. In *Proceedings of 2011 International Conference on Fluid Power and Mechatronics* (pp. 173-178). IEEE.

Sorsa, T., & Koivo, H. N. (1993). Application of artificial neural networks in process fault diagnosis. *Automatica*, *29*(4), 843–849. doi:10.1016/0005-1098(93)90090-G

Stalford, H. L., & Borrás, C. (2002, May). Pattern recognition in hydraulic backlash using neural network. In *Proceedings of the 2002 American Control Conference (IEEE Cat. No. CH37301)* (Vol. 1, pp. 400-405). IEEE. 10.1109/ACC.2002.1024838

Tao, X., Lu, C., Lu, C., & Wang, Z. (2013). An approach to performance assessment and fault diagnosis for rotating machinery equipment. *EURASIP Journal on Advances in Signal Processing*, *2013*(1), 5. doi:10.1186/1687-6180-2013-5

Tao, X., Wang, Z., Ma, J., & Fan, H. (2012, May). Study on fault detection using wavelet packet and SOM neural network. In *Proceedings of the IEEE 2012 Prognostics and System Health Management Conference (PHM-2012 Beijing)* (pp. 1-5). IEEE.

Wang, S., & Xiang, J. (n.d.). A minimum entropy deconvolution-enhanced convolutional neural networks for fault diagnosis of axial piston pumps. *Soft Computing*, 1-15.

Wang, S., Xiang, J., Zhong, Y., & Tang, H. (2018). A data indicator-based deep belief networks to detect multiple faults in axial piston pumps. *Mechanical Systems and Signal Processing*, *112*, 154–170. doi:10.1016/j.ymssp.2018.04.038

Wen, L., Li, X., Gao, L., & Zhang, Y. (2017). A new convolutional neural network-based data-driven fault diagnosis method. *IEEE Transactions on Industrial Electronics*, *65*(7), 5990–5998. doi:10.1109/TIE.2017.2774777

Zhao, W., Wang, S., & Dou, H. (2012, July). Fault simulation analysis of axial piston pump based on the component's failure. In *IEEE 10th International Conference on Industrial Informatics* (pp. 631-634). IEEE.

APPENDIX

Figure 23. P-Welch of pressure signals condition 1 (94,9%)
Source: (Author, 2017)

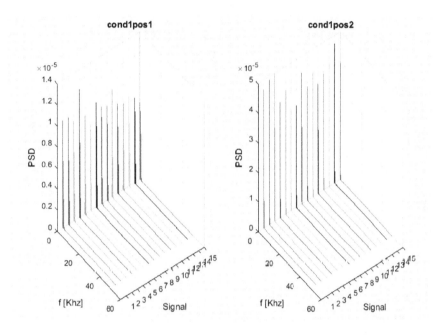

Figure 24. P-Welch of pressure signals condition 1 (64,9%)
Source: (Author, 2017)

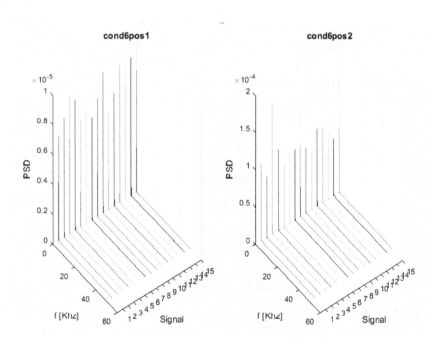

WPT + ANN Methodology Code Developed for Pressure Signals

```
%%
%Auxiliar function used to process Pressure data
% General Signal Parameters
Fs = 100e3; % sample frequency
positions = [1 2]; % must be the respective positions for pressure or vibra-
tion data
%%
%read excel files
raw_data = ReadrawData;
%%
%remove means, trends and cut signals
parameters.Ls = 60e3;          % minimum length signal
parameters.method = 'initial'; % can be 'center' or 'initial'
fData = preProcess(raw_data,parameters);
%%
%plot PSD (Power Spectral Densities)
PSDplot (fData,Fs,positions) % for original frequencies until Fs/2
%%
%plot Low Frequency Signal Contens
% Basic User Parameters (Should be specified)
fPar.Factor = 30;  % downsampled factor (must be 1 to processing original sig-
nal)
fPar.Fp = 10e3;     % Passband frequency in Hz
% Advanced Parameters (optional to change)
fPar.Fst = fPar.Fp + 1e3; % Stopband frequency in Hz
fPar.MinPeakHeight = 0.1; % percentage value of frequency peaks to mantain
fPar.Ap = 1;       % Passband ripple in dB
fPar.Ast = 95;    % Stopband attenuation in dB
[PeaksL, FreqsL, powerL] = PSDplotLower (fData,Fs,fPar,positions); %plot PSD
for lower frequencies below Fp kHz
%%
%plot High Frequency Signal Contens
% Basic User Parameters (Should be specified)
fPar.NPeaks = 3; % percentage value
fPar.Fp = 10e3;    % Passband frequency in Hz
% Advanced Parameters (optional to change)
fPar.Fst = fPar.Fp - 1e3; % Stopband frequency in Hz
fPar.Ap = 1;       % Passband ripple in dB
fPar.Ast = 95;    % Stopband attenuation in dB
```

```
% Plot function
[PeaksH, FreqsH,powerH] = PSDplotHigh (fData,Fs,fPar,positions);
%%
%scatter plots for Frequency data (for pressure data it does not make sense)
PlotGroups(powerL,powerH,FreqsH,FreqsL,PeaksH,PeaksL,positions)
%%
%Function to select mother and level for wavelet analysis
% entropy criterion
CRI = 'log energy'; %'shannon', 'threshold', 'norm','log energy' (or 'logen-
ergy'), 'sure'
[wpM,wpS,Res] = waveletAnalysis(fData,positions,CRI);
%%
%function to compute wavelet features
% mean, rms,std,var,skewness,kurtosis,iqr (aprox y detalle x cada posición)
% wavelet family filter
wname='db2';  %'db1', ...,'db43' or 'coif1',...,'coif5' or 'sym1', ...,
'sym25'
                    %'bior1.1'; 'bior1.3'; 'bior1.5'; 'bior2.2'; 'bior2.4';
'bior2.6'; 'bior2.8';
                    %'bior3.3'; 'bior3.5'; 'bior3.7'; 'bior3.9'; 'bior4.4';
'bior5.5'; 'bior6.8'
Position = [1 2];% can be [1 2 3 4] or [1 3 4] or [1 4] (Positions with useful
information)
wFeat = waveletFeat(fData,wpM,CRI,Position,wname);
%%
%optional wavelet features
% Energy and kurtosis of sub-band wavelet packets (aprox y detalle x cada
posición)
wFeatS = waveletFeats(fData,wpM,CRI,Position,wname);
%%
%This function is used to hold the same number of signals per condition...
num = 15; % It is only necessary if the positions in each condition have dif-
ferent number of signals
wFeatR = cutWfeat(wFeat,num); % wFeat can be replaced with wFeatS if optional
features are used
%%
%Feature Selection using  independent features method with significance thresh-
old
S=2;                %typically, variables are kept when significance S >= 2
[DataF,Targets] = FeatSel(wFeat,S); % wFeat can be replaced with wFeatS if op-
tional features are used
```

315

```
%%
%Results for different types of ANN (Neaural Network Outputs)
Normalizar = true; % or false
[Outputs,performance,errors]= Ann(DataF,Targets,Normalizar);
%%
 %Compare performance of different types of ANN
 [confmat,All_performance]=AnnCompare(DataF,Targets,Normalizar)
```

Compilation of References

Abdi, H., & Williams, L. J. (2010). Principal Component Analysis. *Wiley Interdisciplinary Reviews: Computational Statistics*, *2*(4), 433–459. doi:10.1002/wics.101

Agarwal, G., Belhumeur, P., Feiner, S., Jacobs, D., Kress, J. W., Ramamoorthi, R., ... White, S. (2006). First steps toward an electronic field guide for plants. *Taxon*, *55*(3), 597–610. doi:10.2307/25065637

Aghili, S. F., Mala, H., Shojafar, M., & Peris-Lopez, P. (2019). LACO: Lightweight Three-Factor Authentication, Access Control and Ownership Transfer Scheme for E-Health Systems in IoT. *Future Generation Computer Systems*, *96*, 410–424. doi:10.1016/j.future.2019.02.020

Ahmad, K., Mohammad, O., Atieh, M., & Ramadan, H. (2019). IoT: Architecture, Challenges, and Solutions Using Fog Network and Application Classification. ACIT 2018 - 19th International Arab Conference on Information Technology, 1–7. 10.1109/ACIT.2018.8672696

Ahmad, R. H., & Pathan, A.-S. K. (2016). A study on M2M (machine to machine) system and communication: Its security, threats, and intrusion detection system. Security Solutions and Applied Cryptography in Smart Grid Communications. doi:10.4018/978-1-5225-1829-7.ch010

Ahmad, A., & Dey, L. (2007). A k-mean clustering algorithm for mixed numeric and categorical data. *Data & Knowledge Engineering*, *63*(2), 503–527. doi:10.1016/j.datak.2007.03.016

Ahmad, A., & Hashmi, S. (2016). K-Harmonic means type clustering algorithm for mixed datasets. *Applied Soft Computing*, *48*, 39–49. doi:10.1016/j.asoc.2016.06.019

Ahmad, A., & Khan, S. S. (2018). Survey of state-of-the-art mixed data clustering algorithms. *IEEE Access : Practical Innovations, Open Solutions*, *7*, 1–16. doi:10.1109/ACCESS.2019.2903568

Ahmad, A., & Khan, S. S. (2019). Survey of State-of-the-Art Mixed Data Clustering Algorithms. *IEEE Access : Practical Innovations, Open Solutions*, *7*(i), 31883–31902. doi:10.1109/ACCESS.2019.2903568

Ahmadi, V., & Khorramizadeh, M. (2018). An adaptive heuristic for multi-objective controller placement in software-defined networks. *Computers & Electrical Engineering*, *66*, 204–228. doi:10.1016/j.compeleceng.2017.12.043

Akay, Ö., & Yüksel, G. (2018). Clustering the mixed panel dataset using Gower's distance and k-prototypes algorithms. *Communications in Statistics. Simulation and Computation*, *47*(10), 3031–3041. doi:10.1080/03610918.2017.1367806

Albur, M., Hamilton, F., & MacGowan, A. P. (2016). Early warning score: A dynamic marker of severity and prognosis in patients with Gram-negative bacteraemia and sepsis. *Annals of Clinical Microbiology and Antimicrobials*, *15*(1), 1–10. doi:10.1186/s12941-016-0139-z PubMed

Alcaldía Mayor de Bogotá. (2017). *Logros del plan de gobierno de Peñalosa en 2017*. Obtenido de http://www.bogota.gov.co/logros-del-plan-de-gobierno-de-penalosa-en-2017/mejora-la-seguridad-en-bogota.html

Alchanatis, V., Ridel, L., Hetzroni, A., & Yaroslavsky, L. (2005). Weed detection in multi-spectral images of cotton fields. *Computers and Electronics in Agriculture, 47*(3), 243–260. doi:10.1016/j.compag.2004.11.019

Almusaylim, Z. A., & Zaman, N. (2018). A review on smart home present state and challenges: Linked to context-awareness internet of things (IoT). *Wireless Networks, 5*, 1–12. doi:10.100711276-018-1712-5

Aloulou, N., Ayari, M., Zhani, M. F., Saidane, L., & Pujolle, G. (2017). Effective controller placement in controller-based Named Data Networks. *2017 International Conference on Computing, Networking and Communications, ICNC 2017*, 249–254. 10.1109/ICCNC.2017.7876134

Alqurashi, T., & Wang, W. (2018). Clustering ensemble method. *International Journal of Machine Learning and Cybernetics, 10*(6), 1227–1246. doi:10.1007/s13042-017-0756-7

Al-Sobky, A. S. A., & Hashim, I. H. (2014). A generalized mathematical model to determine the turning movement counts at roundabouts. Alexandria Engineering Journal, 53(3), 669–675. doi:10.1016/j.aej.2014.06.012

Alvarez-Montoya, J., Carvajal-Castrillón, A., & Sierra-Pérez, J. (2020). In-flight and wireless damage detection in a UAV composite wing using fiber optic sensors and strain field pattern recognition. *Mechanical Systems and Signal Processing*, 136.

Alvarez-Montoya, J., & Sierra-Pérez, J. (2018). Fuzzy unsupervised-learning techniques for diagnosis in a composite UAV wing by using fiber optic sensors. *Proceedings of the 7th Asia-Pacific Workshop on Structural Health Monitoring, APWSHM 2018*, 682–690.

Alvarez-Montoya, J., Torres-Arredondo, M., & Sierra-Pérez, J. (2018). Gaussian process modeling for damage detection in composite aerospace structures by using discrete strain measurements. *Proceedings of the 7th Asia-Pacific Workshop on Structural Health Monitoring, APWSHM 2018*, 710–718.

Aly, A. A. (2015). An Artificial Neural Network Flow Control of Variable Displacement Piston Pump with Pressure Compensation. *International Journal of Control, Automation, and Systems, 4*(1), 1–7.

Ambrosio, L., Iglesias, L., Marin, C., & Del Monte, J. P. (2004). Evaluation of sampling methods and assessment of the sample size to estimate the weed seedbank in soil, taking into account spatial variability. *Weed Research, 44*(3), 224–236. doi:10.1111/j.1365-3180.2004.00394.x

Anaya, M. (2016). *Design and validation of a structural health monitoring system based on bio-inspired algorithms* (PhD. Thesis). Universitat Politecnica de Catalunya.

Anaya, M., Tibaduiza, D. A., Forero, E., Castro, R., & Pozo, F. (2015). An acousto-ultrasonics pattern recognition approach for damage detection in wind turbine structures. *20th Symposium on Signal Processing, Images and Computer Vision (STSIVA)*. 10.1109/STSIVA.2015.7330419

Ančić, I., Šestan, A., & Virag, Z. (2014). Determination of Actual Discharge of High-Pressure Low-Discharge Axial Piston Pumps. *Transactions of FAMENA, 38*(2), 1–10.

Andreev, S., Petrov, V., Huang, K., Lema, M. A., & Dohler, M. (2019). Dense Moving Fog for Intelligent IoT: Key Challenges and Opportunities. *IEEE Communications Magazine, 57*(5), 34–41. doi:10.1109/MCOM.2019.1800226

Annamalai, P., Bapat, J., & Das, D. (2019). Emerging Access Technologies and Open Challenges in 5G IoT: From Physical Layer Perspective. *2018 IEEE International Conference on Advanced Networks and Telecommunications Systems (ANTS)*, 1–6. 10.1109/ants.2018.8710133

Anta, J. A. (2012). Análisis verbo-corporal (AVC): Su utilidad en los secuestros. *Revista de Criminologia e Ciencias Penitenciárias*, 9-14.

Arango, G. (2007). *Clasificación de fallas en motores eléctricos utilizando señales de vibración* (Master's thesis). Universidad Tecnológica de Pereira.

Arizmendi, C. J. (2011). *Signal processing techniques for brain tumour diagnosis from magnetic resonance spectroscopy data* (Doctoral dissertation). Universitat Politecnica de Catalunya.

Arora, N., Rastogi, R., Chaturvedi, D. K., Satya, S., Gupta, M., Yadav, V., . . . Sharma, P. (2019b). Chronic TTH Analysis by EMG & GSR Biofeedback on Various Modes and Various Medical Symptoms Using IoT. In Big Data Analytics for Intelligent Healthcare Management. doi:10.1016/B978-0-12-818146-1.00005-2

Arora, N., Trivedi, P., Chauhan, S., Rastogi, R., & Chaturvedi, D. K. (2017a). Framework for Use of Machine Intelligence on Clinical Psychology to study the effects of Spiritual tools on Human Behavior and Psychic Challenges. *Proceedings of NSC-2017*.

Atzori, L., Iera, A., & Morabito, G. (2017). Understanding the Internet of Things: Definition, potentials, and societal role of a fast evolving paradigm. *Ad Hoc Networks*, *56*, 122–140. doi:10.1016/j.adhoc.2016.12.004

Avasalcai, C., & Dustdar, S. (2020). Latency-aware distributed resource provisioning for deploying IoT applications at the edge of the network. Lecture Notes in Networks and Systems, 69, 377–391. doi:10.1007/978-3-030-12388-8_27

Azad, M. A., Bag, S., Hao, F., & Salah, K. (2018). M2M-REP: Reputation system for machines in the internet of things. *Computers & Security*, *79*, 1–16. doi:10.1016/j.cose.2018.07.014

Bakhshipour, A., & Jafari, A. (2018). Evaluation of support vector machine and artificial neural networks in weed detection using shape features. *Computers and Electronics in Agriculture*, *145*, 153–160. doi:10.1016/j.compag.2017.12.032

Banerjee, M. B., Chatterjee, T. N., Roy, R. B., Tudu, B., Bandyopadhyay, R., & Bhattacharyya, N. (2016, April). Multivariate preprocessing techniques towards optimising response of fused sensor from electronic nose and electronic tongue. In *2016 International Conference on Computing, Communication and Automation (ICCCA)* (pp. 949-954). IEEE 10.1109/CCAA.2016.7813875

Banerjee, R., Tudu, B., Bandyopadhyay, R., & Bhattacharyya, N. (2016). A review on combined odor and taste sensor systems. *Journal of Food Engineering*, *190*, 10–21. doi:10.1016/j.jfoodeng.2016.06.001

Bansal, I., Rastogi, R., Chaturvedi, D. K., Satya, S., Arora, N., & Yadav, V. (2018k). Intelligent Analysis for Detection of Complex Human Personality by Clinical Reliable Psychological Surveys on Various Indicators. In *National Conference on 3rd MDNCPDR-2018 at DEI*, Agra.

Barbona Ivana, B. C. (2016). Método de clasificación supervisada support vector machine: Una aplicación a la clasificación automática de textos. *Revista de Epistemología y Ciencias Humanas*, 37-42.

Bari, M. F., Roy, A. R., Chowdhury, S. R., Zhang, Q., Zhani, M. F., Ahmed, R., & Boutaba, R. (2013). Dynamic controller provisioning in software defined networks. *2013 9th International Conference on Network and Service Management, CNSM 2013 and Its Three Collocated Workshops - ICQT 2013, SVM 2013 and SETM 2013*, 18–25.

Barré, P., Stöver, B. C., Müller, K. F., & Steinhage, V. (2017). LeafNet: A computer vision system for automatic plant species identification. *Ecological Informatics*, *40*, 50–56. doi:10.1016/j.ecoinf.2017.05.005

Basatneh, R., Najafi, B., & Armstrong, D. G. (2018). Health Sensors, Smart Home Devices, and the Internet of Medical Things: An Opportunity for Dramatic Improvement in Care for the Lower Extremity Complications of Diabetes. *Journal of Diabetes Science and Technology*, *12*(3), 577–586. doi:10.1177/1932296818768618 PubMed

Bayer, C., & Enge-Rosenblatt, O. (2011, June). Modeling of hydraulic axial piston pumps including specific signs of wear and tear. In *Proceedings of the 8th International Modelica Conference; March 20th-22nd; Technical Univeristy; Dresden; Germany* (pp. 461-466). Linköping University Electronic Press. 10.3384/ecp11063461

Beebe, K. R., Pell, R. J., & Seasholtz, M. B. (1998). *Chemometrics: a practical guide* (Vol. 4). Wiley-Interscience.

Bei, J., Lu, C., & Wang, Z. (2013). Performance Assessment and Fault Diagnosis for Hydraulic Pump Based on WPT and SOM. *Vibroengineering PROCEDIA, 2*, 23–28.

Belouch, M., Hadaj, S. E., & Idhammad, M. (2018). Performance evaluation of intrusion detection based on machine learning using Apache Spark. *Procedia Computer Science, 127*, 1–6. doi:10.1016/j.procs.2018.01.091

Ben Salem, S., Naouali, S., & Chtourou, Z. (2018). A fast and effective partitional clustering algorithm for large categorical datasets using a k-means based approach. *Computers & Electrical Engineering, 68*, 463–483. doi:10.1016/j.compeleceng.2018.04.023

Bhattarai, S., & Wang, Y. (2018). Internet of Things Security and Challenges. IEEE Computer, 78, 544–546. Retrieved from https://www.sciencedirect.com/science/article/pii/S0167739X17316667

Bhondekar, A. P., Kaur, R., Kumar, R., Vig, R., & Kapur, P. (2011). A novel approach using Dynamic Social Impact Theory for optimization of impedance-Tongue (iTongue). *Chemometrics and Intelligent Laboratory Systems, 109*(1), 65–76. doi:10.1016/j.chemolab.2011.08.002

Binder, H., & Blettner, M. (2015). Big Data in Medical Science—a Biostatistical View. *Deutsches Ärzteblatt International, 112*(9), 137.

Bishop, C. M. (2006). *Pattern recognition and machine learning.* New York, NY: Springer. Retrieved from https://www.microsoft.com/en-us/research/publication/pattern-recognition-machine-learning/

Bolaño, Y., & Niño, J. (2016). *Rediseño y construcción de un banco de pruebas para diagnóstico de fallas en una hidrotransmisión bomba variable-motor fijo* (Bachelor Thesis). Universidad Industrial de Santander, Bucaramanga.

Boller, C. (2008). Structural Health Monitoring-An Introduction and Definitions. In Encyclopedia of Structural Health Monitoring. doi:10.1002/9780470061626.shm204

Bolme, D., Beveridge, J. R., Draper, B. A., & Lui, Y. M. (2010). Visual object tracking using adaptive correlation filters. 2010 IEEE Computer Society Conference on Computer Vision and Pattern Recognition, 2544–2550. doi:10.1109/CVPR.2010.5539960

Boongoen, T., & Iam-On, N. (2018). Cluster ensembles: A survey of approaches with recent extensions and applications. *Computer Science Review, 28*, 1–25. doi:10.1016/j.cosrev.2018.01.003

Boresi, A., & Schmidt, R. (2003). *Advanced mechanics of materials.* John Wiley & Sons, Inc.

Boriah, S., Chandola, V., & Kumar, V. (2008). Similarity Measures for Categorical Data: A Comparative Evaluation. Proceedings of the 2008 SIAM International Conference on Data Mining, 243–254. doi:10.1137/1.9781611972788.22

Bossu, J., Gée, C., Jones, G., & Truchetet, F. (2009). Wavelet transform to discriminate between crop and weed in perspective agronomic images. *Computers and Electronics in Agriculture, 65*(1), 133–143. doi:10.1016/j.compag.2008.08.004

Bougrini, M., Tahri, K., Haddi, Z., El Bari, N., Llobet, E., Jaffrezic-Renault, N., & Bouchikhi, B. (2014). Aging time and brand determination of pasteurized milk using a multisensor e-nose combined with a voltammetric e-tongue. *Materials Science and Engineering C, 45*, 348–358. doi:10.1016/j.msec.2014.09.030 PMID:25491839

Boureau, F., Luu, M., & Doubrere, J. F. (2001). Study of experimental pain measures and nociceptive reflex in Chronic pain patients and normal subjects. *Pain*, *44*(3), 131–138. PubMed

Boussaïd, I., Lepagnot, J., & Siarry, P. (2013). A survey on optimization metaheuristics. *Information Sciences*, *237*, 82–117. doi:10.1016/j.ins.2013.02.041

Bouwmans, T., Silva, C., Marghes, C., Zitouni, M. S., Bhaskar, H., & Frelicot, C. (2018). On the role and the importance of features for background modeling and foreground detection. *Computer Science Review*, *28*, 26–91. doi:10.1016/j.cosrev.2018.01.004

Bozorgchami, B., Member, S., Sodagari, S., & Member, S. (n.d.). Spectrally Efficient Telemedicine and In-Hospital Patient Data Transfer. Academic Press.

Br. Medina Olivera, V. J. (2018). *Desarrollo de un Sistema Automatizado Basado en Procesamiento Digital de Imágenes para mejorar el control de Videovigilancia en empresas de Trujillo*. Academic Press.

Bramer, M. (2016). Estimating the Predictive Accuracy of a Classifier. In *Principles of data mining* (3rd ed.; pp. 79–92). London, UK: Springer; doi:10.1007/978-1-4471-7307-6_7

Bramsen, P. (2017). Exploring a New IoT Infrastructure. University of California at Berkley. Retrieved from http://www2.eecs.berkeley.edu/Pubs/TechRpts/2017/EECS-2017-56.html

Brazdil, P. B., Soares, C., & Da Costa, J. P. (2003). Ranking learning algorithms: Using IBL and meta-learning on accuracy and time results. *Machine Learning*, *50*(3), 251–277. doi:10.1023/A:1021713901879

Brinkman, W. F., Haggan, D. E., & Troutman, W. W. (1997). A history of the invention of the transistor and where it will lead us. *IEEE Journal of Solid-State Circuits*, *32*(12), 1858–1865. doi:10.1109/4.643644

Brody, A. A., Arbaje, A. I., DeCherrie, L. V., Federman, A. D., Leff, B., & Siu, A. L. (2019). Starting Up a Hospital at Home Program: Facilitators and Barriers to Implementation. *Journal of the American Geriatrics Society*, *67*(3), 588–595. doi:10.1111/jgs.15782 PubMed

Bull, P., Austin, R., Popov, E., Sharma, M., & Watson, R. (2016). Flow based security for IoT devices using an SDN gateway. Proceedings - 2016 IEEE 4th International Conference on Future Internet of Things and Cloud, FiCloud 2016, 157–163. doi:10.1109/FiCloud.2016.30

Buratti, S., Sinelli, N., Bertone, E., Venturello, A., Casiraghi, E., & Geobaldo, F. (2015). Discrimination between washed Arabica, natural Arabica and Robusta coffees by using near infrared spectroscopy, electronic nose and electronic tongue analysis. *Journal of the Science of Food and Agriculture*, *95*(11), 2192–2200. doi:10.1002/jsfa.6933 PMID:25258213

Burger, W., & Burge, M. J. (2009). Histograms. In *Principles of digital image processing: Fundamental techniques* (pp. 37–54). London, UK: Springer; doi:10.1007/978-1-84800-191-6_3

C.P., V. (2016). Security improvement in IoT based on Software Defined Networking (SDN). *International Journal of Science Engineering and Technology Research*, *5*(1), 291–295.

Caballero Barriga, E. R. (2017). *Aplicación Práctica de la Visión Artificial para el Reconocimiento de Rostros en una Imagen*. Utilizando Redes Neuronales y Algoritmos de Reconocimiento de Objetos de la Biblioteca OPENCV. Obtenido de http://hdl.handle.net/11349/6104

Cámara de Comercio de Bogotá. (2018). *Encuesta de Percepción y Victimización de seguridad en Bogotá*. Bogotá: Author.

Camps-Valls, G., & Bruzzone, L. (2005). Kernel-based methods for hyperspectral image classification. *IEEE Transactions on Geoscience and Remote Sensing*, *43*(6), 1351–1362. doi:10.1109/TGRS.2005.846154

Cao, J., Zhang, Q., & Shi, W. (2018). Challenges and opportunities in edge computing. Í. SpringerBriefs in Computer Science. doi:10.1007/978-3-030-02083-5_5

Carlson, N. (2013). *Physiology of Behavior*. Pearson Education, Inc.

Casadei, R., Fortino, G., Pianini, D., Russo, W., Savaglio, C., & Viroli, M. (2019). A development approach for collective opportunistic Edge-of-Things services. *Information Sciences*, *498*, 154–169. doi:10.1016/j.ins.2019.05.058

Casoli, P., Bedotti, A., Campanini, F., & Pastori, M. (2018). A methodology based on cyclostationary analysis for fault detection of hydraulic axial piston pumps. *Energies*, *11*(7), 1874. doi:10.3390/en11071874

Cassel, R. N. (1997). Biofeedback for developing self-control of tension and stress in one's hierarchy of psychological states. Psychology: A Journal of Human Behavior, 22(2), 50-57.

Castillo, S., & Serrano, J. (2016). *Metodología de mantenimiento predictivo para diagnóstico de fallas en bombas hidráulicas de pistones axiales asociadas con desgaste de los platos rodantes de presión* (Bachelor Thesis). Universidad Industrial de Santander, Bucaramanga.

Cavalcante, E., Pereira, J., Alves, M. P., Maia, P., Moura, R., & Batista, T. … Pires, P. F. (2016). On the interplay of Internet of Things and Cloud Computing: A systematic mapping study. *Computer Communications*. doi:10.1016/j.comcom.2016.03.012

Cetó, X., Céspedes, F., & del Valle, M. (2013). Comparison of methods for the processing of voltammetric electronic tongues data. *Mikrochimica Acta*, *180*(5-6), 319–330. doi:10.100700604-012-0938-7

Cetó, X., Voelcker, N. H., & Prieto-Simón, B. (2016). Bioelectronic tongues: New trends and applications in water and food analysis. *Biosensors & Bioelectronics*, *79*, 608–626. doi:10.1016/j.bios.2015.12.075 PMID:26761617

Cha, S., Ruiz, M. P., Wachowicz, M., Tran, L. H., Cao, H., & Maduako, I. (2016). The role of an IoT platform in the design of real-time recommender systems. 2016 IEEE 3rd World Forum on Internet of Things (WF-IoT), 448–453. 10.1109/WF-IoT.2016.7845469

Chaturvedi, D. K., Rastogi, R., Satya, S., Arora, N., Saini, H., & Verma, H., … Varshney, Y. (2018b). Statistical Analysis of EMG and GSR Therapy on Visual Mode and SF-36 Scores for Chronic TTH. Proceedings of UPCON-2018.

Chaturvedi, D. K., Rastogi, R., Arora, N., Trivedi, P., & Mishra, V. (2017b). Swarm Intelligent Optimized Method of Development of Noble Life in the perspective of Indian Scientific Philosophy and Psychology. *Proceedings of NSC-2017*.

Chauhan, M. A., & Babar, M. A. (2017). Using Reference Architectures for Design and Evaluation of Web of Things Systems: A Case of Smart Homes Domain. Managing the Web of Things: Linking the Real World to the Web. doi:10.1016/B978-0-12-809764-9.00009-3

Chauhan, S., Rastogi, R., Chaturvedi, D. K., Satya, S., Arora, N., Yadav, V., & Sharma, P. (2018d). Analytical Comparison of Efficacy for Electromyography and Galvanic Skin Resistance Biofeedback on Audio-Visual Mode for Chronic TTH on Various Attributes. *Proceedings of the ICCIDA-2018*.

Chen, C., Hu, J., Zhang, J., Sun, C., Zhao, L., & Ren, Z. (2016). Information congestion control on intersections in VANETs: A bargaining game approach. IEEE Vehicular Technology Conference, 2016-July, 1–5. doi:10.1109/VTC-Spring.2016.7504289

Chen, W., Chen, C., Jiang, X., & Liu, L. (2018). Multi-controller placement towards SDN based on louvain heuristic algorithm. *IEEE Access: Practical Innovations, Open Solutions*, *6*, 49486–49497. doi:10.1109/ACCESS.2018.2867931

Chen, W., Zhang, Z., Hong, Z., Chen, C., Wu, J., & Maharjan, S. (2019). *Zhang, Y.* Cooperative and Distributed Computation Offloading for Blockchain-Empowered Industrial Internet of Things. IEEE Internet of Things Journal; doi:10.1109/jiot.2019.2918296

Chen, Y., Liu, T., Chen, J., Li, D., & Wu, M. (2018, November). A Novel Feature Specificity Enhancement for Taste Recognition by Electronic Tongue. In *International Conference on Extreme Learning Machine* (pp. 11-16). Springer.

Cho, S.-H., & Kang, H.-B. (2014). Abnormal behavior detection using hybrid agents in crowded scenes. Pattern Recognition Letters, 44, 64-70. doi:10.1016/j.patrec.2013.11.017

Chu, M. H., Kang, Y., Chang, Y. F., Liu, Y. L., & Chang, C. W. (2003). Model-following controller based on neural network for variable displacement pump. *JSME International Journal. Series C, Mechanical Systems, Machine Elements and Manufacturing*, *46*(1), 176–187. doi:10.1299/jsmec.46.176

Ciosek, P., Brzózka, Z., Wróblewski, W., Martinelli, E., Di Natale, C., & D'amico, A. (2005). Direct and two-stage data analysis procedures based on PCA, PLS-DA and ANN for ISE-based electronic tongue—Effect of supervised feature extraction. *Talanta*, *67*(3), 590–596. doi:10.1016/j.talanta.2005.03.006 PMID:18970211

Ciosek, P., & Wróblewski, W. (2006). The analysis of sensor array data with various pattern recognition techniques. *Sensors and Actuators. B, Chemical*, *114*(1), 85–93. doi:10.1016/j.snb.2005.04.008

Commonwealth. (2017). Hospital at Home" Programs Improve Outcomes, Lower Costs But Face Resistance from Providers and Payers. Author.

Constitución Política de Colombia. (1991). Legis.

Cope, J. S., Corney, D., Clark, J. Y., Remagnino, P., & Wilkin, P. (2012). Plant species identification using digital morphometrics: A review. *Expert Systems with Applications*, *39*(8), 7562–7573. doi:10.1016/j.eswa.2012.01.073

Corporation, I. T. U. (2015). Internet of Things Global Standards Initiative. Internet of Things Global Standards Initiative. Retrieved from http://www.itu.int/en/ITU-T/gsi/iot/Pages/default.aspx

Correa-Morris, J., Martínez-Díaz, Y., Hernández, N., & Méndez-Vázquez, H. (2015). Novel histograms kernels with structural properties. *Pattern Recognition Letters*, *68*, 146–152. doi:10.1016/j.patrec.2015.09.005

Correa-Uribe, A. (2015). *Esquema para la implementación de medición de deformaicones en edificaciones de hormigón*. Universidad EIA.

Corte Constitucional, Sentencia, T - 114 (3 de Abril de 2018).

Costa, C., Taiti, C., Strano, M. C., Morone, G., Antonucci, F., Mancuso, S., ... Menesatti, P. (2016). Multivariate Approaches to Electronic Nose and PTR-TOF-MS Technologies in Agro-Food Products. In *Electronic Noses and Tongues in Food Science* (pp. 73–82). London, UK: Academic Press. doi:10.1016/B978-0-12-800243-8.00008-1

Costa, F. F. (2014). Big Data in Biomedicine. *Drug Discovery Today*, *19*(4), 433–440. doi:10.1016/j.drudis.2013.10.012 PubMed

Crystal, S. C., & Robbins, M. S. (2010). Epidemiology of tension-type headache. *Current Pain and Headache Reports*, *14*(6), 449–45. doi:10.1007/s11916-010-0146-2 PubMed

Cuartero, M., Ruiz, A., Oliva, D. J., & Ortuño, J. A. (2017). Multianalyte detection using potentiometric ionophore-based ion-selective electrodes. *Sensors and Actuators. B, Chemical*, *243*, 144–151. doi:10.1016/j.snb.2016.11.129

Dagamseh, A., Bruinink, C., Kolster, M., Wiegerink, R., Lammerink, T., & Krijnen, G. (2010). Array of biomimetic hair sensor dedicated for flow pattern recognition. *2010 Symposium on Design Test Integration and Packaging of MEMS/MOEMS (DTIP)*, 48-50.

Dalal, N., & Triggs, B. (2005). *Histograms of oriented gradients for human detection.* Obtenido de https://lear.inrialpes.fr/people/triggs/pubs/Dalal-cvpr05.pdf

Dasgupta, D. (1998). An Overview of Artificial Immune Systems and Their Applications, *Artificial Immune Systems and Their Applications.* In D. Dasgupta (Ed.), (pp. 3–21). Berlin, Germany: Springer-Verlag Berlin Heidelberg.

Dasgupta, D. (2012). Immunity-Based Intrusion Detection System: A General Framework. In *Proceedings of Seventh Int. Conf. on Bio-Inspired Computing: Theories and Applications* (pp. 417-428). Springer.

Dasgupta, D., & González, F. (2002). An Immunity-based Technique to Characterize Intrusions in Computer Networks. *IEEE Transactions on Evolutionary Computation, 6*(3), 281–291. doi:10.1109/TEVC.2002.1011541

de la Rocha, E. (2010). Image-processing algorithms for detecting and counting vehicles waiting at a traffic light. *Journal of Electronic Imaging, 19*(4), 043025. doi:10.1117/1.3528465

Del Bratov, A., Abramova, N., & Ipatov, A. (2010). Recent trends in potentiometric sensor arrays—A review. *Analytica Chimica Acta, 678*(2), 149–159. doi:10.1016/j.aca.2010.08.035 PMID:20888446

Del Valle, M. (2007). Potentiometric electronic tongues applied in ion multidetermination. *Comprehensive Analytical Chemistry, 49*, 721–753. doi:10.1016/S0166-526X(06)49030-9

Del Valle, M. (2010). Electronic tongues employing electrochemical sensors. *Electroanalysis, 22*(14), 1539–1555.

Del Valle, M. (2017). Materials for electronic tongues: Smart sensor combining different materials and chemometric tools. In *Materials for Chemical Sensing* (pp. 227–265). Cham: Springer. doi:10.1007/978-3-319-47835-7_9

Despagne, F., & Massart, D. L. (1998). Neural networks in multivariate calibration. *Analyst (London), 123*(11), 157R–178R. doi:10.1039/a805562i PMID:10396805

Diaz, K., Hernández, A., & López, A. (2016). A reduced feature set for driver head pose estimation. Applied Soft Computing, 45, 98 - 107. doi:10.1016/j.asoc.2016.04.027

Díaz, Y. Y. R., Acevedo, C. M. D., & Cuenca, M. (2017). Discriminación De Hidromieles A Través De Una Lengua Electrónica. *Revista Colombiana De Tecnologías De Avanzada (RCTA), 1*(23).

Diaz-Sanchez, A. F. (2014). *Cloud brokering: new value-added services and pricing models.* Retrieved from https://pastel.archives-ouvertes.fr/tel-01276552/

Diaz-Sanchez, F., Al Zahr, S., Gagnaire, M., Laisne, J. P., & Marshall, I. J. (2014). CompatibleOne: Bringing Cloud as a Commodity. *2014 IEEE International Conference on Cloud Engineering*, 397–402. 10.1109/IC2E.2014.62

Dollar, P., Tu, Z., Perona, P., & Belongie, S. (2012). Integral Channel Features. *Proceedings of the British Machine Vision Conference 2009*, 91.1–91.11. 10.5244/c.23.91

Dollar, P., Wojek, C., Schiele, B., & Perona, P. (2012). Pedestrian Detection: An Evaluation of the State of the Art. *IEEE Transactions on Pattern Analysis and Machine Intelligence, 34*(4), 743–761. doi:10.1109/TPAMI.2011.155 PMID:21808091

Dolui, K., & Datta, S. K. (2017). Comparison of edge computing implementations: Fog computing, cloudlet and mobile edge computing. GIoTS 2017 - Global Internet of Things Summit Proceedings. doi:10.1109/GIOTS.2017.8016213

Domínguez, R. B., Moreno-Barón, L., Muñoz, R., & Gutiérrez, J. M. (2014). Voltammetric electronic tongue and support vector machines for identification of selected features in Mexican coffee. *Sensors (Basel)*, *14*(9), 17770–17785. doi:10.3390140917770 PMID:25254303

dos Santos, A., Freitas, D., Silva, G., Pistori, H., & Folhes, M. (2017). Weed detection in soybean crops using ConvNets. *Computers and Electronics in Agriculture*, *143*, 314–324. doi:10.1016/j.compag.2017.10.027

Du, H., Fang, W., Huang, H., & Zeng, S. (2018). MMDBC: Density-Based Clustering Algorithm for Mixed Attributes and Multi-dimension Data. Proceedings - 2018 IEEE International Conference on Big Data and Smart Computing, BigComp 2018, 549–552. 10.1109/BigComp.2018.00093

Du, J., & Wang, S. (2010, January). Hiberarchy clustering fault diagnosis of hydraulic pump. In 2010 Prognostics and System Health Management Conference (pp. 1-7). IEEE.

Duin, R. P. W. (2015). *Distools examples: Classifiers in dissimilarity space*. Available at: http://37steps.com/distools/examples/disspace-classification/

Duin, R. P. W., & Pękalska, E. (2009, Nov). *Datasets and tools for dissimilarity analysis in pattern recognition* (Tech. Rep. No. 2009 9). SIMBAD (EU, FP7, FET). Retrieved from http://simbad-fp7.eu/techreports.php

Duin, R. P. W., Bicego, M., Orozco-Alzate, M., Kim, S.-W., & Loog, M. (2014, Aug). Metric learning in dissimilarity space for improved nearest neighbor performance. In P. Fränti, G. Brown, M. Loog, F. Escolano, & M. Pelillo (Eds.), *Structural, Syntactic and Statistical Pattern Recognition: Proceedings of the joint IAPR international workshop, S+SSPR 2014* (Vol. 8621, p. 183-192). Berlin: Springer. 10.1007/978-3-662-44415-3_19

Duin, R. P. W., & Pękalska, E. (2012). The dissimilarity space: Bridging structural and statistical pattern recognition. *Pattern Recognition Letters*, *33*(7), 826–832. doi:10.1016/j.patrec.2011.04.019

Duin, R. P. W., Pękalska, E., & Loog, M. (2013). Non-Euclidean Dissimilarities: Causes, Embedding and Informativeness. In M. Pelillo (Ed.), *Similarity-based pattern analysis and recognition* (pp. 13–44). Springer London. doi:10.1007/978-1-4471-5628-4_2

Du, J., Wang, S., & Zhang, H. (2013). Layered clustering multi-fault diagnosis for hydraulic piston pump. *Mechanical Systems and Signal Processing*, *36*(2), 487–504. doi:10.1016/j.ymssp.2012.10.020

El-Mougy, A., Al-Shiab, I., & Ibnkahla, M. (2019). Scalable Personalized IoT Networks. *Proceedings of the IEEE*. 10.1109/JPROC.2019.2894515

Fan, J., Zhou, N., Peng, J., & Gao, L. (2015, November). Hierarchical learning of tree classifiers for large-scale plant species identification. *IEEE Transactions on Image Processing*, *24*(11), 4172–4184. doi:10.1109/TIP.2015.2457337 PMID:26353356

Farrar, C. R., & Worden, K. (2007). An introduction to structural health monitoring. *Philosophical Transactions of the Royal Society A: Mathematical, Physical and Engineering Sciences, 365*(1851), 303–315.

Farrar, C. R., & Worden, K. (2012). Structural Health Monitoring. In Structural Health Monitoring: A Machine Learning Perspective.

Farris, I., Taleb, T., Khettab, Y., & Song, J. S. (2018). A survey on emerging SDN and NFV security mechanisms for IoT systems. *IEEE Communications Surveys and Tutorials*, (c): 1–26. doi:10.1109/COMST.2018.2862350

Federal Highway Administration. (2013). Traffic Monitoring Guide. Retrieved from http://www.fhwa.dot.gov/policy-information/tmguide/

Felzenszwalb, P., Girshick, R., McAllester, D., & Ramanan, D. (2010). Object Detection with Discriminatively Trained Part-Based Models. *IEEE Transactions on Pattern Analysis and Machine Intelligence*, 32(9), 1627–1645. doi:10.1109/TPAMI.2009.167 PMID:20634557

Fernández, C., & Baptista, P. (2014). *Metodología de la investigación*. McGraw Hill Education.

Fernández-Lopez, A., Menendez, J. M., & Güemes, A. (2007). Damage Detection in a Stiffened Curved Plate By Measuring Differential Strains. *16th International Conference on Composite Materials*.

Feyaerts, F., & Van Gool, L. (2001). Multi-spectral vision system for weed detection. *Pattern Recognition Letters*, 22(6), 667–674. doi:10.1016/S0167-8655(01)00006-X

Figueiredo, E., & Santos, A. (2018). Machine Learning Algorithms for Damage Detection. In A. S. Nobari & M. H. F. Aliabadi (Eds.), *Vibration-Based Techniques for Damage Detection and Localization in Engineering Structures* (pp. 1–39). Academic Press.

Figueiredo, E., & Santos, A. (2018). Machine Learning Algorithms for Damage Detection. In Computational and Experimental Methods in Structures (Vol. 10, pp. 1–39). doi:10.1142/9781786344977_0001

Forbes. (2018). IoT Market Predicted To Double By 2021, Reaching $520B. Author.

Forrest, S., Perelson, A. S., Allen, L., & Cherukuri, R. (1994). Self-Nonself Discrimination in a Computer. In *Proc. of IEEE Symposium on Research in Security and Privacy* (pp. 202- 212). Academic Press.

Fraley, C., & Raftery, A. E. (2002). Model-Based Clustering, Discriminant Analysis, and Density Estimation. *Journal of the American Statistical Association*, 97(458), 611–631. doi:10.1198/016214502760047131

Freund, Y., & Schapire, R. E. (1997). A Decision-Theoretic Generalization of On-Line Learning and an Application to Boosting. *Journal of Computer and System Sciences*, 55(1), 119–139. doi:10.1006/jcss.1997.1504

Friedman, J., Hastie, T., & Tibshirani, R. (2000). Additive logistic regression: A statistical view of boosting. *Annals of Statistics*, 28(2), 337–407. doi:10.1214/aos/1016218223

Fu, J., Li, G., Qin, Y., & Freeman, W. (2007). A pattern recognition method for electronic noses based on an olfactory neural network. *Sensors and Actuators. B, Chemical*, 125(2), 489–497. doi:10.1016/j.snb.2007.02.058

Fu, K.-S. (1976). Pattern Recognition and Image Processing. *IEEE Transactions on Computers*, C-25(12), 1336–1346. doi:10.1109/TC.1976.1674602

Fumal, A., & Scohnen, J. (2008). Tension-type headache: Current research and clinical management. *Lancet Neurology*, 7(2), 70–83. doi:10.1016/S1474-4422(07)70325-3 PubMed

Galitsky, B., & Parnis, A. (2019). Accessing Validity of Argumentation of Agents of the Internet of Everything. Artificial Intelligence for the Internet of Everything. doi:10.1016/B978-0-12-817636-8.00011-9

Gao, Q., Tang, H., Xiang, J., Zhong, Y., Ye, S., & Pang, J. (2019). A Walsh transform-based teager energy operator demodulation method to detect faults in axial piston pumps. *Measurement*, 134, 293–306. doi:10.1016/j.measurement.2018.10.085

Gao, Y., Kong, X., & Zhang, Q. (2002). Wavelet Analysis for Piston Pump Fault Diagnosis. *SAE Transactions*, 131–136.

Gao, Y., Zhang, Q., & Kong, X. (2003). Wavelet-based pressure analysis for hydraulic pump health diagnosis. *Transactions of the ASAE. American Society of Agricultural Engineers*, 46(4), 969–976.

Gardner, J. W., & Bartlett, P. N. (1994). A brief history of electronic noses. *Sensors and Actuators. B, Chemical, 18*(1-3), 210–211. doi:10.1016/0925-4005(94)87085-3

Garlan, D., & Shaw, M. (1994). An Introduction to Software Architecture. Pittsburgh, PA: School of Computer Science, Carnegie Mellon University.

Gent, P. (2016). *Emotion Recognition With Python, OpenCV and a Face Dataset. A tech blog about fun things with Python and embedded electronics.* Obtenido de http://www.paulvangent.com/2016/04/01/emotion-recognition-with-python-opencv-and-a-face-dataset/

Gerry, S., Birks, J., Bonnici, T., Watkinson, P. J., Kirtley, S., & Collins, G. S. (2017). Early warning scores for detecting deterioration in adult hospital patients: A systematic review protocol. *BMJ Open, 7*(12), 1–5. doi:10.1136/bmjopen-2017-019268 PubMed

Gill, V. (2010). *BBC News.* Obtenido de https://www.bbc.com/mundo/ciencia_tecnologia/2010/05/100528_vigilancia_defensa_pl

Girshick, R. (2015). Fast R-CNN. 2015 IEEE International Conference on Computer Vision (ICCV), 1440–1448. doi:10.1109/ICCV.2015.169

Giurgiutiu, V. (2016). Structural Health Monitoring of Aerospace Composites. In Polymer Composites in the Aerospace Industry (Vol. 16). Academic Press.

Glover, F. (1989). Tabu Search—Part I. *ORSA Journal on Computing, 1*(3), 190–206. doi:10.1287/ijoc.1.3.190

Goffmann, E. (1971). *Conducta en situaciones sociales.* Academic Press.

Gold, H. J., Bay, J., & Wilkerson, G. G. (1996). *Scouting for Weeds, Based on the Negative Binomial Distribution. Weed Science.* Cambridge University PressWeed Science Society of America; doi:10.2307/4045627

Gomes, H. L. (2009). Sensor arrays for liquid sensing (electronic tongue systems). Biossensores, Mestrado Integrado em Eng. Electrónica e Telecomunicações (MIEET-2009/00). Universidade do Algarve, FCT, Campus de Gambelas.

Gómez, C. M. (2016). *Análisis comparativo de los algoritmos fisherfaces y lbph para el reconocimiento facial en diferentes condiciones de iluminación y pose, tacna.* Guia de Protección de Datos Personales de la Superintendencia de Industria y Comercio. Obtenido de http://www.sic.gov.co/sites/default/files/files/Nuestra_Entidad/Guia_Vigilancia_sept16_2016.pdf

Gong, S., Chen, J., Kang, Q., Meng, Q., Zhu, Q., & Zhao, S. (2016). An efficient and coordinated mapping algorithm in virtualized SDN networks. *Frontiers of Information Technology & Electronic Engineering, 17*(7), 701–716. doi:10.1631/FITEE.1500387

Gonzalez, R. C. (2009). *Digital Image Processing.* Pearson Education. Retrieved from https://books.google.com.co/books?id=a62xQ2r_f8wC

Gorji-Chakespari, A., Nikbakht, A. M., Sefidkon, F., Ghasemi-Varnamkhasti, M., Brezmes, J., & Llobet, E. (2016). Performance comparison of Fuzzy ARTMAP and LDA in qualitative classification of Iranian Rosa damascena essential oils by an electronic nose. *Sensors (Basel), 16*(5), 636. doi:10.339016050636 PMID:27153069

Grabner, H., Grabner, M., & Bischof, H. (2006). Real-Time Tracking via On-line Boosting. Procedings of the British Machine Vision Conference 2006, 6.1–6.10. 10.5244/C.20.6

Grigoryan, G., Njilla, L., Kamhoua, C., & Kwiat, K. (2017). *Enabling Cooperative IoT Security via Software Defined Networks.* SDN; doi:10.1109/ICC.2018.8423017

Grinblat, G. L., Uzal, L. C., Larese, M. G., & Granitto, P. M. (2016). Deep learning for plant identification using vein morphological patterns. *Computers and Electronics in Agriculture*, *127*, 418–424. doi:10.1016/j.compag.2016.07.003

Grün, B. (2018). Model-Based Clustering. doi:10.100700357-016-9211-9

Gualdrón, O., Llobet, E., Brezmes, J., Vilanova, X., & Correig, X. (2006). Coupling fast variable selection methods to neural network-based classifiers: Application to multisensor systems. *Sensors and Actuators. B, Chemical*, *114*(1), 522–529. doi:10.1016/j.snb.2005.04.046

Güemes, A., & Menendez, J. M. (2010). Fiber-Optic Sensors. In Structural Health Monitoring (pp. 225–285). Academic Press.

Güemes, A., Fernández-Lopez, A., & Díaz-Maroto, P. (2016). A permanent inspection system for damage detection at composite laminates, based on distributed fiber optics sensing. *Proceedings of the 8th International Symposium on NDT in Aerospace*.

Güemes, A., Fernández-López, A., Díaz-Maroto, P., Lozano, A., & Sierra-Perez, J. (2018). Structural Health Monitoring in Composite Structures by Fiber-Optic Sensors. *Sensors (Basel)*, *18*(4), 1094. doi:10.339018041094 PMID:29617345

Guillén, M. L., Paredes, J. L., & Quintero, O. E. C. (2004). Detección y diagnóstico de fallas utilizando la transformada wavelet. *Ciencia e Ingeniería*, *25*(1), 35–42.

Gulati, M., Rastogi, R., Chaturvedi, D. K., Sharma, P., Yadav, V., Chauhan, S., & Singhal, P. (2019e). Statistical Resultant Analysis of Psychosomatic Survey on Various Human Personality Indicators: Statistical Survey to Map Stress and Mental Health. In *Handbook of Research on Learning in the Age of Transhumanism*. Hershey, PA: IGI Global; doi:10.4018/978-1-5225-8431-5.ch022

Guo, H., Xiao, G., Mrad, N., & Yao, J. (2011). Fiber optic sensors for structural health monitoring of air platforms. *Sensors (Basel)*, *11*(4), 3687–3705. doi:10.3390110403687 PMID:22163816

Gupta, M., Rastogi, R., Chaturvedi, D. K., Satya, S. A., Verma, H., Singhal, P., & Singh, A. (2019a). Comparative Study of Trends Observed During Different Medications by Subjects under EMG & GSR Biofeedback. IJITEE, 8(6S), 748-756. Retrieved from https://www.ijitee.org/download/volume-8-issue-6S/

Gupta, M., & Shrivastava, S. K. (2015). Intrusion Detection System based on SVM and Bee Colony. *International Journal of Computers and Applications*, *111*(10), 27–32. doi:10.5120/19576-1377

Gutiérrez, J. M., Haddi, Z., Amari, A., Bouchikhi, B., Mimendia, A., Cetó, X., & del Valle, M. (2013). Hybrid electronic tongue based on multisensor data fusion for discrimination of beers. *Sensors and Actuators. B, Chemical*, *177*, 989–996. doi:10.1016/j.snb.2012.11.110

Haar, A. (1910). Zur Theorie der orthogonalen Funktionensysteme—Erste Mitteilung. *Mathematische Annalen*, *69*(3), 331–371. doi:10.1007/BF01456326

Haddock, C. K., Rowan, A. B., Andrasik, F., Wilson, P. G., Talcott, G. W., & Stein, R. J. (1997). Home- based behavioral Treatments for chronic benign headache: A meta-analysis of controlled trials. *Cephalalgia*, *17*(1), 113–118. doi:10.1046/j.1468-2982.1997.1702113.x PubMed

Haldurai, Madhubala, Rajalakshmi, & Anderson. (1980). Computer Security Threat Monitoring and Surveillance. Technical Report, Fort Washington.

Haldurai, L., Madhubala, T., & Rajalakshmi, R. (2016). A Study on Genetic Algorithm and its Applications. *International Journal on Computer Science and Engineering*, *4*(10), 139–143.

Hand, D. J. (2006). Classifier technology and the illusion of progress. *Statistical Science*, *21*(1), 1–14. doi:10.1214/088342306000000060 PMID:17906740

Han, J., & Kamber, M. (2000). *Data Mining Concepts and Techniques*. Elsevier.

Haroon, M. (2009). Free and Forced Vibration Models. In C. Boller, F.-K. Chang, & Y. Fujino (Eds.), Encyclopedia of Structural Health Monitoring (pp. 1–28). Academic Press.

Harvey, D. (2010). *Analytical Chemistry 2.0*. Retrieved from: https://chem.libretexts.org/Bookshelves/Analytical_Chemistry/Book%3A_Analytical_Chemistry_2.0_(Harvey)

Hastie, T., Tibshirani, R., & Friedman, J. (2009). *The Elements of Statistical Learning: Data Mining, Inference, and Prediction* (Vol. 64). New York: Springer; doi:10.1007/978-0-387-84858-7

Haynes, S. N., Griffin, P., Mooney, D., & Parise, M. (2005). Electromyographic BF and Relaxation Instructions in the Treatment of Muscle Contraction Headaches. *Behavior Therapy*, *6*(1), 672–678.

He, Z., Wang, S., Wang, K., & Li, K. (2012, July). Prognostic analysis based on hybrid prediction method for axial piston pump. In *IEEE 10th International Conference on Industrial Informatics* (pp. 688-692). IEEE. 10.1109/INDIN.2012.6301185

Held, D., Thrun, S., & Savarese, S. (2016). Learning to track at 100 FPS with deep regression networks. *Lecture Notes in Computer Science*, *9905*, 749–765. doi:10.1007/978-3-319-46448-0_45

Heller, B., Sherwood, R., & Mckeown, N. (2012). *The Controller Placement Problem*. Academic Press.

Henriques, J. F., Caseiro, R., Martins, P., & Batista, J. (2015). High-speed tracking with kernelized correlation filters. *IEEE Transactions on Pattern Analysis and Machine Intelligence*, *37*(3), 583–596. doi:10.1109/TPAMI.2014.2345390 PubMed

He, Z., Deng, S., & Xu, X. (2006, August). Approximation algorithms for k-modes clustering. In *International Conference on Intelligent Computing* (pp. 296-302). Springer.

He, Z., Xu, X., & Deng, S. (2005). Scalable algorithms for clustering large datasets with mixed type attributes. *International Journal of Intelligent Systems*, *20*(10), 1077–1089. doi:10.1002/int.20108

Hindman, J., Burton, R., & Schoenau, G. (2002). Condition monitoring of fluid power systems: A survey. *SAE Transactions*, 69–75.

Hitesh, K. A. C. (2018). Feature Selection Optimization in SPL using Genetic Algorithm. *Procedia Computer Science*, *132*, 1477-1486.

Holmberg, M., Eriksson, M., Krantz-Rülcker, C., Artursson, T., Winquist, F., Lloyd-Spetz, A., & Lundström, I. (2004). 2nd workshop of the second network on artificial olfactory sensing (NOSE II). *Sensors and Actuators. B, Chemical*, *101*(1-2), 213–223. doi:10.1016/j.snb.2004.02.054

Holzinger, A. (2018, Aug). From machine learning to explainable AI. In *2018 World Symposium on Digital Intelligence for Systems and Machines (DISA)* (p. 55-66). 10.1109/DISA.2018.8490530

Hong, X., & Wang, J. (2014). Detection of adulteration in cherry tomato juices based on electronic nose and tongue: Comparison of different data fusion approaches. *Journal of Food Engineering*, *126*, 89–97. doi:10.1016/j.jfoodeng.2013.11.008

Hoogendoorn, S. P., & Bovy, P. H. L. (2001). State-of-the-art of vehicular traffic flow modelling. *Proceedings of the Institution of Mechanical Engineers. Part I, Journal of Systems and Control Engineering*, *215*(4), 283–303. doi:10.1177/095965180121500402

Hosseinpour, F., Amoli, P. V., Farahnakian, F., Plosila, J., & Hämäläinen, T. (2014). Artificial Immune System Based Intrusion Detection: Innate Immunity using an Unsupervised Learning Approach. *Int. J. Digit. Content Technol. and its Appl.*, 8(5), 1-12.

Hsieh, H.-C., Lee, C.-S., & Chen, J.-L. (2018). Mobile Edge Computing Platform with Container-Based Virtualization Technology for IoT Applications. *Wireless Personal Communications, 1*. doi:10.100711277-018-5856-5

Hu, W., Wang, S., He, Z., & Zhao, S. (2010, June). Prognostic analysis based on updated grey model for axial piston pump. In *2010 5th IEEE Conference on Industrial Electronics and Applications* (pp. 571-575). IEEE. 10.1109/ICIEA.2010.5517068

Huang, Z. (1998). Extensions to the k-means algorithm for clustering large data sets with categorical values. *Data Mining and Knowledge Discovery*, 2(2(3)), 283–304. doi:10.1023/A:1009769707641

Hu, M. (1962). Visual pattern recognition by moment invariants. *I.R.E. Transactions on Information Theory*, 8(2), 179–187. doi:10.1109/TIT.1962.1057692

Hu, W., Xie, N., Hu, R., Ling, H., Chen, Q., Yan, S., & Maybank, S. (2014, December). Bin ratio-based histogram distances and their application to image classification. *IEEE Transactions on Pattern Analysis and Machine Intelligence*, 36(12), 2338–2352. doi:10.1109/TPAMI.2014.2327975 PMID:26353143

Iannacci, J. (2018). Internet of things (IoT); internet of everything (IoE); tactile internet; 5G – A (not so evanescent) unifying vision empowered by EH-MEMS (energy harvesting MEMS) and RF-MEMS (radio frequency MEMS). *Sensors and Actuators. A, Physical*, 272, 187–198. doi:10.1016/j.sna.2018.01.038

Ibba, A., Duin, R. P. W., & Lee, W.-J. (2010). A study on combining sets of differently measured dissimilarities. In *20th International Conference on Pattern Recognition (ICPR 2010)* (p. 3360-3363). 10.1109/ICPR.2010.820

Igbe, O., Ajayi, O., & Saadawi, T. (2017). Denial of Service Attack Detection using Dendritic Cell Algorithm. In *2017 IEEE 8th Annual Ubiquitous Computing, Electronics and Mobile Communication Conference (UEMCON)* (pp. 294-299). New York, NY: IEEE. 10.1109/UEMCON.2017.8249054

Ikävalko, H., Turkama, P., & Smedlund, A. (2018a). Enabling the Mapping of Internet of Things Ecosystem Business Models Through Roles and Activities in Value Co-creation. Proceedings of the 51st Hawaii International Conference on System Sciences. doi:10.24251/HICSS.2018.620

Ikävalko, H., Turkama, P., & Smedlund, A. (2018b). *Value Creation in the Internet of Things: Mapping Business Models and Ecosystem Roles*. Technology Innovation Management Review; doi:10.22215/timreview/1142

Inaudi, D. (2010). Overview of fiber optic sensing technologies for structural health monitoring. *Informacije MIDEM*, 40(4), 263–272.

Instituto Nacional de Vías (INVIAS). (2013). Volúmenes de Tránsito 2010-2011. Retrieved from https://www.invias.gov.co/index.php/archivo-y-documentos/documentos-tecnicos/volumenes-de-transito-2008/1921-volumenes-de-transito-2010-2011

Intel Software. (2018). *Developer Reference for Intel® Integrated Performance Primitives 2019*. Obtenido de https://software.intel.com/en-us/ipp-dev-reference-histogram-of-oriented-gradients-hog-descriptor

Irias Tejeda, A. J., & Castro Castro, R. (2019). Algorithm of Weed Detection in Crops by Computational Vision. In *2019 International Conference on Electronics, Communications and Computers (CONIELECOMP)* (pp. 124–128). IEEE. 10.1109/CONIELECOMP.2019.8673182

Jabbar, S., Ullah, F., Khalid, S., Khan, M., & Han, K. (2017). Semantic {Interoperability} in {Heterogeneous} {IoT} {Infrastructure} for {Healthcare}. *Wireless Communications and Mobile Computing, 2017*, e9731806. doi:10.1155/2017/9731806

Jadhav, B. D., & Patil, P. M. (2014). Hyperspectral Remote Sensing For Agricultural Management: A Survey. *International Journal of Computers and Applications, 106*(7).

Jain, D., & Singh, V. (2018). An Efficient Hybrid Feature Selection model for Dimensionality Reduction. *Procedia Computer Science, 132*, 333-341.

Jalili, A., Keshtgari, M., Akbari, R., & Javidan, R. (2018, July). Multi criteria analysis of controller placement problem in software defined networks. *Computer Communications*.

Jalili, A., Keshtgari, M., & Akbari, R. (2019). A new framework for reliable control placement in software-defined networks based on multi-criteria clustering approach. *Soft Computing*, 1–20.

Jamaica Alejandro, D., & ... (2013). *Dinámica espacial y temporal de poblaciones de malezas en cultivos de papa, espinaca y caña de azúcar y su relación con propiedades del suelo en dos localidades de Colombia. Weed Research.* Universidad Nacional de Colombia.

Janjanam, P., & Reddy, C. P. (2019). Text Summarization: An Essential Study. *2019 International Conference on Computational Intelligence in Data Science (ICCIDS)*, 1-6.

Janjua, Z. H., Vecchio, M., Antonini, M., & Antonelli, F. (2019). IRESE: An intelligent rare-event detection system using unsupervised learning on the IoT edge. *Engineering Applications of Artificial Intelligence, 84*(May), 41–50. doi:10.1016/j.engappal.2019.05.011

Ji, J., Bai, T., Zhou, C., Ma, C., & Wang, Z. (2013). An improved k-prototypes clustering algorithm for mixed numeric and categorical data. *Neurocomputing, 120*, 590–596. doi:10.1016/j.neucom.2013.04.011

Jin, C., Lin, L., & Xiang, M. (2009). An Improved ID3 Decision Tree Algorithm. In *Proceedings of International Conference on Computer Science and Education* (pp. 127-130). Academic Press.

Jinyin, C., Huihao, H., Jungan, C., Shanqing, Y., & Zhaoxia, S. (2017). Fast Density Clustering Algorithm for Numerical Data and Categorical Data. *Mathematical Problems in Engineering, 2017*, 1–15. doi:10.1155/2017/6393652

Johnson, R., & Wichern, D. (2008). *Applied Multivariate Statistical Analysis (6th ed.)*. Pearson.

Jolliffe, I. T. (2011). Principal Component Analysis. In International Encyclopedia of Statistical Science. Springer.

Journal, I., Engineering, A., & Kumar, S. (2019). An Extensive Review on Sensing as a Service Paradigm in IoT : Architecture, Research Challenges. *Lessons Learned and Future Directions, 14*(6), 1220–1243.

Junguito, R., Perfetti, J. J., & Becerra, A. (2014). *Desarrollo de la agricultura colombiana*. Retrieved from https://www.repository.fedesarrollo.org.co/handle/11445/151

Kabugo, J. C., Jämsä-Jounela, S.-L., Schiemann, R., & Binder, C. (2020). Industry 4.0 based process data analytics platform: A waste-to-energy plant case study. *International Journal of Electrical Power & Energy Systems, 115*. doi:10.1016/j.ijepes.2019.105508

Kalal, Z., Matas, J., & Mikolajczyk, K. (2009). Online learning of robust object detectors during unstable tracking. In *IEEE 12th International Conference on Computer Vision Workshops* (pp. 1417–1424). IEEE.

Kalal, Z., Mikolajczyk, K., & Matas, J. (2010). Tracking-Learning-Detection. *IEEE Transactions on Pattern Analysis and Machine Intelligence, 6*(1). Retrieved from http://info.ee.surrey.ac.uk/Personal/Z.Kalal/

Kalal, Z., Mikolajczyk, K., Matas, J., & Republic, C. (2010). Forward-Backward Error: Autonomous Identification of Tracking Failures. *Proceedings of the 20th International Conference on Pattern Recognition*, 2756–2759. 10.1109/ICPR.2010.675

Kane, A. (2012). Determining the number of clusters for a k-means clustering algorithm. *Indian Journal of Computer Science and Engineering, 3*(5), 670–672.

Karagiannis, V., Chatzimisios, P., Vazquez-Gallego, F., & Alonso-Zarate, J. (2015). *A Survey on Application Layer Protocols for the Internet of Things*. Transaction on IoT and Cloud Computing; doi:10.5281/ZENODO.51613

Karkoub, M. A., Gad, O. E., & Rabie, M. G. (1999). Predicting axial piston pump performance using neural networks. *Mechanism and Machine Theory, 34*(8), 1211–1226. doi:10.1016/S0094-114X(98)00086-X

Kerautret, B., Colom, M., Lopresti, D., Monasse, P., & Talbot, H. (Eds.). (2019). *Reproducible Research in Pattern Recognition - Second International Workshop, RRPR 2018, Beijing, China, August 20, 2018, revised selected papers* (Vol. 11455). Springer. doi: 10.1007/978-3-030-23987-9

Kerschen, G., & Golinval, J. (2009). Dimensionality Reduction Using Linear and Nonlinear Transformation. In C. Boller, F.-K. Chang, & Y. Fujino (Eds.), *Encyclopedia of Structural Health Monitoring* (pp. 1–13). John Wiley & Sons, Ltd.

Khalili, R., Despotovic, Z., & Hecker, A. (2018). Flow Setup Latency in SDN Networks. *IEEE Journal on Selected Areas in Communications, 36*(12), 2631–2639. doi:10.1109/JSAC.2018.2871291

Killi, B. P. R., & Rao, S. V. (2016). Optimal Model for Failure Foresight Capacitated Controller Placement in Software-Defined Networks. *IEEE Communications Letters, 20*(6), 1108–1111. doi:10.1109/LCOMM.2016.2550026

Kim, D., Lee, K. H., & Lee, D. (2004). Fuzzy clustering of categorical data using fuzzy centroids. *Pattern Recognition, 25*(11), 1263–1271. doi:10.1016/j.patrec.2004.04.004

Kim, K. H., Kim, H. J., & Chung, Y. S. (2019). Case Study: Cost-effective Weed Patch Detection by Multi-Spectral Camera Mounted on Unmanned Aerial Vehicle in the Buckwheat Field. *Hangug Jagmul Haghoeji, 64*(2), 159–164. doi:10.7740/kjcs.2019.64.2.159

Kimmel, D. W., LeBlanc, G., Meschievitz, M. E., & Cliffel, D. E. (2011). Electrochemical sensors and biosensors. *Analytical Chemistry, 84*(2), 685–707. doi:10.1021/ac202878q PMID:22044045

Kivelä, T., & Mattila, J. (2013, October). Internal leakage fault detection for variable displacement axial piston pump. In *ASME/BATH 2013 Symposium on Fluid Power and Motion Control*. American Society of Mechanical Engineers Digital Collection. 10.1115/FPMC2013-4445

Knight, S., Nguyen, H. X., Falkner, N. R. B., & Roughan, M. (2011). *The Internet Topology Zoo*. Retrieved January 18, 2019, from http://www.topology-zoo.org/

Koc, L., Mazzuchi, T. A., & Sarkani, S. (2012). A Network Intrusion Detection System Based on a Hidden Naive Bayes Multiclass Classifier. *Expert Systems with Applications, 39*(18), 13492–13500. doi:10.1016/j.eswa.2012.07.009

Kohavi, R., & John, G. H. (1997). Wrappers for feature subset selection. *Artificial Intelligence, 97*(1-2), 273–324. doi:10.1016/S0004-3702(97)00043-X

Kotsiantis, S. B., Zaharakis, I., & Pintelas, P. (2007). Supervised machine learning: A review of classification techniques. *Emerging Artificial Intelligence Applications in Computer Engineering, 160*, 3-24.

Kramer, M. A. (1991). Nonlinear principal component analysis using autoassociative neural networks. *AIChE Journal. American Institute of Chemical Engineers, 37*(2), 233–243. doi:10.1002/aic.690370209

Krishnan, P., Najeem, J. S., & Achuthan, K. (2018). SDN framework for securing IoT networks. Lecture Notes of the Institute for Computer Sciences. *Social-Informatics and Telecommunications Engineering, LNICST, 218*, 116–129. doi:10.1007/978-3-319-73423-1_11

Kropotov, J. D. (2009). *Quantitative EMG, event-related potentials and neurotherapy.* San Diego, CA: Academic Press.

Kuhl, M., Neugebauer, R., & Mickel, P. (2007). Methods for a multisensorsystem for in-line process- and quality monitoring of welding seams using fuzzy pattern recognition. *2007 IEEE Conference on Emerging Technologies and Factory Automation (EFTA 2007)*, 908-911. 10.1109/EFTA.2007.4416879

Kulkarni, N. J., & Bakal, J. W. (2019). E-Health: IoT Based System and Correlation of Vital Stats in Identification of Mass Disaster Event. Proceedings - 2018 4th International Conference on Computing, Communication Control and Automation, ICCUBEA 2018, 1–6. 10.1109/ICCUBEA.2018.8697529

Kumar, N. M., & Mallick, P. K. (2018). The Internet of Things: Insights into the building blocks, component interactions, and architecture layers. Í. *Procedia Computer Science.* doi:10.1016/j.procs.2018.05.170

Kuncheva, L. I., & Vetrov, D. P. (2006). Evaluation of the stability of k-means cluster ensembles with respect to the random initialization. *IEEE Transactions on Pattern Analysis and Machine Intelligence, 28*(11), 1798–1808. doi:10.1109/TPAMI.2006.226 PubMed

Lam, D., Wei, M., & Wunsch, D. (2015). Clustering Data of Mixed Categorical and Numerical Type With Unsupervised Feature Learning. *IEEE Access: Practical Innovations, Open Solutions, 3*, 1605–1616. doi:10.1109/ACCESS.2015.2477216

Lan, Y., Hu, J., Huang, J., Niu, L., Zeng, X., Xiong, X., & Wu, B. (2018). Fault diagnosis on slipper abrasion of axial piston pump based on extreme learning machine. *Measurement, 124*, 378–385. doi:10.1016/j.measurement.2018.03.050

Larese, M. G., Namías, R., Craviotto, R. M., Arango, M. R., Gallo, C., & Granitto, P. M. (2014). Automatic classification of legumes using leaf vein image features. *Pattern Recognition, 47*(1), 158–168. doi:10.1016/j.patcog.2013.06.012

Lázaro, J. G. M., Pinilla, C. B., & Prada, S. R. (2016, November). A Survey of Approaches for Fault Diagnosis in Axial Piston Pumps. In *ASME 2016 International Mechanical Engineering Congress and Exposition* (pp. V04AT05A052-V04AT05A052). American Society of Mechanical Engineers.

Leach, M., Sparks, E., & Robertson, N. (2014). Contextual anomaly detection in crowded surveillance scenes. *Pattern Recognition Letters, 44*, 71–79. doi:10.1016/j.patrec.2013.11.018

Lee, S., Bae, M., & Kim, H. (2017). Future of IoT Networks: A Survey. Applied Sciences, 7(10), 1072. doi:10.3390/app7101072

Lee, C. H., & Yoon, H. J. (2017). Medical big data: Promise and challenges. *Kidney Research and Clinical Practice, 36*(1), 461–475. doi:10.23876/j.krcp.2017.36.1.3

Lee, I., & Lee, K. (2015). The Internet of Things (IoT): Applications, investments, and challenges for enterprises. *Business Horizons, 58*(4), 431–440. doi:10.1016/j.bushor.2015.03.008

Lehoczky, É., Riczu, P., Mazsu, N., Viktória Dellaszéga, L., & Tamás, J. (2018). Applicability of remote sensing in weed detection. *20th EGU General Assembly, EGU2018, Proceedings from the Conference Held 4-13 April, 2018 in Vienna, Austria, p.14644, 20*, 14644. Retrieved from http://adsabs.harvard.edu/abs/2018EGUGA.2014644L

Lehtola, V., Huttunen, H., Christophe, F., & Mikkonen, T. (2017). Evaluation of visual tracking algorithms for embedded devices. Lecture Notes in Computer Science, 10269, 88–97. doi:10.1007/978-3-319-59126-1_8

Leonardo, D., & Dominguez, A. J. (2016). Herramientas para la detección seguimiento de personas a partir de cámaras de seguridad. *Congreso Argentino de Ciencias de la Computación*, 22, 251–260.

Leon-Medina, J. X., Cardenas-Flechas, L. J., & Tibaduiza, D. A. (2019). A data-driven methodology for the classification of different liquids in artificial taste recognition applications with a pulse voltammetric electronic tongue. *International Journal of Distributed Sensor Networks*, 15(10). doi:10.1177/1550147719881601

Levine, D. M., Ouchi, K., Blanchfield, B., Diamond, K., Licurse, A., Pu, C. T., & Schnipper, J. L. (2018). Hospital-Level Care at Home for Acutely Ill Adults: A Pilot Randomized Controlled Trial. Journal of General Internal Medicine, 1–8. doi:10.100711606-018-4307-z

Ley 1581 (Congreso de la República 17 de Octubre de 2012).

Li, H., Beck, F., Dupouy, O., Herszberg, I., Stoddart, P. R., Davis, C. E., & Mouritz, A. P. (2006a). Strain-based health assessment of bonded composite repairs. *Composite Structures*, 76(3), 234–242. doi:10.1016/j.compstruct.2006.06.032

Li, H., Herszberg, I., Davis, C. E., Mouritz, A. P., & Galea, S. C. (2006b). Health monitoring of marine composite structural joints using fibre optic sensors. *Composite Structures*, 75(1), 321–327. doi:10.1016/j.compstruct.2006.04.054

Lin, T.-Y., Goyal, P., Girshick, R., He, K., & Dollar, P. (2017). Focal Loss for Dense Object Detection. 2017 IEEE International Conference on Computer Vision (ICCV), 2999–3007. doi:10.1109/ICCV.2017.324

Lin, Z., Zhang, Q., Liu, R., Gao, X., Zhang, L., Kang, B., ... Li, X. (2016). Evaluation of the bitterness of traditional Chinese medicines using an e-tongue coupled with a robust partial least squares regression method. *Sensors (Basel)*, 16(2), 151. doi:10.339016020151 PMID:26821026

Liu, J., Liu, J., & Xie, R. (2016). Reliability-based controller placement algorithm in software defined networking. *Computer Science and Information Systems*, 13(2), 547–560. doi:10.2298/CSIS160225014L

Liu, J., Shi, Y., Zhao, L., Cao, Y., Sun, W., & Kato, N. (2018). Joint Placement of Controllers and Gateways in SDN-Enabled 5G-Satellite Integrated Network. *IEEE Journal on Selected Areas in Communications*, 36(2), 221–232. doi:10.1109/JSAC.2018.2804019

Liu, M., Wang, M., Wang, J., & Li, D. (2013). Comparison of random forest, support vector machine and back propagation neural network for electronic tongue data classification: Application to the recognition of orange beverage and Chinese vinegar. *Sensors and Actuators. B, Chemical*, 177, 970–980. doi:10.1016/j.snb.2012.11.071

Liu, N., Liang, Y., Bin, J., Zhang, Z., Huang, J., Shu, R., & Yang, K. (2014). Classification of green and black teas by PCA and SVM analysis of cyclic voltammetric signals from metallic oxide-modified electrode. *Food Analytical Methods*, 7(2), 472–480. doi:10.100712161-013-9649-x

Liu, T., Chen, Y., Li, D., & Wu, M. (2018). An active feature selection strategy for DWT in artificial taste. *Journal of Sensors*.

Liu, W., Wang, Z., Liu, X., Zeng, N., Liu, Y., & Alsaadi, F. E. (2017). A survey of deep neural network architectures and their applications. *Neurocomputing*, 234, 11–26. doi:10.1016/j.neucom.2016.12.038

Liu, X., & Ansari, N. (2019). Toward Green IoT: Energy Solutions and Key Challenges. *IEEE Communications Magazine*, 57(3), 104–110. doi:10.1109/MCOM.2019.1800175

Li, W., & Wang, P. (2019). Two-factor authentication in industrial Internet-of-Things: Attacks, evaluation and new construction. *Future Generation Computer Systems*, 101, 694–708. doi:10.1016/j.future.2019.06.020

Li, Z., Xu, D., & Guo, X. (2014). Remote sensing of ecosystem health: Opportunities, challenges, and future perspectives. *Sensors (Basel)*, *14*(11), 21117–21139. doi:10.3390141121117 PMID:25386759

Lojek, B. (2007). Shockley Semiconductor Laboratories. In History of Semiconductor Engineering. Springer.

Lopez Garcia, J. C. (2009). *Algoritmos y Programación para Docentes*. Fundación Gabriel Piedrahita Uribe.

López-Granados, F. (2011). Weed detection for site-specific weed management: Mapping and real-time approaches. *Weed Research*, *51*(1), 1–11. doi:10.1111/j.1365-3180.2010.00829.x

Lopez, H. (2011). *Detección y seguimiento de objetos con cámaras en movimiento*. Madrid: Universidad Autónoma de Madrid.

Lopez, I., & Sarigul-Klijn, N. (2010). A review of uncertainty in flight vehicle structural damage monitoring, diagnosis and control: Challenges and opportunities. *Progress in Aerospace Sciences*, *46*(7), 247–273. doi:10.1016/j.paerosci.2010.03.003

Lorieul, T., Pearson, K. D., Ellwood, E. R., Goëau, H., Molino, J.-F., Sweeney, P. W., ... Joly, A. (2019). Toward a large-scale and deep phenological stage annotation of herbarium specimens: Case studies from temperate, tropical, and equatorial floras. *Applications in Plant Sciences*, *7*(3), e01233. doi:10.1002/aps3.1233 PMID:30937225

Lottes, P., Khanna, R., Pfeifer, J., Siegwart, R., & Stachniss, C. (2017). UAV-based crop and weed classification for smart farming. In *Robotics and Automation (ICRA), 2017 IEEE International Conference on* (pp. 3024–3031). 10.1109/ICRA.2017.7989347

Louargant, M., Jones, G., Faroux, R., Paoli, J.-N., Maillot, T., Gée, C., & Villette, S. (2018). Unsupervised Classification Algorithm for Early Weed Detection in Row-Crops by Combining Spatial and Spectral Information. *Remote Sensing*, *10*(5), 761. doi:10.3390/rs10050761

Lowe, D. G. (1999). Object recognition from local scale-invariant features. Proceedings of the Seventh IEEE International Conference on Computer Vision, 1150–1157. doi:10.1109/ICCV.1999.790410

Lu, J., Zhang, Z., Hu, T., Yi, P., & Lan, J. (2019). A Survey of Controller Placement Problem in Software-Defined Networking. *IEEE Access: Practical Innovations, Open Solutions*, *7*, 24290–24307. doi:10.1109/ACCESS.2019.2893283

Lukežič, A., Vojíř, T., Čehovin Zajc, L., Matas, J., & Kristan, M. (2018). Discriminative Correlation Filter Tracker with Channel and Spatial Reliability. *International Journal of Computer Vision*, *126*(7), 671–688. doi:10.1007/s11263-017-1061-3

Lu, L., Hu, X., & Zhu, Z. (2019). Joint Voltammetry Technology with a Multi-electrode Array for Four Basic Tastes. *Current Analytical Chemistry*, *15*(1), 75–83. doi:10.2174/1573411014666180522100504

Luo, J., Ma, Y., Takikawa, E., Lao, S., Kawade, M., & Lu, B. L. (2007). Person-specific SIFT features for face recognition. ICASSP, IEEE International Conference on Acoustics, Speech and Signal Processing - Proceedings, 2, 593–596. 10.1109/ICASSP.2007.366305

Lurette, C., & Lecoeuche, S. (2003). Unsupervised and auto-adaptive neural architecture for on-line monitoring. Application to a hydraulic process. *Engineering Applications of Artificial Intelligence*, *16*(5-6), 441–451. doi:10.1016/S0952-1976(03)00064-2

Madakam, S., Ramaswamy, R., & Tripathi, S. (2015). Internet of Things (IoT): A Literature Review. Journal of Computer and Communications. doi:10.4236/jcc.2015.35021

Mallah, C., Cope, J., & Orwell, J. (2013, Feb). Plant leaf classification using probabilistic integration of shape, texture and margin features. In L. Linsen, & M. Kampel (Eds.), *Proceedings of the 14th IASTED international conference on computer graphics and imaging (CGIM 2013) / track 798: Signal processing, pattern recognition and applications* (p. 1-8). Calgary, Canada: Acta Press. 10.2316/P.2013.798-098

Mannini, A., & Sabatini, A. M. (2010). *Machine Learning Methods for Classifying Human Physical Activity from On-Body Accelerometers.* Obtenido de mdpi.com: https://www.mdpi.com/1424-8220/10/2/1154/htm

Marco, S., & Gutierrez-Galvez, A. (2012). Signal and data processing for machine olfaction and chemical sensing: A review. *IEEE Sensors Journal, 12*(11), 3189–3214. doi:10.1109/JSEN.2012.2192920

Marolla, C. (2018). Information and Communication Technology for Sustainable Development. Information and Communication Technology for Sustainable Development. Springer Singapore. doi:10.1201/9781351045230

Marx, Í., Rodrigues, N., Dias, L. G., Veloso, A. C., Pereira, J. A., Drunkler, D. A., & Peres, A. M. (2017). Sensory classification of table olives using an electronic tongue: Analysis of aqueous pastes and brines. *Talanta, 162*, 98–106. doi:10.1016/j.talanta.2016.10.028 PMID:27837890

Mathew, S. S., Atif, Y., & El-Barachi, M. (2017). From the Internet of Things to the web of things-enabling by sensing as-A service. *Proceedings of the 2016 12th International Conference on Innovations in Information Technology, IIT 2016.* 10.1109/INNOVATIONS.2016.7880055

McCrory, D., Penzien, D. B., Hasselblad, V., & Gray, R. (2001). *Behavioral and physical treatments for tension Type and cervocogenic headaches.* Des Moines, IA: Foundation for Chiropractic Education and Research.

McCurdy, B. R. (2012). hospital-at-home programs for patients with acute exacerbations of chronic obstructive pulmonary disease (COPD): An evidence-based analysis. Ontario Health Technology Assessment Series.

McDaniel, N. L., Novicoff, W., Gunnell, B., & Cattell Gordon, D. (2018). Comparison of a Novel Handheld Telehealth Device with Stand-Alone Examination Tools in a Clinic Setting. Telemedicine Journal and e-Health. doi:10.1089/tmj.2018.0214

Mckeown, N., Anderson, T., Peterson, L., Rexford, J., Shenker, S., & Louis, S. (2008). Sigcomm08_Openflow. *Pdf., 38*(2), 69–74.

McParland, D., & Gormley, I. C. (2016). Model-based clustering for mixed data: clustMD. *Advances in Data Analysis and Classification, 10*(2), 155–169. doi:10.1007/s11634-016-0238-x

Médini, L., Mrissa, M., Khalfi, E. M., Terdjimi, M., Le Sommer, N., Capdepuy, P., … Touseau, L. (2017). Building a Web of Things with Avatars: A comprehensive approach for concern management in WoT applications. Managing the Web of Things: Linking the Real World to the Web. doi:10.1016/B978-0-12-809764-9.00007-X

Mehner, S. (n.d.). Secure and Flexible Internet of Things using Software Defined Networking. Academic Press.

Menendez, J. M., & Güemes, A. (2006). SHM Using Fiber Sensors in Aerospace Applications. *Optical Fiber Sensors*, MF1.

Millea, J. P., & Brodie, J. J. (2002). Tension type headache. *American Family Physician, 66*(5), 797–803.

Mimendia, A., Gutiérrez, J. M., Leija, L., Hernández, P. R., Favari, L., Muñoz, R., & del Valle, M. (2010). A review of the use of the potentiometric electronic tongue in the monitoring of environmental systems. *Environmental Modelling & Software, 25*(9), 1023–1030. doi:10.1016/j.envsoft.2009.12.003

Minerva, R., Biru, A., & Rotondi, D. (2015). Towards a definition of the Internet of Things (IoT). Retrieved from https://iot.ieee.org/definition.html

Mirón Rubio, M., Ceballos Fernández, R., Parras Pastor, I., Palomo Iloro, A., Fernández Félix, B. M., Medina Miralles, J., ... Alonso-Viteri, S. (2018). Telemonitoring and home hospitalization in patients with chronic obstructive pulmonary disease: Study TELEPOC. *Expert Review of Respiratory Medicine, 12*(4), 335–343. doi:10.1080/17476348.2018.144 2214 PubMed

Mishalanie, E. A., Lesnik, B., Araki, R., & Segall, R. (2005). *Validation and peer review of US Environmental Protection Agency chemical methods of analysis.* Washington, DC: Environmental Protection Agency.

Mostafa, S. A., Gunasekaran, S. S., Mustapha, A., Mohammed, M. A., & Abduallah, W. M. (2020). Modelling an adjustable autonomous multi-agent internet of things system for elderly smart home. Advances in Intelligent Systems and Computing, 953, 301–311. doi:10.1007/978-3-030-20473-0_29

MOTChallenge. (n.d.). Obtenido de Multiple Object Tracking Benchmark: https://motchallenge.net/

Mountrakis, G., Im, J., & Ogole, C. (2011). Support vector machines in remote sensing: A review. *ISPRS Journal of Photogrammetry and Remote Sensing, 66*(3), 247–259. doi:10.1016/j.isprsjprs.2010.11.001

Mourad, M. H., Nassehi, A., Schaefer, D., & Newman, S. T. (2020). Assessment of interoperability in cloud manufacturing. *Robotics and Computer-integrated Manufacturing, 61*(June), 101832. doi:10.1016/j.rcim.2019.101832

Mujica, L., Tibaduiza-Burgos, D. A., & Rodellar, J. (2010). Data-driven multiactuator piezoelectric system for structural damage localization. *Fifth World Conference on Structural Control and Monitoring (5WCSCM).*

Mujica, L. E., Rodellar, J., Fernández, A., & Güemes, A. (2011). Q statistic and T2-statistic PCA-based measures for damage assessment in structures. *Structural Health Monitoring: An International Journal, 10*(5), 539–553. doi:10.1177/1475921710388972

Mullaly, J. W., Hall, K., & Goldstein, R. (2009). Efficacy of BF in the Treatment of Migraine and Tension Type Headaches. *Pain Physician, 12*(1), 1005–1011. PubMed

Murawwat, S., Qureshi, A., Ahmad, S., & Shahid, Y. (2018). *Weed Detection Using SVMs. Technology & Applied Science Research* (Vol. 8). Retrieved from www.etasr.com

Naeem, R. Z., Bashir, S., Amjad, M. F., Abbas, H., & Afzal, H. (2019). Fog computing in internet of things: Practical applications and future directions. *Peer-to-Peer Networking and Applications, 12*(5), 1236–1262. doi:10.1007/s12083-019-00728-0

Nguyen Thanh, T. K., Truong, Q. B., Truong, Q. D., & Huynh, H. X. (2018). Depth learning with convolutional neural network for leaves classifier based on shape of leaf vein. In N. T. Nguyen, D. H. Hoang, T.-P. Hong, H. Pham, & B. Trawiński (Eds.), *Intelligent information and database systems - 10th Asian Conference, ACIIDS 2018, Dong Hoi city, Vietnam, March 19-21, 2018, Proceedings, part I* (p. 565-575). Cham: Springer International Publishing. 10.1007/978-3-319-75417-8_53

Nguyen, H. A., & Choi, D. (2008). Application of Data Mining to Network Intrusion Detection: Classifier Selection Model. In *Asia-Pacific Network Operations and Management Symposium.* Springer-Verlag.

Nixon, O., & Nathan, A. (1995). Magnetic Pattern Recognition Sensor Arrays using CCD Readout. *ESSDERC '95: Proceedings of the 25th European Solid State Device Research Conference*, 273-276.

Nkoa, R., Owen, M. D. K., & Swanton, C. J. (2015). Weed Abundance, Distribution, Diversity, and Community Analyses. *Weed Science, 63*(SP1), 64–90. doi:10.1614/WS-D-13-00075.1

Nomikos, P., & MacGregor, J. F. (1994). Monitoring batch processes using multiway principal component analysis. *AIChE Journal. American Institute of Chemical Engineers, 40*(8), 1361–1375. doi:10.1002/aic.690400809

Norza, E., Peñalosa, M. J., & Rodríguez, J. D. (2016). *Exégesis de los registros de criminalidad y actividad operativa de la Policía Nacional en Colombia.* Academic Press.

Nunna, S., & Ganesan, K. (2017). Mobile edge computing. Health 4.0: How Virtualization and Big Data are Revolutionizing Healthcare. doi:10.1007/978-3-319-47617-9_9

Oerke, E. (2006). Crop losses to pests. *The Journal of Agricultural Science, 144*(1), 31–43. doi:10.1017/S0021859605005708

Oficina de Análisis de Información y Estudios Estratégicos (OAIEE). (2017). *Boletín mensual de indicadores de seguridad y convivencia.* Bogotá: Author.

Oficina de análisis de información y estudios estratégicos (OAIEE). (2019). *Boletín mensual de indicadores de seguridad y convivencia.* Bogotá: Author.

Oficina de Naciones Unidas contra la Droga y el Delito. (2010). *12° Congreso de las Naciones Unidas sobre Prevención del Delito y Justicia Penal.* Obtenido de https://www.un.org/es/events/crimecongress2010/pdf/factsheet_ebook_es.pdf

Oliveri, P., Casolino, M. C., & Forina, M. (2010). Chemometric brains for artificial tongues. *Advances in Food and Nutrition Research, 61,* 57–117. doi:10.1016/B978-0-12-374468-5.00002-7 PMID:21092902

Omer, G., Mutanga, O., Abdel-Rahman, E. M., & Adam, E. (2016). Empirical prediction of Leaf Area Index (LAI) of endangered tree species in intact and fragmented indigenous forests ecosystems using WorldView-2 data and two robust machine learning algorithms. *Remote Sensing, 8*(4), 1–26. doi:10.3390/rs8040324

Open Networking Foundation. (2013). SDN Architecture Overview. *Onf,* (1), 1–5.

Open Networking Foundation. (2016). *SDN Architecture.* Author.

Orozco-Alzate, M., Villegas-Jaramillo, E.-J., & Uribe-Hurtado, A.-L. (2017, Jun). A block-separable parallel implementation for the weighted distribution matching similarity measure. In S. Omatu, S. Rodr'ıguez, G. Villarrubia, P. Faria, P. Sitek, & J. Prieto (Eds.), *Distributed computing and artificial intelligence, 14th international conference, DCAI 2017, Porto, Portugal, 21-23 June, 2017* (Vol. 620, p. 239-246). Cham, Switzerland: Springer. doi: 10.1007/978-3-319-62410-5_29

Ostachowicz, W., Soman, R., & Malinowski, P. (2019). Optimization of sensor placement for structural health monitoring: A review. *Structural Health Monitoring, 18*(3), 963–988. doi:10.1177/1475921719825601

Otsu, N. (1979). A threshold selection method from gray-level hostgrams. *IEEE Transactions on Systems, Man, and Cybernetics, 9*(1), 62–66. doi:10.1109/TSMC.1979.4310076

Özmen, Ö., Sınanoğlu, C., Batbat, T., & Güven, A. (2019). Prediction of Slipper Pressure Distribution and Leakage Behaviour in Axial Piston Pumps Using ANN and MGGP. *Mathematical Problems in Engineering.*

Palit, M., Tudu, B., Bhattacharyya, N., Dutta, A., Dutta, P. K., Jana, A., ... Chatterjee, A. (2010). Comparison of multivariate preprocessing techniques as applied to electronic tongue based pattern classification for black tea. *Analytica Chimica Acta, 675*(1), 8–15. doi:10.1016/j.aca.2010.06.036 PMID:20708109

Panda, C. K., & Bhatnagar, R. (2020a). Social Internet of Things in Agriculture: An Overview and Future Scope. doi:10.1007/978-3-030-24513-9_18

Panda, C. K., & Bhatnagar, R. (2020b). Toward Social Internet of Things (SIoT): Enabling Technologies, Architectures and Applications (B. 846). doi:10.1007/978-3-030-24513-9

Panda, M., & Patra, M. R. (2007). Network Intrusion Detection Using Naïve Bayes. *Int. Journal of Computer Science and Network Security, 7*(12), 258–263.

Pandey, H. M. (2016). Performance Evaluation of Selection Methods of Genetic Algorithm and Network Security Concerns. In *Int. Conf. on Information Security and Privacy* (no. 78, pp. 13-18). 10.1016/j.procs.2016.02.004

Pękalska, E., Harol, A., Duin, R. P. W., Spillmann, B., & Bunke, H. (2006). Non-Euclidean or non-metric measures can be informative. In D.-Y. Yeung, J. Kwok, A. Fred, F. Roli, & D. de Ridder (Eds.), Structural, Syntactic, and Statistical Pattern Recognition Joint IAPR International Workshops, SSPR 2006 and SPR 2006 (Vol. 4109, p. 871-880). Springer. doi:10.1007/11815921_96

Pękalska, E., & Duin, R. P. W. (2005). *The dissimilarity representation for pattern recognition: Foundations and applications* (Vol. 64). Singapore: World Scientific. doi:10.1142/5965

Peng, Y., Fan, M., Song, J., Cui, T., & Li, R. (2018). Assessment of plant species diversity based on hyperspectral indices at a fine scale. *Scientific Reports*, *8*(4776), 1–11. doi:10.103841598-018-23136-5 PMID:29555982

Peng, Z. K., & Chu, F. L. (2004). Application of the wavelet transform in machine condition monitoring and fault diagnostics: A review with bibliography. *Mechanical Systems and Signal Processing*, *18*(2), 199–221. doi:10.1016/S0888-3270(03)00075-X

Perafan, J. (2018). *An unsupervised clustering methodology by means of an improved dbscan algorithm for operational conditions classification in a structure.* Universidad Pontificia Bolivariana.

Perez, A. J., Lopez, F., Benlloch, J. V., & Christensen, S. (2000). Colour and shape analysis techniques for weed detection in cereal fields. *Computers and Electronics in Agriculture*, *25*(3), 197–212. doi:10.1016/S0168-1699(99)00068-X

Personería de Bogotá, D. C. (2018). *Alarmante situación de inseguridad en Bogotá.* Bogotá: Author.

Piccardi, M. (2004). Background subtraction techniques: a review. *2004 IEEE International Conference on Systems, Man and Cybernetics.* 10.1109/ICSMC.2004.1400815

Pico-Valencia, P., Holgado-Terriza, J. A., Herrera-Sánchez, D., & Sampietro, J. (2018). Towards the internet of agents: An analysis of the internet of things from the intelligence and autonomy perspective. *Ingenieria e Investigacion*, *38*(1), 121–129. doi:10.15446/ing.investig.v38n1.65638

Pinto, J. (2014). *Análisis e interpretación de señales cardiorrespiratorias para determinar el momento óptimo de desconexión de un paciente asistido mediante ventilación* (Bachelor Thesis). Universidad Autónoma de Bucaramanga, Bucaramanga.

Pires, I.M., Garcia, N.M., Pombo, N., Flórez-Revuelta, F., Spinsante, S., Canavarro Teixeira, M., & Zdravevski, E. (2019). *Pattern recognition techniques for the identification of Activities of Daily Living using mobile device accelerometer.* PeerJ Preprints 7:e27225v2.

Piron, A., Leemans, V., Lebeau, F., & Destain, M.-F. (2009). Improving in-row weed detection in multispectral stereoscopic images. *Computers and Electronics in Agriculture*, *69*(1), 73–79. doi:10.1016/j.compag.2009.07.001

Plastria, F., De Bruyne, S., & Carrizosa, E. (2008, October). Dimensionality reduction for classification. In *International Conference on Advanced Data Mining and Applications* (pp. 411-418). Springer. 10.1007/978-3-540-88192-6_38

Pławiak, P. (2014). An estimation of the state of consumption of a positive displacement pump based on dynamic pressure or vibrations using neural networks. *Neurocomputing*, *144*, 471–483. doi:10.1016/j.neucom.2014.04.026

Poongodi, T., Krishnamurthi, R., Indrakumari, R., Suresh, P., & Balusamy, B. (2020). Wearable Devices and IoT. doi:10.1007/978-3-030-23983-1_10

Popov, E. (1998). *Engineering Mechanics of Solids (2nd ed.).* Pearson.

Potthast, J. (2011). *Sense and Security. A Comparative View on Access Control at Airports.* Science, Technology & Innovation Studies.

Pravdová, V., Pravda, M., & Guilbault, G. G. (2002). Role of chemometrics for electrochemical sensors. *Analytical Letters*, *35*(15), 2389–2419. doi:10.1081/AL-120016533

Prieto, N., Oliveri, P., Leardi, R., Gay, M., Apetrei, C., Rodriguez-Méndez, M. L., & De Saja, J. A. (2013). Application of a GA–PLS strategy for variable reduction of electronic tongue signals. *Sensors and Actuators. B, Chemical*, *183*, 52–57. doi:10.1016/j.snb.2013.03.114

Puchi-Gómez, C., Paravic-Klijn, T., & Salazar, A. (2018). Indicators of the quality of health care in home hospitalization: An integrative review. *Aquichan*, *18*(2), 186–197. doi:10.5294/aqui.2018.18.2.6

Pyšek, P., Hulme, P. E., Meyerson, L. A., Smith, G. F., Boatwright, J. S., Crouch, N. R., ... Wilson, J. R. U. (2013). Hitting the right target: taxonomic challenges for, and of, plant invasions. *AoB Plants*, *5*(plt042), 1-25. doi:10.1093/aobpla/plt042

Qing, X., Li, W., Wang, Y., & Sun, H. (2019). Piezoelectric transducer-based structural health monitoring for aircraft applications. Sensors, 19(3).

Rahmani, M. E., Amine, A., & Hamou, R. M. (2016). Supervised machine learning for plants identification based on images of their leaves. *International Journal of Agricultural and Environmental Information Systems*, *7*(4), 17–31. doi:10.4018/IJAEIS.2016100102

Ramdén, T., Krus, P., & Palmberg, J. O. (1996). Reliability and sensitivity analysis of a condition monitoring technique. In *Proceedings of the JFPS International Symposium on Fluid Power* (pp. 567-572). The Japan Fluid Power System Society. 10.5739/isfp.1996.567

Ramirez Lopez, L. J., Puerta Aponte, G., & Rodriguez Garcia, A. (2019). Internet of Things Applied in Healthcare Based on Open Hardware with Low-Energy Consumption. *Healthcare Informatics Research*, *25*(3), 230. doi:10.4258/hir.2019.25.3.230 PubMed

Ramya, G., & Manoharan, R. (2018). Enhanced Multi-Controller Placements in SDN. *International Conference on Wireless Communications*.

Rastogi, R., Chaturvedi, D. K., Satya, S., Arora, N., & Chauhan, S. (2018a). An Optimized Biofeedback Therapy for Chronic TTH between Electromyography and Galvanic Skin Resistance Biofeedback on Audio, Visual and Audio Visual Modes on Various Medical Symptoms. In the National Conference on 3rd MDNCPDR-2018 at DEI, Agra.

Rastogi, R., Chaturvedi, D. K., Satya, S., Arora, N., Sirohi, H., Singh, M., . . . Singh, V. (2018i). Which One is Best: Electromyography Biofeedback Efficacy Analysis on Audio, Visual and Audio-Visual Modes for Chronic TTH on Different Characteristics. Proceedings of ICCIIoT- 2018. Retrieved from https://ssrn.com/abstract=3354375

Rastogi, R., Chaturvedi, D. K., Satya, S., Arora, N., Singhal, P., & Gulati, M. (2018f). Statistical Resultant Analysis of Spiritual & Psychosomatic Stress Survey on Various Human Personality Indicators. *The International Conference Proceedings of ICCI 2018*. Doi:10.1007/978-981-13-8222-2_25

Rastogi, R., Chaturvedi, D. K., Satya, S., Arora, N., Yadav, V., Chauhan, S., & Sharma, P. (2018c). 'F-36 Scores Analysis for EMG and GSR Therapy on Audio, Visual and Audio Visual Modes for Chronic TTH. *Proceedings of the ICCIDA-2018*.

Rath, H. K., Revoori, V., Nadaf, S. M., & Simha, A. (2014). Optimal controller placement in Software Defined Networks (SDN) using a non-zero-sum game. *Proceeding of IEEE International Symposium on a World of Wireless, Mobile and Multimedia Networks 2014, WoWMoM 2014*.

Rayes, A., & Salam, S. (2016). Internet of things-from hype to reality: The road to digitization. doi:10.1007/978-3-319-44860-2

Real Academia de Ingeniería-Raing. (2018). *Electroanálisis*. Retrieved from: http://diccionario.raing.es/es/lema/electroan%C3%A1lisis

Rebel, G. (2009). *El lenguaje corporal. Lo que expresanlas actitudes, las posturas, los gestos y su interpretación*. Madrid: EDAF, S.L.

Redmon, J., & Farhadi, A. (2018). YOLOv3: An Incremental Improvement. Retrieved from http://arxiv.org/abs/1804.02767

Redmon, J., Divvala, S., Girshick, R., & Farhadi, A. (2016). You Only Look Once: Unified, Real-Time Object Detection. 2016 IEEE Conference on Computer Vision and Pattern Recognition (CVPR), 779–788. doi:10.1109/CVPR.2016.91

Ren, W., Sun, Y., Luo, H., & Guizani, M. (2019). A novel control plane optimization strategy for important nodes in SDN-IoT networks. *IEEE Internet of Things Journal*, *6*(2), 3558–3571. doi:10.1109/JIOT.2018.2888504

Rew, L. J., & Cousens, R. D. (2001). Spatial distribution of weeds in arable crops: Are current sampling and analytical methods appropriate? *Weed Research*, *41*(1), 1–18. doi:10.1046/j.1365-3180.2001.00215.x

Röbler, C. (2008). Design of a fuel cell powered UAV wing and V-tail building instruction. Munich: Academic Press.

Robusto, C. C. (1957). The cosine-haversine formula. *The American Mathematical Monthly*, *64*(1), 38–40. doi:10.2307/2309088

Rocha, B., Silva, C., Keulen, C., Yildiz, M., & Suleman, A. (2013). Structural Health Monitoring of Aircraft Structures. In W. Ostachowicz & A. Güemes (Eds.), *New Trends in Structural Health Monitoring*. Udine: Springer. doi:10.1007/978-3-7091-1390-5_2

Rodriguez, M., Laptev, I., & Audibert, J. (2011). Density-aware person detection and tracking in crowds. In *IEEE International Conference on Computer Vision* (pp. 2423-2430). IEEE. 10.1109/ICCV.2011.6126526

Rodríguez, M., Plaza, G., & Gil, R. (2008). Reconocimiento y fluctuación poblacional arvense en el cultivo de espinaca (Spinacea oleracea L.) para el municipio de Cota, Cundinamarca. *Agronomia Colombiana*, *26*(1), 87–96.

Roger, S., & Pressman, P. (2010). *Ingeniería del software un enfoque práctico*. McGraw Hill.

Roopa, M. S., Pattar, S., Buyya, R., Venugopal, K. R., Iyengar, S. S., & Patnaik, L. M. (2019). Social Internet of Things (SIoT): Foundations, thrust areas, systematic review and future directions. *Computer Communications*, *139*, 32–57. doi:10.1016/j.comcom.2019.03.009

Ros, F. J., & Ruiz, P. M. (2014). Five nines of southbound reliability in software-defined networks. *Proceedings of the Third Workshop on Hot Topics in Software Defined Networking - HotSDN '14*, 31–36. 10.1145/2620728.2620752

Rubin, A. (1999). Biofeedback and binocular vision. *Journal of Behavioral Optometry*, *3*(4), 95–98.

Sabokrou, M., Fayyaz, M., Fathy, M., Moayed, Z., & Klette, R. (2018). Deep-anomaly: Fully convolutional neural network for fast anomaly detection in crowded scenes. *Computer Vision and Image Understanding*, *172*, 88–97. doi:10.1016/j.cviu.2018.02.006

Saini, H., Rastogi, R., Chaturvedi, D. K., Satya, S., Arora, N., Verma, H., & Mehlyan, K. (2018j). Comparative Efficacy Analysis of Electromyography and Galvanic Skin Resistance Biofeedback on Audio Mode for Chronic TTH on Various Indicators. Proceedings of ICCIIoT- 2018. Retrieved from https://ssrn.com/abstract=3354371

Saini, H., Rastogi, R., Chaturvedi, D. K., Satya, S., Arora, N., Gupta, M., & Verma, H. (2019c). An Optimized Biofeedback EMG and GSR Biofeedback Therapy for Chronic TTH on SF-36 Scores of Different MMBD Modes on Various Medical Symptoms. In *Hybrid Machine Intelligence for Medical Image Analysis*. Springer Nature Singapore Pte Ltd.; doi:10.1007/978-981-13-8930-6_8

Saini, M., Alelaiwi, A., & Saddik, A. E. (2015). How Close are We to Realizing a Pragmatic VANET Solution? A Meta-Survey. *ACM Computing Surveys*, *48*(2), 1–40. doi:10.1145/2817552

Sakthi, P., & Yuvarani, P. (2018). Detection and Removal of Weed between Crops in Agricultural Field using Image Processing. *International Journal of Pure and Applied Mathematics*, *118*, 201–206. Retrieved from http://www.ijpam.eu

Salamatian, K. (2011). Toward a polymorphic future internet: A networking science approach. *IEEE Communications Magazine*, *49*(10), 174–178. doi:10.1109/MCOM.2011.6035832

Salimi, E., & Niromandfam, B. (2015). *Condition Monitoring of Hydrolic Pump by Wavelet Transform and Artificial Neural Network*. Academic Press.

Sallahi, A., & St-Hilaire, M. (2017). Expansion model for the controller placement problem in software defined networks. *IEEE Communications Letters*, *21*(2), 274–277. doi:10.1109/LCOMM.2016.2621746

Salles, M. O., & Paixão, T. R. (2014). Application of Pattern Recognition Techniques in the Development of Electronic Tongues. *Advanced Synthetic Materials in Detection Science*, (3), 197.

Salman, O., Elhajj, I., Chehab, A., & Kayssi, A. (2017). Software Defined IoT security framework. 2017 Fourth International Conference on Software Defined Systems (SDS), 75–80. doi:10.1109/SDS.2017.7939144

Salman, T., & Jain, R. (2017). Networking protocols and standards for internet of things. Internet of Things and Data Analytics Handbook. doi:10.1002/9781119173601.ch13

Salman, O., Elhajj, I., Chehab, A., & Kayssi, A. (2018). IoT survey: An SDN and fog computing perspective. *Computer Networks*, *143*, 221–246. doi:10.1016/j.comnet.2018.07.020

Santos, J., Wauters, T., Volckaert, B., & de Turck, F. (2018). Fog computing: Enabling the management and orchestration of smart city applications in 5G networks. *Entropy (Basel, Switzerland)*, *20*(1). doi:10.3390/e20010004

Satya, S., Arora, N., Trivedi, P., Singh, A., Sharma, A., Singh, A., ... Chaturvedi, D. K. (2019a). Intelligent Analysis for Personality Detection on Various Indicators by Clinical Reliable Psychological TTH and Stress Surveys. *Proceedings of CIPR 2019 at Indian Institute of Engineering Science and Technology*.

Satya, S., Rastogi, R., Chaturvedi, D. K., Arora, N., Singh, P., & Vyas, P. (2018e). Statistical Analysis for Effect of Positive Thinking on Stress Management and Creative Problem Solving for Adolescents. *Proceedings of the 12th INDIACom*, 245-251.

Schauerte, B. (2013). *Histogram distances*. Available at: https://www.mathworks.com/matlabcentral/fileexchange/39275-histogram-distances

Scheirer, W. J., Wilber, M. J., Eckmann, M., & Boult, T. E. (2014). Good recognition is non-metric. *Pattern Recognition*, *47*(8), 2721–2731. doi:10.1016/j.patcog.2014.02.018

Schoenau, G. J., Stecki, J. S., & Burton, R. T. (2000). Utilization of Artificial Neural Networks in the Control, Identification and Condition Monitoring of Hydraulic Systems—An Overview. *SAE Transactions*, 205–212.

Scholz, M., Fraunholz, M., & Selbig, J. (2008). Nonlinear Principal Component Analysis: Neural Network Models and Applications. In A. N. Gorban, B. Kégl, D. C. Wunsch, & A. Y. Zinovyev (Eds.), *Principal Manifolds for Data Visualization and Dimension Reduction* (pp. 44–67). Berlin: Springer Berlin Heidelberg. doi:10.1007/978-3-540-73750-6_2

Scott, D. S., & Lundeen, T. F. (2004). Myofascial pain involving the masticatory muscles: An experimental model. *Pain*, *8*(2), 207–215. doi:10.1016/0304-3959(88)90008-5 PubMed

Seeland, M., Rzanny, M., Boho, D., Wäldchen, J., & Mäder, P. (2019, January). Image-based classification of plant genus and family for trained and untrained plant species. *BMC Bioinformatics*, *20*(4), 1–13. doi:10.118612859-018-2474-x PMID:30606100

Sermanet, P., Eigen, D., Zhang, X., Mathieu, M., Fergus, R., & LeCun, Y. (2013). OverFeat: Integrated Recognition, Localization and Detection using Convolutional Networks. Retrieved from http://arxiv.org/abs/1312.6229

Serrato, N., & Castillo, C. (2018). Colombia Land of Opportunities to Apply Precision Agriculture: An Overview. *International Journal of Agricultural Sciences*, *3*. Retrieved from http://iaras.org/iaras/journals/ijas

Sethi, P., & Sarangi, S. R. (2017a). Internet of Things: Architectures, Protocols, and Applications. *Journal of Electrical and Computer Engineering*, *2017*, 1–25. doi:10.1155/2017/9324035

Sethi, P., & Sarangi, S. R. (2017b). *Review Article Internet of Things : Architectures, Protocols, and Applications*. Academic Press.

Shala, B., Trick, U., Lehmann, A., Ghita, B., & Shiaeles, S. (2019). Novel Trust Consensus Protocol and Blockchain-based Trust Evaluation System for M2M Application Services. Internet of Things. doi:10.1016/j.iot.2019.100058

Shang, W., Yu, Y., Zhang, L., & Droms, R. (2016). Challenges in IoT Networking via TCP/IP Architecture. NDN Project, Tech. Rep. NDN-0038.

Sharma, S., Rastogi, R., Chaturvedi, D. K., Bansal, A., & Agrawal, A. (2018g). Audio Visual EMG & GSR Biofeedback Analysis for Effect of Spiritual Techniques on Human Behavior and Psychic Challenges. *Proceedings of the 12th INDIACom*, 252-258.

Shi, Q., Guo, T., Yin, T., Wang, Z., Li, C., Sun, X., ... Yuan, W. (2018). Classification of Pericarpium Citri Reticulatae of Different Ages by Using a Voltammetric Electronic Tongue System. *International Journal of Electrochemical Science*, *13*(12), 11359–11374. doi:10.20964/2018.12.45

Sierra, J., Frövel, M., Del Olmo, E., Pintado, J. M., & Güemes, A. (2014). A robust procedure for damage identification in a lattice spacecraft structural element by mean of Strain field pattern recognition techniques. *16th European Conference on Composite Materials, ECCM 2014*.

Sierra, J., Güemes, A., & Mujica, E. (2013). Damage detection by using FBGs and strain field pattern recognition techniques. *Smart Materials and Structures*, *22*(2).

Sierra-Pérez, J. (2014). *Smart aeronautical structures: development and experimental validation of a structural health monitoring system for damage detection*. Universidad Politécnica de Madrid.

Sierra-Pérez, J., & Alvarez-Montoya, J. (2018). Damage detection in composite aerostructures from strain and telemetry data fusion by means of pattern recognition techniques. *9th European Workshop on Structural Health Monitoring, EWSHM 2018*, 1–12.

Sierra-Pérez, J., & Güemes, A. (2016). Damage detection in aerostructures from strain measurements. *Aircraft Engineering and Aerospace Technology*, *88*(3), 441–451. doi:10.1108/AEAT-11-2013-0210

Sierra-Pérez, J., Güemes, A., Mujica, L. E., & Ruiz, M. (2015). Damage detection in composite materials structures under variable loads conditions by using fiber Bragg gratings and principal component analysis, involving new unfolding and scaling methods. *Journal of Intelligent Material Systems and Structures, 26*(11), 1346–1359. doi:10.1177/1045389X14541493

Sierra-Pérez, J., Torres-Arredondo, M. A., & Güemes, A. (2016). Damage and nonlinearities detection in wind turbine blades based on strain field pattern recognition. FBGs, OBR and strain gauges comparison. *Composite Structures, 135*, 156–166. doi:10.1016/j.compstruct.2015.08.137

Sierra-Pérez, J., Torres-Arredondo, M.-A., & Alvarez-Montoya, J. (2018). Damage detection methodology under variable load conditions based on strain field pattern recognition using FBGs, nonlinear principal component analysis, and clustering techniques. *Smart Materials and Structures, 27*(1), 015002. doi:10.1088/1361-665X/aa9797

Silva, C., Bouwmans, T., & Frélicot, C. (2015). An eXtended Center-Symmetric Local Binary Pattern for Background Modeling and Subtraction in Videos. Proceedings of the 10th International Conference on Computer Vision Theory and Applications, 395–402. doi:10.5220/0005266303950402

Singh, A. K., & Srivastava, S. (2018). A survey and classification of controller placement problem in SDN. *International Journal of Network Management, 28*(3), e2018. doi:10.1002/nem.2018

Singh, A., Rastogi, R., Chaturvedi, D. K., Satya, S., Arora, N., Sharma, A., & Singh, A. (2019d). Intelligent Personality Analysis on Indicators in IoT-MMBD Enabled Environment. In *Multimedia Big Data Computing for IoT Applications: Concepts, Paradigms, and Solutions.* Springer; doi:10.1007/978-981-13-8759-3_7

Singhal, P., Rastogi, R., Chaturvedi, D. K., Satya, S., Arora, N., Gupta, M., . . . Gulati, M. (2019b). Statistical Analysis of Exponential and Polynomial Models of EMG & GSR Biofeedback for Correlation between Subjects Medications Movement & Medication Scores. IJITEE, 8(6S), 625-635. Retrieved from https://www.ijitee.org/download/volume-8-issue-6S/

Siyuan, L., Linlin, D., & Wanlu, J. (2011, August). Study on application of Principal Component Analysis to fault detection in hydraulic pump. In *Proceedings of 2011 International Conference on Fluid Power and Mechatronics* (pp. 173-178). IEEE.

Skopal, T., & Bustos, B. (2011, Oct). On nonmetric similarity search problems in complex domains. *ACM Computing Surveys, 43*(4), 34:1-34:50. doi:10.1145/1978802.1978813

Sliwinska, M., Wisniewska, P., Dymerski, T., Namiesnik, J., & Wardencki, W. (2014). Food analysis using artificial senses. *Journal of Agricultural and Food Chemistry, 62*(7), 1423–1448. doi:10.1021/jf403215y PMID:24506450

Söderkvist, O. J. O. (2001). *Computer vision classification of leaves from Swedish trees* (Master's thesis). Linköping University, Linköping, Sweden. Retrieved from http://www.cvl.isy.liu.se/en/research/datasets/swedish-leaf/

Sohn, H., & Oh, C. K. (2009). Statistical Pattern Recognition. In C. Boller, F.-K. Chang, & Y. Fujino (Eds.), Encyclopedia of Structural Health Monitoring (pp. 1–18). Academic Press.

Somoza Castro, O. (2004). *La muerte violenta. Inspección ocular y cuerpo del delito.* Madrid: Grefol, S.L.

Son, J. (2018). Integrated Provisioning of Compute and Network Resources in Software-Defined Cloud Data Centers. Retrieved from https://minerva-access.unimelb.edu.au/bitstream/handle/11343/212287/thesis.pdf?sequence=1&isAllowed=y

Soriano, R. R. (2002). *Investigación social teoría y praxis.* Editorial Plaza y Valdés, S.A. de C.V.

Sorsa, T., & Koivo, H. N. (1993). Application of artificial neural networks in process fault diagnosis. *Automatica, 29*(4), 843–849. doi:10.1016/0005-1098(93)90090-G

Soursos, S., Zarko, I. P., Zwickl, P., Gojmerac, I., Bianchi, G., & Carrozzo, G. (2016). Towards the cross-domain interoperability of IoT platforms. EUCNC 2016 - European Conference on Networks and Communications, 398–402. 10.1109/EuCNC.2016.7561070

Stalford, H. L., & Borrás, C. (2002, May). Pattern recognition in hydraulic backlash using neural network. In *Proceedings of the 2002 American Control Conference (IEEE Cat. No. CH37301)* (Vol. 1, pp. 400-405). IEEE. 10.1109/ACC.2002.1024838

Stańczyk, U. (2015). Feature evaluation by filter, wrapper, and embedded approaches. In *Feature Selection for Data and Pattern Recognition* (pp. 29–44). Berlin: Springer. doi:10.1007/978-3-662-45620-0_3

Suda, J. (2017). Misstatements of GPS- location of public transport vehicles in Warsaw. Transportation Overview - Przeglad Komunikacyjny, 2017(2), 37–45. doi:10.35117/A_ENG_17_02_05

Sujaritha, M., Annadurai, S., Satheeshkumar, J., Sharan, S. K., & Mahesh, L. (2017). Weed detecting robot in sugarcane fields using fuzzy real time classifier. *Computers and Electronics in Agriculture, 134*, 160–171. doi:10.1016/j.compag.2017.01.008

Sujitha, B. B., & Ramani, R. R., & Parameswari. (2012). Intrusion Detection System using Fuzzy Genetic Approach. *Int. J. Adv. Res. Comp. Commun. Eng., 1*(10), 827–831.

Suliman, S. I., Abd Shukor, M. S., Kassim, M., Mohamad, R., & Shahbudin, S. (2018). Network Intrusion Detection System Using Artificial Immune System (AIS). In *Proc. of 2018 3rd International Conference on Computer and Communication Systems (ICCCS)* (pp. 178-182). Nagoya, Japan: Academic Press. 10.1109/CCOMS.2018.8463274

SuperDataScience. (2017). *SuperDataScience.* Obtenido de https://www.superdatascience.com/blogs/opencv-face-recognition

Superintendencia de Industria y Comercio. (n.d.). *Oficina Asesora Jurídica.* Obtenido de http://www.sic.gov.co/sites/default/files/files/Nuestra_Entidad/Guia_Vigilancia_sept16_2016.pdf

Swaminathan, M., Yadav, P. K., Piloto, O., Sjöblom, T., & Cheong, I. (2017). A new distance measure for non-identical data with application to image classification. *Pattern Recognition, 63*, 384–396. doi:10.1016/j.patcog.2016.10.018

Szeliski, R. (2010). *Computer vision: algorithms and applications.* Springer Science & Business Media.

Szöllősi, D., Dénes, D. L., Firtha, F., Kovács, Z., & Fekete, A. (2012). Comparison of six multiclass classifiers by the use of different classification performance indicators. *Journal of Chemometrics, 26*(3-4), 76–84. doi:10.1002/cem.2432

Tabatabaefar, M., Miriestahbanati, M., & Grégoire, J.-C. (2017). Network Intrusion Detection through Artificial Immune System. In *2017 Annual IEEE International Systems Conference (SysCon).* Montreal, QC, Canada: IEEE. 10.1109/SYSCON.2017.7934751

Takiddeen, N., & Zualkernan, I. (2019). Smartwatches as IoT Edge Devices: A Framework and Survey. 2019 Fourth International Conference on Fog and Mobile Edge Computing (FMEC), 216–222. doi:10.1109/FMEC.2019.8795338

Tanha, M., Sajjadi, D., Ruby, R., & Pan, J. (2018). Capacity-Aware and Delay-Guaranteed Resilient Controller Placement for Software-Defined WANs. *IEEE eTransactions on Network and Service Management, 15*(3), 991–1005. doi:10.1109/TNSM.2018.2829661

Tao, F., Zuo, Y., Da Xu, L., & Zhang, L. (2014). IoT-Based intelligent perception and access of manufacturing resource toward cloud manufacturing. *IEEE Transactions on Industrial Informatics.* doi:10.1109/TII.2014.2306397

Tao, X., Lu, C., Lu, C., & Wang, Z. (2013). An approach to performance assessment and fault diagnosis for rotating machinery equipment. *EURASIP Journal on Advances in Signal Processing, 2013*(1), 5. doi:10.1186/1687-6180-2013-5

Tao, X., Wang, Z., Ma, J., & Fan, H. (2012, May). Study on fault detection using wavelet packet and SOM neural network. In *Proceedings of the IEEE 2012 Prognostics and System Health Management Conference (PHM-2012 Beijing)* (pp. 1-5). IEEE.

Tavallaee, M., Bagheri, E., Lu, W., & Ghorbani, A. A. (2009). A Detailed Analysis of the KDD Cup 99 Data Set. *IEEE Int. Symposium on Computational Intelligence for Security and Defense Applications.* 10.1109/CISDA.2009.5356528

Tayyaba, S. K., Shah, M. A., Khan, O. A., & Ahmed, A. W. (2017). Software defined network (SDN) based internet of things (IoT): A road ahead. ACM International Conference Proceeding Series, Part F1305. doi:10.1145/3102304.3102319

Theodoridis, S., & Koutroumbas, K. (2003). *Pattern Recognition* (2nd ed.). Elsevier Acedemic press.

Theodoridis, S., & Koutroumbas, K. (2009). *Pattern recognition* (4th ed.). Burlington, MA: Academic Press. doi:10.1016/B978-1-59749-272-0.X0001-2

Thorp, K. R., & Tian, L. F. (2004). A review on remote sensing of weeds in agriculture. *Precision Agriculture, 5*(5), 477–508. doi:10.100711119-004-5321-1

Tian, S. Y., Deng, S. P., & Chen, Z. X. (2007). Multifrequency large amplitude pulse voltammetry: A novel electrochemical method for electronic tongue. *Sensors and Actuators. B, Chemical, 123*(2), 1049–1056. doi:10.1016/j.snb.2006.11.011

Tibaduiza, D. A. (2013). *Design and validation of a structural health monitoring system for aeronautical structures* (PhD thesis). Universitat Politecnica de Catalunya.

Tibaduiza, D. A., Mujica, L. E., & Rodellar, J. (2011). Comparison of several methods for damage localization using indices and contributions based on PCA. *Journal of Physics: Conference Series, 305,* 12013. doi:10.1088/1742-6596/305/1/012013

Tsolakidis, D. G., Kosmopoulos, D. I., & Papadourakis, G. (2014). Plant leaf recognition using Zernike moments and histogram of oriented gradients. In *Hellenic Conference on Artificial Intelligence* (pp. 406–417). 10.1007/978-3-319-07064-3_33

Turk, D. C., Swanson, K. S., & Tunks, E. R. (2008). Psychological approaches in the treatment of chronic pain patients- -When pills, scalpels, and needles are not enough. *Canadian Journal of Psychiatry, 53*(4), 213–223. doi:10.1177/070674370805300402 PubMed

Tuydes-Yaman, H., Altintasi, O., & Sendil, N. (2015). Better estimation of origin–destination matrix using automated intersection movement count data. *Canadian Journal of Civil Engineering, 42*(7), 490–502. doi:10.1139/cjce-2014-0555

Umamaheswari, S., Arjun, R., & Meganathan, D. (2018). Weed Detection in Farm Crops using Parallel Image Processing. In *2018 Conference on Information and Communication Technology (CICT)* (pp. 1–4). IEEE. 10.1109/INFOCOM-TECH.2018.8722369

Universidad de California & Departamento de Policía de los Ángeles. (2010). *Predpol*. Obtenido de Predpol: https://www.predpol.com/technology/

Valdivieso, A., Peral, A., Barona, L., & García, L. (2014). Evolution and Opportunities in the Development IoT Applications. Retrieved from Http://Journals.Sagepub.Com/Doi/Full/10.1155/2014/735142

Valera, M., & Velastin, S. (2005). Intelligent distributed surveillance systems: a review. *IEE Proceedings Vision, Image and Signal Processing.*

van Oorschot, P. C., & Smith, S. W. (2019). The Internet of Things: Security Challenges. *IEEE Security and Privacy*, *17*(5), 7–9. doi:10.1109/MSEC.2019.2925918

Vasan, K. K., & Surendiran, B. (2016). Dimensionality reduction using principal component analysis for network intrusion detection. *Perspectives on Science*, *8*, 510–512. doi:10.1016/j.pisc.2016.05.010

Vilela, P. H., Rodrigues, J. J. P. C., Solic, P., Saleem, K., & Furtado, V. (2019). Performance evaluation of a Fog-assisted IoT solution for e-Health applications. *Future Generation Computer Systems*, *97*, 379–386. doi:10.1016/j.future.2019.02.055

Viola, P., & Jones, M. (2001). *Rapid Object Detection using a Boosted Cascade of Simple*. Obtenido de https://www.cs.cmu.edu/~efros/courses/LBMV07/Papers/viola-cvpr-01.pdf

Viola, P., & Jones, M. J. (2004). Robust Real-Time Face Detection. *International Journal of Computer Vision*, *57*(2), 137–154. doi:10.1023/B:VISI.0000013087.49260.fb

Vitola, J., Pozo, F., Tibaduiza, D. A., & Anaya, M. (2017). Distributed Piezoelectric Sensor System for Damage Identification in Structures Subjected to Temperature Changes. *Sensors*, *17*(6), 1252. doi:10.339017061252 PMID:28561786

Vlasov, Y. G., Legin, A. V., Rudnitskaya, A. M., D'amico, A., & Di Natale, C. (2000). «Electronic tongue»—New analytical tool for liquid analysis on the basis of non-specific sensors and methods of pattern recognition. *Sensors and Actuators. B, Chemical*, *65*(1-3), 235–236. doi:10.1016/S0925-4005(99)00323-8

Vögler Matrikelnummer, M. (2016a). Efficient IoT Application Delivery and Management in Smart City Environments. Universität Wien. Retrieved from http://www.Infosys.tuwien.ac.at/Staff/sd/papers/Diss_Voegler_Michael.pdf

Vögler Matrikelnummer, M. (2016b). *Efficient IoT Application Delivery and Management in Smart City Environments*. Academic Press.

Vyas, P., Rastogi, R., Chaturvedi, D. K., Arora, N., Trivedi, P., & Singh, P. (2018h). Study on Efficacy of Electromyography and Electroencephalography Biofeedback with Mindful Meditation on Mental health of Youths. *Proceedings of the 12th INDIA-Com*, 84-89.

Wäldchen, J., & Mäder, P. (2018). Machine learning for image based species identification. *Methods in Ecology and Evolution*, *9*(11), 2216–2225. doi:10.1111/2041-210X.13075

Wäldchen, J., Rzanny, M., Seeland, M., & Mäder, P. (2018, April). Automated plant species identification— Trends and future directions. *PLoS Computational Biology*, *14*(4), 1–19. doi:10.1371/journal.pcbi.1005993 PMID:29621236

Wang, G., Zhao, Y., & Huang, J. (2017). *The controller placement problem in software defined networking: a survey*. Homepages.Dcc.Ufmg.Br.

Wang, J., & Li, D. (2018). Adaptive computing optimization in software-defined network-based industrial internet of things with fog computing. Sensors (Switzerland), 18(8), 2509. doi:10.339018082509

Wang, S., & Xiang, J. (n.d.). A minimum entropy deconvolution-enhanced convolutional neural networks for fault diagnosis of axial piston pumps. *Soft Computing*, 1-15.

Wang, X., Han, T. X., & Yan, S. (2009). An HOG-LBP human detector with partial occlusion handling. 2009 IEEE 12th International Conference on Computer Vision, 32–39. 10.1109/ICCV.2009.5459207

Wang, A., Zhang, W., & Wei, X. (2019). A review on weed detection using ground-based machine vision and image processing techniques. *Computers and Electronics in Agriculture*, *158*, 226–240. doi:10.1016/j.compag.2019.02.005

Wang, L., Niu, Q., Hui, Y., & Jin, H. (2015). Discrimination of rice with different pretreatment methods by using a voltammetric electronic tongue. *Sensors (Basel)*, *15*(7), 17767–17785. doi:10.3390150717767 PMID:26205274

Wang, R., Li, M., Peng, L., Hu, Y., Hassan, M. M., & Alelaiwi, A. (2020). Cognitive multi-agent empowering mobile edge computing for resource caching and collaboration. *Future Generation Computer Systems*, *102*, 66–74. doi:10.1016/j.future.2019.08.001

Wang, S., Xiang, J., Zhong, Y., & Tang, H. (2018). A data indicator-based deep belief networks to detect multiple faults in axial piston pumps. *Mechanical Systems and Signal Processing*, *112*, 154–170. doi:10.1016/j.ymssp.2018.04.038

Wang, X., Kruger, U., Irwin, G. W., McCullough, G., & McDowell, N. (2008). Nonlinear PCA With the Local Approach for Diesel Engine Fault Detection and Diagnosis. *IEEE Transactions on Control Systems Technology*, *16*(1), 122–129. doi:10.1109/TCST.2007.899744

Wang, X., Mueen, A., Ding, H., Trajcevski, G., Scheuermann, P., & Keogh, E. (2013). Experimental comparison of representation methods and distance measures for time series data. *Data Mining and Knowledge Discovery*, *26*(2), 275–309. doi:10.100710618-012-0250-5

Wang, Z., & Xue, X. (2014). Multi-Class Support Vector Machine. In *Support Vector Machines Applications* (pp. 23–48). Cham, Switzerland: Springer. doi:10.1007/978-3-319-02300-7_2

Wang, Z., Yoon, S., Xie, S. J., Lu, Y., & Park, D. S. (2016). Visual tracking with semi-supervised online weighted multiple instance learning. *The Visual Computer*, *32*(3), 307–320. doi:10.1007/s00371-015-1067-1

Wei, M., Chow, T. W. S., & Chan, R. H. M. (2015). Clustering heterogeneous data with k-means by mutual information-based unsupervised feature transformation. *Entropy (Basel, Switzerland)*, *17*(3), 1535–1548. doi:10.3390/e17031535

Wei, Z., Wang, J., & Jin, W. (2013). Evaluation of varieties of set yogurts and their physical properties using a voltammetric electronic tongue based on various potential waveforms. *Sensors and Actuators. B, Chemical*, *177*, 684–694. doi:10.1016/j.snb.2012.11.056

Wei, Z., Yang, Y., Wang, J., Zhang, W., & Ren, Q. (2018). The measurement principles, working parameters and configurations of voltammetric electronic tongues and its applications for foodstuff analysis. *Journal of Food Engineering*, *217*, 75–92. doi:10.1016/j.jfoodeng.2017.08.005

Wenk-Sormaz, H. (2005). Meditation can reduce habitual responding. *Advances in Mind-Body*, *3*(4), 34–39.

Wen, L., Li, X., Gao, L., & Zhang, Y. (2017). A new convolutional neural network-based data-driven fault diagnosis method. *IEEE Transactions on Industrial Electronics*, *65*(7), 5990–5998. doi:10.1109/TIE.2017.2774777

West, J., & Bhattacharya, M. (2016). Intelligent financial fraud detection: A comprehensive review. *Computers & Security*, *57*, 47–66. doi:10.1016/j.cose.2015.09.005

Westwood, J. H., Charudattan, R., Duke, S. O., Fennimore, S. A., Marrone, P., Slaughter, D. C., ... Zollinger, R. (2018). Weed Management in 2050: Perspectives on the Future of Weed Science. *Weed Science*, *66*(3), 275–285. doi:10.1017/wsc.2017.78

Whiting, M. L., Ustin, S. L., Zarco-Tejada, P., Palacios-Orueta, A., & Vanderbilt, V. C. (2006). Hyperspectral mapping of crop and soils for precision agriculture. In *SPIE Optics* (pp. 62980B–62980B). Photonics. doi:10.1117/12.681289

Wilken, A., Kraft, V., Girod, S., Winter, M., & Nowak, S. (2016). A fluoride-selective electrode (Fse) for the quantification of fluoride in lithium-ion battery (Lib) electrolytes. *Analytical Methods*, *8*(38), 6932–6940. doi:10.1039/C6AY02264B

Wilmer Rivas Asanza, B. M. (2018). *Redes neuronales artificiales aplicadas al reconocimiento de patrones*. Editorial UTMACH.

Wilson, A. D., & Baietto, M. (2011). Advances in electronic-nose technologies developed for biomedical applications. *Sensors (Basel)*, *11*(1), 1105–1176. doi:10.3390110101105 PMID:22346620

Winquist, F., Wide, P., & Lundström, I. (1997). An electronic tongue based on voltammetry. *Analytica Chimica Acta*, *357*(1-2), 21–31. doi:10.1016/S0003-2670(97)00498-4

Wold, S., Kettaneh, N., Fridén, H., & Holmberg, A. (1998). Modelling and diagnostics of batch processes and analogous kinetic experiments. *Chemometrics and Intelligent Laboratory Systems*, *44*(1), 331–340. doi:10.1016/S0169-7439(98)00162-2

Wu, S., Chen, X., Yang, L., Fan, C., & Zhao, Y. (2018). Dynamic and static controller placement in Software-Defined Satellite Networking. *Acta Astronautica*, *152*, 49–58. doi:10.1016/j.actaastro.2018.07.017

Xia, Q., Zhu, H.-D., Gan, Y., & Shang, L. (2014). Plant leaf recognition using histograms of oriented gradients. In *International Conference on Intelligent Computing* (pp. 369–374). 10.1007/978-3-319-09339-0_38

Xu, X., Xu, S., Jin, L., & Song, E. (2011). Characteristic analysis of Otsu threshold and its applications. *Pattern Recognition Letters*, *32*(7), 956–961. doi:10.1016/j.patrec.2011.01.021

Yadav, V., Rastogi, R., Chaturvedi, D. K., Satya, S., Arora, N., Yadav, V., ... Chauhan, S. (2018j). Statistical Analysis of EMG & GSR Biofeedback Efficacy on Different Modes for Chronic TTH on Various Indicators. *Int. J. Advanced Intelligence Paradigms*, *13*(1), 251–275. doi:10.1504/IJAIP.2019.10021825

Yang, R., Wen, Z., Mckee, D., Lin, T., Xu, J., & Garraghan, P. (2018). *Chapter #: Fog Orchestration and Simulation for IoT Services*. Academic Press.

Yan, R., Chen, X., & Mukhopadhyay, S. C. C. (2017). Advanced Signal Processing for Structural Health Monitoring. In *Smart Sensors* (Vol. 26). Measurement and Instrumentation.

Yaqoob, I., Ahmed, E., Hashem, I. A. T., Ahmed, A. I. A., Gani, A., Imran, M., & Guizani, M. (2017). Internet of Things Architecture: Recent Advances, Taxonomy, Requirements, and Open Challenges. *IEEE Wireless Communications*, *24*(3), 10–16. doi:10.1109/MWC.2017.1600421

Yazıcı, B., Yaslı, F., Gürleyik, H. Y., Turgut, U. O., Aktas, M. S., & Kalıpsız, O. (2015). Veri Madenciliğinde Özellik Seçim Tekniklerinin Bankacılık Verisine Uygulanması Üzerine Araştırma ve Karşılaştırmalı Uygulama, Ulusal Yazılım Mühendisliği Sempozyumu (a national conference). İzmir.

You, T. (2016). *Toward the future of Internet architecture for IoE*. Academic Press.

Yu, Z. (2013). Hybrid Fuzzy Cluster Ensemble Framework for Tumor Clustering from Biomolecular Data. Computational Biology and Bioinformatics, IEEE/ACM Transactions On, 10(3), 657–670.

Yu, W., Liang, F., He, X., Hatcher, W. G., Lu, C., Lin, J., & Yang, X. (2017). A Survey on the Edge Computing for the Internet of Things. *IEEE Access : Practical Innovations, Open Solutions*. doi:10.1109/ACCESS.2017.2778504

Yu, Z., Li, L., Liu, J., Zhang, J., & Han, G. (2015). Adaptive noise immune cluster ensemble using affinity propagation. *IEEE Transactions on Knowledge and Data Engineering*, *27*(12), 3176–3189. doi:10.1109/TKDE.2015.2453162

Zanella, A., Bui, N., Castellani, A., Vangelista, L., & Zorzi, M. (2004). Internet of things for smart cities. IEEE Internet Things J., 1(1), 22–32. doi:10.1109/JIOT.2014.2306328

Zhang, B., Wang, X., & Huang, M. (2018). Multi-objective optimization controller placement problem in internet-oriented software defined network. *Computer Communications*, *123*, 24–35. doi:10.1016/j.comcom.2018.04.008

Zhang, L., & Tian, F. C. (2014). A new kernel discriminant analysis framework for electronic nose recognition. *Analytica Chimica Acta*, *816*, 8–17. doi:10.1016/j.aca.2014.01.049 PMID:24580850

Zhang, L., Wang, X., Huang, G. B., Liu, T., & Tan, X. (2018). Taste recognition in e-tongue using local discriminant preservation projection. *IEEE Transactions on Cybernetics*, *49*(3), 947–960. doi:10.1109/TCYB.2018.2789889 PMID:29994190

Zhao, Chellappa, Rosenfeld, & Phillips. (2003). *Face Recognition: A Literature Survey.* Academic Press.

Zhao, W., Wang, S., & Dou, H. (2012, July). Fault simulation analysis of axial piston pump based on the component's failure. In *IEEE 10th International Conference on Industrial Informatics* (pp. 631-634). IEEE.

Zhao, W., Bhushan, A., Santamaria, A. D., Simon, M. G., & Davis, C. E. (2008). Machine learning: A crucial tool for sensor design. *Algorithms*, *1*(2), 130–152. doi:10.3390/a1020130 PMID:20191110

Zhao, W., & Davis, C. E. (2009). Swarm intelligence based wavelet coefficient feature selection for mass spectral classification: An application to proteomics data. *Analytica Chimica Acta*, *651*(1), 15–23. doi:10.1016/j.aca.2009.08.008 PMID:19733729

Zhao, X., Cao, F., & Liang, J. (2018). A sequential ensemble clusterings generation algorithm for mixed data. *Applied Mathematics and Computation*, *335*, 264–277. doi:10.1016/j.amc.2018.04.035

Zhao, Z.-Q., Ma, L.-H., Cheung, Y., Wu, X., Tang, Y., & Chen, C. L. P. (2015). ApLeaf: An efficient android-based plant leaf identification system. *Neurocomputing*, *151*, 1112–1119. doi:10.1016/j.neucom.2014.02.077

Zhong, Y. H., Zhang, S., He, R., Zhang, J., Zhou, Z., Cheng, X., ... Zhang, J. (2019). A Convolutional Neural Network Based Auto Features Extraction Method for Tea Classification with Electronic Tongue. *Applied Sciences*, *9*(12), 2518. doi:10.3390/app9122518

Zhuang, L., Guo, T., Cao, D., Ling, L., Su, K., Hu, N., & Wang, P. (2015). Detection and classification natural odors with an in vivo bioelectronic nose. *Biosensors & Bioelectronics*, *67*, 694–699. doi:10.1016/j.bios.2014.09.102 PMID:25459058

Zhu, S., & Xu, L. (2018). Many-objective fuzzy centroids clustering algorithm for categorical data. *Expert Systems with Applications*, *96*, 230–248. doi:10.1016/j.eswa.2017.12.013

Zikria, Y., Kim, S. W., Hahm, O., Afzal, M. K., & Aalsalem, M. Y. (2019). Internet of Things (IoT) Operating Systems Management: Opportunities, Challenges, and Solution. *Sensors (Basel)*, *19*(8), 1793. doi:10.3390/s19081793 PubMed

Zimdahl, R. (2007). *Fundamentals of weed science.* Elsevier. doi:10.1016/0378-4290(95)90065-9

About the Contributors

Diego Alexander Tibaduiza Burgos received in 2003 his Electronic Engineer Degree from the Industrial University of Santander in Colombia, the M.Sc. degree in 2011 in electronic engineering from the same university and his Doctoral Degree in 2013 from the "Universitat Politècnica de Catalunya" in Barcelona-Spain. He is the author and coauthor of more than 60 publications in journals, conferences, books and chapter books. He is currently a professor in the department of Electrics and Electronics engineering in the "Universidad Nacional de Colombia" Bogotá-Colombia. His research interest includes Structural Health Monitoring, Artificial Intelligence, Control Systems, Robotics and Signal Processing.

Maribel Anaya Vejar received in 2005 her Electronic Engineer Degree from the Industrial University of Santander in Colombia, the M.Sc. degree in 2010 in electronic engineering from the same university. Her Doctoral Degree was obtained in 2016 from the "Universitat Politècnica de Catalunya" in Barcelona-Spain, where she worked in the Structural Health Monitoring area focused on pattern recognition techniques. Her research interest includes Structural Health Monitoring, Artificial Intelligence, Pattern Recognition and Signal Processing.

Francesc Pozo received the degree in mathematics from the University of Barcelona, Barcelona, Spain, in 2000, and the Ph.D. degree in applied mathematics from the Universitat Politècnica de Catalunya, Barcelona, in 2005. Since 2000, he has been with the Department of Mathematics and the Barcelona East School of Engineering (EEBE), Universitat Politècnica de Catalunya, where he is currently an Associate Professor with the Control, Modeling, Identification and Applications Research Group (CoDAlab, codalab.upc.edu/en). He is also a Teaching Collaborator at the Open University of Catalonia, Barcelona. His research interests include wind turbine control, semiactive vibration mitigation in civil engineering structures (buildings and bridges), automotive and aeronautic systems, and offshore support structures, structural health monitoring (SHM) and condition monitoring (CM) for wind turbines and, in general, the application of applied mathematics in engineering problems. Dr. Pozo serves as a Secretary of the Spanish Joint Chapter of the IEEE Control Systems Society (CSS) and the IEEE Industrial Applications Society (IAS). He is also a member of the European Association for the Control of Structures (EACS) and an Editorial Board Member for international journals, such as Sensors, Structural Control and Health Monitoring, Journal of Vibration and Control, International Journal of Distributed Sensor Networks and Mathematical Problems in Engineering. Dr. Pozo is author of more than 55 research papers, 13 book chapters and 120 conference papers.

* * *

Richard Abuabara was born in Mompox, Colombia in 1988. He holds a B.S. degree in Electronic Engineering from Universidad de Pamplona in 2010 and a M.Sc. degree in Telecommunications and ICT Regulation from Universidad Santo Tomás in 2019. He has more than 4 years of experience working in digital business transformation projects. Richard is currently working for the biggest mobile operators in Colombia such as Claro Colombia and Telefonica, adding award-winning value to their BSS operations by leading continuous upgrades through holistic knowledge.

Joham Alvarez-Montoya is currently a lecturer and master student in Universidad Pontificia Bolivariana. His research interests are related to smart materials and structural health monitoring.

Isabel Amigo is an associate professor at IMT Atlantique. She obtained a PhD (2013) in computer science from Telecom Bretagne, France, and Universidad de la Repblica, Uruguay, and an electrical engineering diploma (2007) from Universidad de la Republica. Her main research interests are traffic engineering, interdomain QoS, New architectures and networking paradigms, network economics, game theory, and network performance.

Juliana Arevalo received a B.S. degree in Telecommunications Engineering from Universidad Santo Tomás in 2006, and an M.Sc. degree in IT security from Universitat Oberta de Catalunya in 2012. Currently, her field of study is the optimization of controller placement problem and security in software-defined networks. She worked in telecommunication companies in Colombia such as Huawei Technologies until 2016, when she joined the faculty of Telecommunications Engineering Department of Universidad Santo Tomas as full-time Professor.

Pragathi C. H. is currently Pursuing her Ph.D. in Computer Science and Engineering from VIT University, Vellore, India. She received her B.Tech in Electrical and Electronics Engineering and M. Tech in Computer Science & Engineering from JNT University, Andhra Pradesh, India. Her research interests include Clustering, Health Care Analytics, Data Analytics, and Data Classifications.

D. K. Chaturvedi is working in Dept. of Elect. Engg, Faculty of Engg, D.E.I., Dayalbagh, Agra since 1989. Presently he is Professor. He did his B.E. from Govt. Engineering College Ujjain, M.P. then he did his M.Tech. and Ph.D. from D.E.I. Dayalbagh. He is gold medalist and received Young Scientists Fellowship from DST, Government of India in 2001-2002 for post doctorial research at Univ. of Calgary, Canada. He is fellow of The Institution of Engineers (India), Aeronautical Society of India, IETE, SMIEEE, USA and Member of IET, U.K., ISTE, Delhi, ISCE, Roorkee, IIIE, Mumbai and SSI etc. and The IEE, U.K.

Felipe Díaz-Sánchez was born in Bogotá, Colombia. He received the B.S. degree in telecommunications engineering from Universidad Santo Tomás, Bogotá, Colombia, in 2008. He received the M.S.degree in design and engineering of convergent networks from Telecom Bretagne, Rennes, France in 2010 and the Ph.D. degree in networks and computer science from Telecom ParisTech, Paris, France, 2015. He worked as a research assistant at NEC Europe Ltd during a 6-month internship in 2010. He was a Full-time Professor at the Telecommunications Engineering Department, Universidad Santo Tomas from 2016 to 2018 and currently works as Innovation Manager in a private company. He is author of the book "Cloud brokering: new value-added services and pricing policies" (Ediciones USTA, 2016). His research interests include resources allocation in cloud computing, evolutionary algorithms, and IoT.

Tolga Ensari completed his undergraduate and graduate studies at Istanbul Technical University, Istanbul, TURKEY, and completed his Ph.D. study at department of Computer Engineering, Istanbul University, TURKEY. He is an assistant professor at Istanbul University-Cerrahpasa. His research interests lie in the area of machine learning, data mining and pattern recognition. He has been visiting scholar between 2011-2012, 2014 and 2019-2020 Department of Electrical and Computer Engineering, Computational Intelligence Lab at University of Louisville, USA. His works have been cited more than 300 times. He has served for several international conferences as committee member and also serves as reviewer in several journals. He is member of IEEE and also completed several research projects. He was awarded the international successful researcher by Istanbul University.

Muskan Gulati is B.Tech. Second Year student of CSE in ABESEC, Ghaziabad. She is working presently working on DataMining and Machine Learning. Her hobbies are singing and reading books. She is young, talented and dynamic.

Mayank Gupta is acting as System and IT Analyst in Tata Consultancy services, Noida and expert of Data sciences and Business Analytics. He has skill to visualize the situations from different perspectives and explore the real facts through critical Analysis. He has deep interest in Human health domains.

Jersson Leon is an Electromechanical Engineering that have worked in several research projects related with electronics and software development. Particularly in his master thesis he developed a bioinspired optimization algorithm to solve problems of topology optimization, besides he worked in multi objective optimization. Nowadays Jersson is student in the Doctoral programme in Mechanical and Mechatronics Engineering at National University of Colombia in Bogotá,his thesis is related with the pattern recognition and signal processing in sensor arrays type electronic noses and tongues.

Santiago Morales received a M.S. degree from Universidad Nacional de Colombia and a bachelor degree in Electronics Engineering. He is a developer of Intelligent Transportation Systems and Embedded Systems at Universidad Nacional de Colombia, Bogotá. His current research are Intelligent Transportation Systems and embedded systems.

Agastyaraju Nagaraja Rao is an Associate Professor from the School of Computer Science and Engineering at VIT University Vellore. He is an experienced expert faculty at VIT University and under his Supervision, many Research Scholars have successfully completed their PhDs. His commitment towards work resulted in publishing many National papers, International Scopus Journals and International Conference research articles. His expertise in the field has made him a reviewer and a chairperson for many Journals. His research interests include Clustering, Health Care Analytics, Data Analytics, and Data Classifications.

Mauricio Orozco-Alzate received his undergraduate degree in Electronic Engineering, his M.Eng. degree in Industrial Automation and his Dr.Eng. degree in Automatics from Universidad Nacional de Colombia - Sede Manizales, in 2003, 2005 and 2008 respectively. In 2007, he served as a Research Fellow at the Information and Communication Theory Group - Pattern Recognition Laboratory (PRLab) of Delft University of Technology (TU Delft), The Netherlands. Since 2008 he has been with Universidad Nacional de Colombia - Sede Manizales, where he is currently a Full Professor in the Department of Informatics and Computing. Mauricio regularly teaches undergraduate courses on Data Structures, Computer Architecture and Research Methodology and, occasionally, other undergraduate courses on Computer Programming and a graduate course on Pattern Recognition. He is also a researcher of the Group on Adaptive-Intelligent Environments and the Group on Signal Processing and Recognition. His main research interests encompass pattern recognition, digital signal processing and their applications to analysis and classification of seismic, bioacoustic and hydro-meteorological signals.

Cesar Pedraza received the Ph.D. degree in Informatics Engineering from the Rey Juan Carlos University in Madrid, Spain. Also received a M.S. degree from the Universidad de Los Andes, Bogota Colombia. He is currently a professor of parallel computing and operating systems at the Universidad Nacional de Colombia. His current research interests are digital signal processing applied to intelligent transportation systems and agriculture, and approximate computing techniques.

Pratyusha is B.Tech. Second Year student of CSE in ABESEC, Ghaziabad. She is working presently working on Machine Learning and blockchain. Her hobbies are doodling and bibliography. She is young, talented and dynamic.

Leonardo Ramirez Lopez was born on February 4, 1969 in Bogotá - Colombia, has a PhD in Biomedical Engineering from the University of Mogi das Cruzes in Brazil, MSc in Computer Science Engineering from the National University of Colombia, an Electronic Engineer of the Antonio Nariño University of Colombia. He has published on topics related to communication systems, telemedicine architectures, telemedicine security, routing algorithms and medical applications, currently researches on applications and sensors on the internet of things. Prof. Ramírez is a member of the IEEE Society Medicine and Biology Society.

Rohit Rastogi received his B.E. degree in Computer Science and Engineering from C.C.S.Univ. Meerut in 2003, the M.E. degree in Computer Science from NITTTR-Chandigarh (National Institute of Technical Teachers Training and Research-affiliated to MHRD, Govt. of India), Punjab Univ. Chandigarh in 2010. Currently he is pursuing his Ph.D. In computer science from Dayalbagh Educational Institute, Agra under renowned professor of Electrical Engineering Dr. D.K. Chaturvedi in area of spiritual consciousness. Dr. Santosh Satya of IIT-Delhi and dr. Navneet Arora of IIT-Roorkee have happily consented him to co supervise. He is also working presently with Dr. Piyush Trivedi of DSVV Hardwar, India in center of Scientific spirituality. He is a Associate Professor of CSE Dept. in ABES Engineering. College, Ghaziabad (U.P.-India), affiliated to Dr. A.P. J. Abdul Kalam Technical Univ. Lucknow (earlier Uttar Pradesh Tech. University). Also, He is preparing some interesting algorithms on Swarm Intelligence approaches like PSO, ACO and BCO etc. Rohit Rastogi is involved actively with Vichaar Krnati Abhiyaan and strongly believe that transformation starts within self.

Nelson Rosas Jimenez is an Electronic Engineer of the National University of Colombia, magister in engineering - Telecommunications of the National University. He is currently the dean of the Faculty of Engineering of Unipanamericana - Compensar. He has specialized in embedded systems, telecommunications and telemedicine.

Julian Sierra-Pérez pursued his doctoral studies in Universidad Politécnica de Madrid under Dr. Güemes supervision. He is also Master of Science in Aerospace Engineering and Master of Science in Engineering. His research fields include smart materials, smart structures, structural health monitoring, pattern recognition and wind energy. He currently is full professor in Universidad Pontificia Bolivariana.

Heena Sirohi is B.Tech. Second Year student of CSE in ABESEC, Ghaziabad. She is working presently on Raspberry pi automation and internet of things (IoT). Her areas of interest are data structure, networking and android development. She is young and dynamic personality.

Felix Vega (M' 12) received the Ph.D. in electrical engineering (Hons.) from the Universidad Nacional de Colombia, Bogota, Colombia, in 2011, and a second Ph.D. degree in electrical engineering from the Swiss Federal Institute of Technology of Lausanne (EPFL), Lausanne, Switzerland, in 2013. He is currently a Professor of Antennas and Electromagnetism at the Universidad Nacional de Colombia

Derya Yiltas-Kaplan received the BSc, MSc, and PhD degrees in computer engineering from Istanbul University, Istanbul, Turkey, in 2001, 2003 and 2007, respectively. She was a post-doctorate researcher at the North Carolina State University and she received post-doctorate research scholarship from The Scientific and Technological Research Council of Turkey during the period of April 2008-April 2009. She is currently working as a faculty member in the Department of Computer Engineering at Istanbul University.

Index

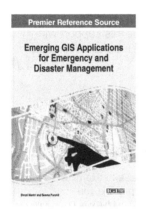
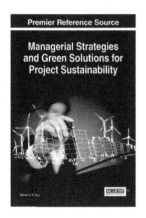

Printed in the United States
By Bookmasters